Jan Drugowitsch

Design and Analysis of Learning Classifier Systems

Studies in Computational Intelligence, Volume 139

Editor-in-Chief

Prof. Janusz Kacprzyk
Systems Research Institute
Polish Academy of Sciences
ul. Newelska 6
01-447 Warsaw
Poland
E-mail: kacprzyk@ibspan.waw.pl

Further volumes of this series can be found on our homepage: springer.com

Vol. 118. Tsau Young Lin, Ying Xie, Anita Wasilewska and Churn-Jung Liau (Eds.)
Data Mining: Foundations and Practice, 2008
ISBN 978-3-540-78487-6

Vol. 119. Slawomir Wiak, Andrzej Krawczyk and Ivo Dolezel (Eds.)
Intelligent Computer Techniques in Applied Electromagnetics, 2008
ISBN 978-3-540-78489-0

Vol. 120. George A. Tsihrintzis and Lakhmi C. Jain (Eds.)
Multimedia Interactive Services in Intelligent Environments, 2008
ISBN 978-3-540-78491-3

Vol. 121. Nadia Nedjah, Leandro dos Santos Coelho and Luiza de Macedo Mourelle (Eds.)
Quantum Inspired Intelligent Systems, 2008
ISBN 978-3-540-78531-6

Vol. 122. Tomasz G. Smolinski, Mariofanna G. Milanova and Aboul-Ella Hassanien (Eds.)
Applications of Computational Intelligence in Biology, 2008
ISBN 978-3-540-78533-0

Vol. 123. Shuichi Iwata, Yukio Ohsawa, Shusaku Tsumoto, Ning Zhong, Yong Shi and Lorenzo Magnani (Eds.)
Communications and Discoveries from Multidisciplinary Data, 2008
ISBN 978-3-540-78732-7

Vol. 124. Ricardo Zavala Yoe
Modelling and Control of Dynamical Systems: Numerical Implementation in a Behavioral Framework, 2008
ISBN 978-3-540-78734-1

Vol. 125. Larry Bull, Bernadó-Mansilla Ester and John Holmes (Eds.)
Learning Classifier Systems in Data Mining, 2008
ISBN 978-3-540-78978-9

Vol. 126. Oleg Okun and Giorgio Valentini (Eds.)
Supervised and Unsupervised Ensemble Methods and their Applications, 2008
ISBN 978-3-540-78980-2

Vol. 127. Régie Gras, Einoshin Suzuki, Fabrice Guillet and Filippo Spagnolo (Eds.)
Statistical Implicative Analysis, 2008
ISBN 978-3-540-78982-6

Vol. 128. Fatos Xhafa and Ajith Abraham (Eds.)
Metaheuristics for Scheduling in Industrial and Manufacturing Applications, 2008
ISBN 978-3-540-78984-0

Vol. 129. Natalio Krasnogor, Giuseppe Nicosia, Mario Pavone and David Pelta (Eds.)
Nature Inspired Cooperative Strategies for Optimization (NICSO 2007), 2008
ISBN 978-3-540-78986-4

Vol. 130. Richi Nayak, Nikhil Ichalkaranje and Lakhmi C. Jain (Eds.)
Evolution of the Web in Artificial Intelligence Environments, 2008
ISBN 978-3-540-79139-3

Vol. 131. Roger Lee and Haeng-Kon Kim (Eds.)
Computer and Information Science, 2008
ISBN 978-3-540-79186-7

Vol. 132. Danil Prokhorov (Ed.)
Computational Intelligence in Automotive Applications, 2008
ISBN 978-3-540-79256-7

Vol. 133. Manuel Graña and Richard J. Duro (Eds.)
Computational Intelligence for Remote Sensing, 2008
ISBN 978-3-540-79352-6

Vol. 134. Ngoc Thanh Nguyen and Radoslaw Katarzyniak (Eds.)
New Challenges in Applied Intelligence Technologies, 2008
ISBN 978-3-540-79354-0

Vol. 135. Hsinchun Chen and Christopher C. Yang (Eds.)
Intelligence and Security Informatics, 2008
ISBN 978-3-540-69207-2

Vol. 136. Carlos Cotta, Marc Sevaux and Kenneth Sörensen (Eds.)
Adaptive and Multilevel Metaheuristics, 2008
ISBN 978-3-540-79437-0

Vol. 137. Lakhmi C. Jain, Mika Sato-Ilic, Maria Virvou, George A. Tsihrintzis, Valentina Emilia Balas and Canicious Abeynayake (Eds.)
Computational Intelligence Paradigms, 2008
ISBN 978-3-540-79473-8

Vol. 138. Bruno Apolloni, Witold Pedrycz, Simone Bassis and Dario Malchiodi
The Puzzle of Granular Computing, 2008
ISBN 978-3-540-79863-7

Vol. 139. Jan Drugowitsch
Design and Analysis of Learning Classifier Systems, 2008
ISBN 978-3-540-79865-1

Jan Drugowitsch

Design and Analysis of Learning Classifier Systems

A Probabilistic Approach

Springer

Dr. Jan Drugowitsch
Department of Brain & Cognitive Sciences
Meliora Hall
University of Rochester
Rochester, NY 14627
USA
Email: jdrugowitsch@bcs.rochester.edu

ISBN 978-3-540-79865-1 e-ISBN 978-3-540-79866-8

DOI 10.1007/978-3-540-79866-8

Studies in Computational Intelligence ISSN 1860949X

Library of Congress Control Number: 2008926082

Typeset & Cover Design: Scientific Publishing Services Pvt. Ltd., Chennai, India.

Printed in acid-free paper

9 8 7 6 5 4 3 2 1

springer.com

Foreword

This book is probably best summarized as providing a principled foundation for Learning Classifier Systems. Something is happening in LCS, and particularly XCS and its variants that clearly often produces good results. Jan Drugowitsch wishes to understand this from a broader machine learning perspective and thereby perhaps to improve the systems. His approach centers on choosing a statistical definition – derived from machine learning – of "a good set of classifiers", based on a model according to which such a set represents the data. For an illustration of this approach, he designs the model to be close to XCS, and tests it by evolving a set of classifiers using that definition as a fitness criterion, seeing if the set provides a good solution to two different function approximation problems. It appears to, meaning that in some sense his definition of "good set of classifiers" (also, in his terms, a good model structure) captures the essence, in machine learning terms, of what XCS is doing.

In the process of designing the model, the author describes its components and their training in clear detail and links it to currently used LCS, giving rise to recommendations for how those LCS can directly gain from the design of the model and its probabilistic formulation. The seeming complexity of evaluating the quality of a set of classifiers is alleviated by giving an algorithmic description of how to do it, which is carried out via a simple Pittsburgh-style LCS. A final chapter on sequential decision tasks round out the model-centered formulation that has until then focused on function approximation and classification, by providing criteria for method stability and insights into new developments.

The link provided between LCS on a theoretical level and machine learning work in general is important, especially since the latter has a more developed theory and may in part stand off from LCS because of LCS's relative lack thereof (I stress "relative"). Also the problem addressed is important because out of greater theoretical understanding may result better classifier systems, as already demonstrated in this work by the improvements suggested for current LCS.

A particularly appealing feature of Drugowitsch's novel approach is its universal applicability to any kind of LCS that seeks to perform function approximation, classification, or handle sequential decision tasks by means of dynamic programming or reinforcement learning. Its close relation to XCS in this book results from the authors commitment to an LCS model structure that relates to XCS, but nothing speaks against applying the same approach to greatly different model types, resulting in different, potentially novel, LCS.

While its connection to Pittsburgh-style LCS is straightforward and clearly established in this work, using the same approach for the design of Michigan-style LCS remains a significant future challenge. Also, it will be interesting to see how the author's theoretical basis for reinforcement learning is built upon in future LCS, and how LCS designed by the model-based approach perform in comparison to currently existing LCS.

Overall, the work is elegant and approaches LCS from a refreshingly different perspective. It's also stylistically pretty novel for LCS work - but that's surely healthy!

Concord, MA, USA
March, 2008

Stewart W. Wilson

Preface

I entered the world of Learning Classifier Systems (LCS) through their introduction by Will Browne as part of a lecture series on "Advanced Artificial Intelligence" at the University of Reading, UK. Their immediate appeal as as a flexible architecture that combines the power of evolutionary computation with machine learning by splitting larger problems into tractable sub-problems made me decide to pursue them further, for which I got the opportunity during my Ph.D., supervised by Alwyn Barry, at the University of Bath.

Modest dissatisfaction followed my initial euphoria when I had to discover that their theoretical basis that I planned to rest my work upon did not live up to my initial expectation. Indeed, despite being generally referred to as Genetic-based Machine Learning, their formal development had little in common with machine learning itself. Their loose definition, ad-hoc design, complex structure of interwoven sub-components, and yet surprisingly competitive performance made me comprehend why David Goldberg referred to them as "a glorious, wondrous, and inventing quagmire, but a quagmire nonetheless."

The work presented in this book is an attempt to "clean up" on LCS and lay the foundations for a principled approach to their design by pragmatically following the road of machine learning, in order to bridge the gap between LCS and machine learning. Their design is approached from first principles, based on the question "What is a classifier system supposed to learn?". As presented here, the work is intended for researchers in LCS, genetic-based machine learning, and machine learning, but also for anyone else who is interested in LCS. The content is in most parts based on work performed during my Ph.D., but also includes extensions to it, most notably a complete formulation for classification tasks rather than only regression tasks. The content of this book is not to be understood as the development of a new LCS, but rather as the groundwork for a new approach to their design that I and hopefully others will build upon.

Numerous people have supported me in performing this work, and I am grateful for their constant encouragement. Most notably, I would not have been able to fully focus on my work without the generous financial support of my parents, Elsbeth and Knut Drugowitsch, during my Ph.D. time. Also, my Ph.D.

supervisor, Alwyn Barry, helped me to stay focused on the main questions, and his guidance, his constructive comments, and his initiative were essential to the completion of this work. Many people in an around Bath, UK, have helped me with comments, discussions, or equally valuable moral support: Dan Richardson, Marelee Hurn, Hagen Lehmann, Tristan Caulfield, Mark Price, Jonty Needham, Joanna Bryson, and especially Will Lowe for emphasising the model behind each method. Various researchers in LCS and machine learning have offered their support thought constructive discussions at conferences or per e-mail: Pier Luca Lanzi, Daniele Loiacono, Martin Butz, Stewart Wilson, Will Browne, Tim Kovacs, Gavin Brown, James Marshall, Lashon Booker, Xavier Llorà, Gavin Brown, Christopher Bishop, Markus Svensén, Matthew Beal, Tommi Jaakkola, Lei Xu, Peter Grünwald, Arta Doci, and Michael Littman. Special thanks go to Larry Bull for not giving me a too hard time at my Ph.D. viva, and for encouraging me to publish my work as a book, therefore taking full responsibility for it. Last, but certainly not least, I am deeply grateful for the moral support and patience of Odali Sanhueza throughout the years that I was working on what resulted in this book.

Rochester, NY, USA
March, 2008

Jan Drugowitsch

Contents

1 Introduction

The work in this book shows how acquiring a model-centred view to reformulating Learning Classifier Systems (LCS), a rule-based method for machine learning, provides an holistic approach to their design, analysis and understanding. This results in a new methodology for their design and analysis, a probabilistic model of their structure that reveals their underlying assumptions, a formal definition of when they perform optimally, new approaches to their analysis, and strong links to other machine learning methods that have not been available before. The work opens up the prospects of advances in several areas, such as the development of new LCS implementations that have formal performance guarantees, the derivation of representational properties of the solutions that they aim for, and improved performance.

Introducing the work start with a short overview of machine learning, its applications, and the most common problem types that it is concerned with. An example that follows highlights the difference between ad-hoc and model-centred approaches to designing machine learning algorithms and emphasises the advantages of the latter. This is followed by a short introduction to LCS, their applications and current issues. Thereafter, the research question of this work is introduced, together with the approach that is used to approach this question, and a short overview of the chapters that are to follow.

1.1 Machine Learning

Machine learning (ML) is a sub-field of artificial intelligence (AI) that is concerned with methods and algorithms that allow machines to learn. Thus, rather than instructing a computer explicitly with regards to which aspects certain data is to be classified, about relations between entities, or with which sequence of actions to achieve certain goals, machine learning algorithms allow this knowledge to be inferred from a limited number of observations, or a description of the task and its goal.

Their use is manifold, including speech and handwriting recognition, object recognition, fraud detection, path planning for robot locomotion, game playing,

J. Drugowitsch: Des. & Anal. of Learn. Class. Sys.: A Prob. Approach, SCI 139, pp. 1–11 2008.
springerlink.com

natural language processing, medical diagnosis, and many more [19, 168]. There is no universal method to handle all of these tasks, but a large set of different approaches exists, each of which is specialised in particular problem classes.

Probably the most distinct differences between the numerous machine learning methods is the type of task that they can handle, the approach that they are designed with, and the assumptions that they are based upon. Describing firstly a set of common machine learning task types, let us then, based on a simple example, consider two common approaches to how one can develop machine learning algorithms.

1.1.1 Common Machine Learning Tasks

The most common problem types of tasks that machine learning deals with are:

Supervised Learning. In such tasks a set of input/output pairs are available, and the function between the inputs and the associated outputs is to be learned. Given a new input, the learned relation can be used to predict the corresponding output. An example for a supervised learning task is a classification task: given several examples of a set of object properties and the type of this object, a supervised learning approach can be taken to find the relation between the properties and the associated type, which subsequently allows us to predict the object type for a set of properties.

Unsupervised Learning. Unsupervised learning is similar to supervised learning, with the difference that no outputs are available. Thus, rather than learning the relationship between inputs and associated outputs, the learner builds a model of the inputs. Consider a clustering task where several examples of the properties of some object are given and we want to group the objects by the similarity of their properties: this is an unsupervised learning task because the given examples only contain the object properties, but not the group assignment of these objects.

Sequential Decision Tasks. Such tasks are characterised by a set of states, and a set of actions that can be performed in these states, causing a transition to another state. The transitions are mediated by a scalar reward and the aim of the learner is to find the action for each state that maximises the reward in the long run. An example for such a task is in a labyrinth to find the shortest path the goal by assigning each step (that is, each transition) a reward of -1. As the aim is to maximise the reward, the number of steps is minimised. The most common approach to sequential decision tasks is that of dynamic programming and reinforcement learning: to learn the optimal value of a state, which is the expected sum of rewards when always performing the optimal actions from that state, and subsequently to derive the optimal actions from these values.

There exists a wide range of different machine learning methods that deal with each of the problem types. As we are interested in their design, let us consider two possible design approaches to an unsupervised learning task.

1.1.2 Designing an Unsupervised Learning Algorithm

Let us consider the well-known Iris data-set [86] that contains 150 instances of four scalar attribute values and a class assignment each. Each of the four attributes refer to a particular measure of the physical appearance of the flower. Each instance belongs to one of the three possible classes of the plant.

Assume that it is unknown which class each instance belongs to and that we want to design an algorithm that groups the instances into three classes, based on their similarity of appearance that is inferred from the similarity of their attribute values. This task is an unsupervised learning task with the inputs given by the attribute values of each instance.

Ad-Hoc Design of an Algorithm

Let us firstly approach the task intuitively by designing an algorithm that aims at grouping the instances such that the similarity of any two instances within the same group or *cluster* is maximised, and between different clusters is minimised. The similarity between two instances is measured by the inverse squared Euclidean distance[1] between the points that represent these instances in the four-dimensional attribute space, spun by the attribute values.

Starting by randomly assigning each instance to one of the three clusters, the centre of each of these clusters is computed by the average attribute values of all instances assigned to that cluster. To group similar instances into the same cluster, each instance is now re-assigned to the cluster to whose centre it is closest. Subsequently, the centres of the clusters are recomputed. Iterating these two steps causes the distance between instances within the same cluster to be minimised, and between clusters to be maximised. Thus, we have reached our goal. The concept of clustering by using the inverse distance between the data points as a measure of their similarity is illustrated in Figure 1.1(a).

This clustering algorithm is the well-known *K-means* algorithm, which is guaranteed to converge to a stable solution, which is, however, not always optimal [160, 19]. While it is a functional algorithm, it leaves open many question: is the squared Euclidean distance indeed the best distance measure to use? What are the implicit assumptions that are made about the data? How should we handle data where the number of classes is unknown? In which cases would the algorithm fail?

Design of Algorithm by Modelling the Data

Let us approach the same problem from a different perspective: assume that for each Iris class there is a virtual standard instance — something like a prototypical Iris — and that all instances of a class are just noisy instantiations of the

[1] The squared Euclidean distance between two equally-sized vectors $\boldsymbol{a} = (a_1, a_2, \dots)^T$ and $\boldsymbol{b} = (b_1, b_2, \dots)^T$ is given by $\sum_i (a_i - b_i)^2$ and is thus proportional to the sum of squared differences between the vectors' elements (see also Section 5.2). Therefore, two instances are considered as being similar if the squared differences between their attribute values is small.

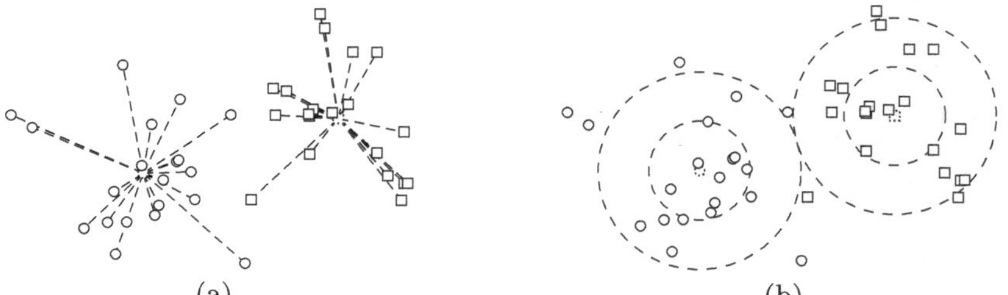

Fig. 1.1. Two different interpretations for clustering a set of data points into two distinct clusters. The circles and squares are data points that are assigned to different clusters. The dashed circle and square represent the centres of the identified clusters. (a) Identifying clusters by minimising the distance between the data points within a cluster, and reassigning data points to the cluster to whose centre they are closest to. The dashed lines indicate the assignment of data points to cluster centres, given by the mean of all data points within the cluster. (b) Interpreting the data points as being generated by Gaussians that are centred on the cluster centres. The two dashed circles around the centres represent the first and the second standard deviation of the generating Gaussian.

standard instance. In other words, assume the attribute values of each instance of a particular class to be *generated* by sampling from a Gaussian that is centred on the attribute values of the standard instance of this class, where this Gaussian models the noisy instantiation process (for an illustration see Figure 1.1(b)). Furthermore, let us assume that each class has generated *all* instances with a certain probability.

This model is completely specified by it *parameters*, which are the centres of the Gaussians and their covariance matrices, and the probability that is assigned to each class. It can be trained by the principle of maximum likelihood by adjusting its parameters such that the probability of having generated all observed instances is maximised; that is, we want to find the model parameters that best explain the data. This can be achieved by using a standard machine learning algorithm known as the *expectation-maximisation (EM)* algorithm [71]. In fact, assuming that each dimension of each Gaussians is independent and has equal variance in each of the dimensions, the resulting algorithm provides the same results as the K-means algorithm [19]; so why take effort of specifying a model rather than using K-means directly?

Reconsidering the questions that were posed in the previous section makes the benefit of having a model clear: it makes explicit the assumptions that are made about the data. This also allows us to specify when the method is likely to fail, which is when we apply it to data that does not conform to the assumptions that the model makes. Furthermore, in this particular example, instances are not assigned to single clusters, but their probability of belonging to either cluster is given. Also, the best number of clusters can be found by facilitating techniques from the field of *model selection* that select the number of clusters that are most suitable to explain the data. Additional advantages are that if Gaussians do not describe the data well, they can be easily replaced by other distributions, while retaining the same techniques to train the model; and if new training methods for that model type become available, they can be used as a drop-in replacement for the ones that are currently used.

Clearly, due to the many advantages of the model-based approach, it should always be preferred to the ad-hoc approach, as the example in this section has demonstrated.

1.2 Learning Classifier Systems

Learning Classifier Systems are a family of machine learning algorithms that are usually designed by the ad-hoc approach. Generally, they can be characterised by handling sequential decision tasks with a rule-based representation and by the use of evolutionary computation methods (for example, [167, 95]), although some variants also perform supervised learning (for example, [161]) or unsupervised learning (for example, [211]), or do not rely on evolutionary computation (for example, [89]).

1.2.1 A Brief Overview

Based on initial ideas by Holland [109, 110, 111, 109] to handle sequential decision tasks and to escape the brittleness of expert systems of that time, LCS initially did not provide the required operational stability that was hoped for [88, 196, 133], until Wilson introduced the simplified versions ZCS [236] and XCS [237], which solved most of the problems of earlier LCS and caused most of the LCS community to concentrate on these two systems and their variants.

Learning Classifier Systems are based on a population of rules (also called the *classifiers*) formed by a condition/action pair, that compete and cooperate to provide the desired solution. In sequential decision tasks, classifiers whose condition *matches* the current states are activated and promote their action. One or several of these classifiers are selected, their promoted action is performed, and the received reward is assigned to these classifiers, and additionally propagated to previously active classifiers that also contributed to receiving the current reward. Occasionally, classifiers of low quality are removed from the current population, and new ones are induced, with their condition and action based on current high-quality classifiers. The aim of replacing classifiers is to improve the overall quality of the classifiers in the population.

Different LCS differ in how they select classifiers, in how they distribute the reward, in whether they additionally maintain an internal state, and in how they evaluate the quality of classifiers. The latter is the most significant difference between early LCS, which based the quality of a classifier on the reward that it contributed to receiving, and the currently most popular LCS, XCS [237], that evaluates the quality of a classifier by how accurate it is at predicting its contribution to the reward.

Shifting from *strength-based* to *accuracy-based* LCS also allowed them to be directly applied to regression tasks [240, 241], which are supervised learning tasks where the output is of interval scale. That also changed the perspective of how LCS handle sequential decision tasks: they act as function approximators for the value function that map the states and actions into the long-run reward that can be expected to be received when performing the action in this state, where the value function estimate is updated by reinforcement learning. By replacing classifiers in the population, LCS aim at finding the best representation of this value function [138].

1.2.2 Applications and Current Issues

Learning Classifier Systems are applied in many areas, such as autonomous robotics (for example, [75, 100]), multi-agent systems (for example, [87, 61]), economics (for example, [221, 169, 3]), and even traffic light control [39]. Particularly in classification tasks, which are supervised learning tasks where the output is of nominal scale, their performance has been found to be competitive with other state-of-the-art machine learning algorithms [98, 152, 7].

Nonetheless, even modern LCS are not free of problems, the most significant being the following:

- Even though initially designed for such tasks, LCS are still not particularly successful in handling sequential decision tasks [11, 12]. This is unfortunate, as "there is a lot of commonality in perspective between the RL community and the LCS community" and more communication between the two communities would be welcome [149].
- Most LCS feature a high number of system parameters, and while the effect of some of them is ill-understood, setting others requires a specialised knowledge of the system. XCS, for example, has 20 partially interacting system parameters [57].
- No LCS features any formal performance guarantees, and even if such guarantees might not always seem particularly important in applications, the choice between a method with such guarantees and an equally powerful method without them will be for the one that features such guarantees.
- There is no knowledge about the assumptions made about the data, and as a result there is also hardly any knowledge about when some LCS might fail.
- Very few direct links between LCS and other machine learning methods are established, which makes the transfer of knowledge for mutual gain hard, if not impossible.
- The general lack of rigour in the design of LCS leads to a lack of their acceptance in the field of machine learning. Together with the previous point this inhibits the exchange of ideas between possibly closely related methods.

These problems concern both practitioners and theoreticians, and solving them should be a top priority in LCS research. Many of them are caused by designing LCS by an ad-hoc approach, with all the disadvantages that we have described before. This was justified when insufficient links were drawn between LCS and other approaches, and in particular when the formalisms were insufficiently developed within other machine learning methods, but now such a position is difficult to argue for.

1.3 About the Model-Centred Approach to LCS

This work arises from the lack of theoretical understanding of LCS, and the missing formality when developing them. Its objective is to develop a formal framework for LCS that lets us design, analyse, and interpret LCS. In that process it focuses on related machine learning approaches and techniques to gain from their understanding and their relation to LCS.

The immediate aim of this work is not to develop a new LCS. Rather it is to give a different perspective on LCS, to increase the understanding and performance of current LCS, and to lay the foundations for a more formal approach to developing new LCS. Neither is the introduced model to be taken as *the* LCS model. It was chosen for demonstrative purposes, due to its similarity to the popular XCS. Other LCS model types can be constructed and analysed by the same approach, to represent other LCS types, such as ZCS.

1.3.1 The Initial Approach

The initial approach was to concentrate on an LCS structure similar to XCSF [240] and to split it conceptually into its function approximation, reinforcement learning and classifier replacement component. Each of these was to be analysed separately but with subsequent integration in mind, and resulted in some studies [78, 83, 155] for the function approximation component and others [79, 80, 81] for the reinforcement learning component.

When analysing these components, the goal-centred approach was followed both pragmatically and successfully: firstly, a formal definition of what is to be learned was given, followed by applying methods from machine learning that reach that goal. The algorithms resulting from this approach are equivalent or improve over those of XCSF, with the additional gain of having a goal definition, a derivation of the method from first principles, and a strong link to associated machine learning methods from which their theoretical analysis was borrowed.

When concentrating on classifier replacement, however, taking this approach was hindered by the lack of a formal definition of what set of classifiers the process of classifier replacement should aim at. Even though some studies aimed at defining the optimal set for limited classifier representations [130, 133, 135], the was still no general definition available. But without having a formally expressible definition of the goal it was impossible to define a method that reaches it.

1.3.2 Taking a Model-Centred View

The definition of the optimal set of classifiers is at the core of LCS: given a certain problem, most LCS aim at finding the set of classifiers that provides the most compact competent solution to the problem.

Fortunately, taking the model-centred view to finding such a definition simplifies its approach significantly: a set of classifiers can be interpreted as a model for the data. With such a perspective, the aim of finding the best set of classifiers becomes that of finding the model that explains the data best. This is the core problem of the field of *model selection*, and many methods have been developed to handle it, such as structural risk minimisation (SRM) [218], minimum description length (MDL) [101], or Bayesian model selection [159].

The advantage of taking the model-centred approach is not only to be able to provide a formal definition for the optimal classifier set. It also reveals the assumptions made about the data, and hence gives us hints about the cases in which the method might excel the performance of other related methods. Also, the model is independent of the method to train it, and therefore we can choose amongst several to perform this task and also acquire their performance guarantees. Furthermore, it makes LCS directly comparable to other machine learning methods that explicitly identify their underlying model.

The probabilistic formulation of the model underlying a set of classifiers was inspired by the related Mixtures-of-Experts model [120, 121], which was extended such that it can describe such a set. This process was simplified by having already analysed the function approximation and reinforcement learning component which allowed the integration of related LCS concepts into the description of the model. In fact, the resulting model allows for expressing both function approximation and reinforcement learning, which makes the model-centred approach for LCS holistic — it integrates function approximation, reinforcement learning and classifier replacement.

1.3.3 Summarising the Approach

In summary, the taken approach is the following: firstly, the relevant problem types are described formally, followed by a probabilistic formulation of a set of classifiers, and how such a model can be trained by methods from adaptive filter theory [105] and statistical machine learning [19, 165], given some data.

The definition of the optimal set of classifiers that is to be sought for is based on Bayesian model selection [19, 119], which requires a Bayesian LCS model. Adding priors to the probabilistic LCS model results in such a Bayesian model. It can be trained by variational Bayesian inference, and two methods of searching the space of classifier sets are introduced. These are then used to demonstrate that defining the best set of classifiers as the one that describes the data best leads to viable results, as preliminary studies have already shown [82].

As handling sequential decision tasks requires the merger of the introduced LCS model with methods from reinforcement learning, it is shown how such a combination can be derived from first principles. One of the major issues of such combinations is their algorithmic stability, and so we discuss how this can be analysed. In addition, some further issues, such as learning tasks that require long action sequences, and the exploration/exploitation dilemma, are discussed in the light of the model.

1.3.4 Novelties

The main novelty of this work are a new methodology for the design and analysis of LCS, a probabilistic model of their structure that reveals their underlying assumptions, a formal definition of when they perform optimally, new approaches to their analysis, and strong links to other machine learning methods that have not been available before.

The methodology is based on taking the model-centred approach to describing the model underlying LCS, and applying standard machine learning methods to train it. It supports the development of new LCS by modifying their model and adjusting the training methods such that they conform to the new model structure. Thus, the introduced approach, if widely adopted, will ensure a formal as well as empirical comparability between approaches. In that sense, it defines a reusable framework for the development of LCS.

1.4 How to Read This Book

Many concepts that are frequently used in this work are introduced throughout the text whenever they are required. Therefore, this work is best read sequentially, in the order that the chapters are presented. However, this might not be an option for all readers, and so some chapters will be emphasised that might be of particular interest for people with a background in LCS and/or ML.

Anyone new to both LCS and ML might want to first do some introductory reading on LCS (for example, [42, 133]) and ML (for example, [19, 102]) before reading this work from cover to cover. LCS workers who are particularly interested in the definition of the optimal set of classifiers should concentrate on Chapters 3 and 4 for the LCS model, Chapter 7 for its Bayesian formulation and the optimality criterion, and Chapter 8 for its application. Those who want to know how the introduced model relates to currently used LCS should read Chapters 3 and 4 for the definition of the model, Chapters 5 and 6 for training the classifiers and how they are combined, and Chapter 9 for reinforcement learning with LCS. People who know ML and are most interested in the LCS model itself should concentrate on the second half of Chapter 3, Chapter 4, and Chapter 7 for its Bayesian formulation.

1.4.1 Chapter Overview

Chapter 2 gives an overview of the initial LCS idea, the general LCS framework, and the problems of early LCS. It also describes how the role of classifiers changed with the introduction of XCS, and how this influences the structure of the LCS model. As our objective is also to advance the theoretical understanding of LCS, the chapter gives a brief introduction to previous attempts that analyse the inner workings of LCS and compares them with the approach that is taken here.

Chapter 3 begins with a formal definition of the problem types, interleaved with what it means to build a model to handle these problems. It then gives a high-level overview of the LCS model by characterising it as a parametric ML model, continuing by discussing how such a model can be trained, and relating it back to the initial LCS idea.

Chapter 4 concentrates on formulating a probabilistic basis for the LCS model by first introducing the Mixture-of-Experts model [121], and subsequently modifying it such that it can describe a set of classifiers in LCS. Certain training issues are resolved by training the classifiers independently. The consequences of this independent training and its relation to current LCS and other LCS model types are discussed at the end of this chapter.

Chapter 5 is concerned with the training of a single classifier, either when all data is available at once, or when it is acquired incrementally. For both cases it is defined what it means for a classifier to perform optimally, based on training the LCS model with respect to the principle of maximum likelihood. For regression models, methods from adaptive filter theory that either are based on the gradient of the cost function, or that directly track the optimum, are derived and discussed, together with a new incremental approach to track the variance estimate of the classifier model. It is also shown how to perform batch and incremental learning with classification models.

Chapter 6 shows how the local model of several classifiers can be combined to a global model, based on maximum likelihood training of the LCS model from Chap. 4. As the approach turns out to be computationally expensive, a set of heuristics are introduced, which are shown to feature competitive performance in a set of experiments. How the content of this chapter differs from closely related previous work [83] is also discussed.

Chapter 7 deals with the core question of LCS: what is the best set of classifiers for a given problem? Relating this question to model selection, a Bayesian LCS model for use within Bayesian model selection is introduced. The model is based on the one elaborated in Chap. 4, but is again discussed in detail with special emphasis on the assumptions that are made about the data. To provide an approach to evaluate the optimality criterion, the second half of this chapter is concerned with deriving an analytical solution to the Bayesian model selection criterion by the use of variational Bayesian inference. Throughout this derivation, obvious similarities to the methods used in Chaps. 5 and 6 are highlighted.

Chapter 8 describes two simple prototype algorithms for using the optimality criterion to find the optimal set of classifiers, one based on Markov Chain Monte Carlo (MCMC) methods, and the other based on GA's. Their core is formed by evaluating the quality of a set of classifiers, for a detailed algorithmic description based on the variational Bayesian inference approach from Chap. 7 is given. Based on these algorithms, the viability of the optimality criterion is demonstrated on a set of regression tasks that highlight some of its features and how they relate to current LCS.

Chapter 9 returns to the treatment of sequential decision tasks after having exclusively dealt with regression and classification tasks in Chaps. 4 to 8. It firstly gives a formal definition of these tasks and their goal, together with an introduction to methods from dynamic programming and reinforcement learning. Then, the exact role of LCS in handling such tasks is defined, and a possible method is partially derived from first principles. This derivation clarifies some of the current issues of how to correctly perform RL with XCS(F), which is discussed in more detail. Based on the LCS model, it is also shown how the stability of LCS with RL can be studied, together with how to handle learning long action sequences and the trade-off between exploring the space and exploiting current knowledge.

Chapter 10 summarises the work and puts it into the perspective of the initial objective.

2 Background

To give the reader a perspective on what characterises LCS exactly, and to which level they are theoretically understood, this chapter gives some background on the initial ideas behind designing LCS, and describes what can be learned from their development over the years and the existing theoretical descriptions. As an example of a current LCS we will concentrate on XCS [237] — not only because it is at the time of this writing the most used and best understood LCS, but also because it is in its structure similar to the LCS model that is developed in this book. Therefore, when discussing the theoretical understanding of LCS, special emphasis is put on XCS and its variants, in addition to describing general approaches that have been used to analyse LCS.

Even though the presented work borrows numerous concepts and methods from statistical machine learning, these methods and their background are not described in this chapter, as this would deviate too much from the main topic of interest. However, whenever using new concepts and applying new methods, a short discussion about their underlying ideas is given at adequate places throughout the text. A more thorough description of the methods used in this work can be found in a wide range of textbooks [17, 19, 102, 105, 164, 165], of which the ones by Bishop [19] and Bertsekas and Tsitsiklis [17] are particularly relevant to the content of this book.

In general, LCS describe a very flexible framework that differs from other machine learning methods in its generality. It can potentially handle a large number of different problem types and can do so by using a wide range of different representations. In particular, LCS have the potential of handling the complex problem class of POMDPs (as described below) that even the currently most powerful machine learning algorithms still struggle with. Another appealing feature is the possible use of human-readable representations that simplify the introspection of found solutions without the requirement of converting them into a different format. Their flexibility comes from the use of evolutionary computation techniques to search for adequate substructures of potential solutions. In combination, this makes LCS an interesting target for theoretical investigation, in particularly to promote a more principled approach to their design.

J. Drugowitsch: Des. & Anal. of Learn. Class. Sys.: A Prob. Approach, SCI 139, pp. 13–28, 2008.
springerlink.com

This chapters begins with a general overview of the problems that were the prime motivator for the development of LCS. This is followed by a review of the ideas behind LCS, describing the motivation and structure of Holland's first LCS, the CS-1 [116]. Many of the LCS that followed had a similar structure and so instead of describing them in detail, Sect. 2.2.5 focuses on some of the problems that they struggled with. With the introduction of XCS [237] many of these problems disappeared and the role of the classifier within the population was redefined, as discussed in Sect. 2.3. However, as our theoretical understanding even of XCS is still insufficient, and as this work aims at advancing the understanding of XCS and LCS in general, Sect. 2.4 gives an overview over recent significant approaches to the theoretical analysis of LCS, before Sect. 2.5 puts the model-based design approach into the general LCS context.

2.1 A General Problem Description

Consider an agent that interacts with an environment. At each discrete time step the environment is in a particular *hidden state* that is not observable by the agent. Instead, the agent senses the *observable state* of the environment that is stochastically determined by its hidden state. Based on this observed state, the agent performs an action that changes the hidden state of the environment and consequently also the observable state. The hidden state transitions conform to the Markov property, such that the current hidden state is completely determined

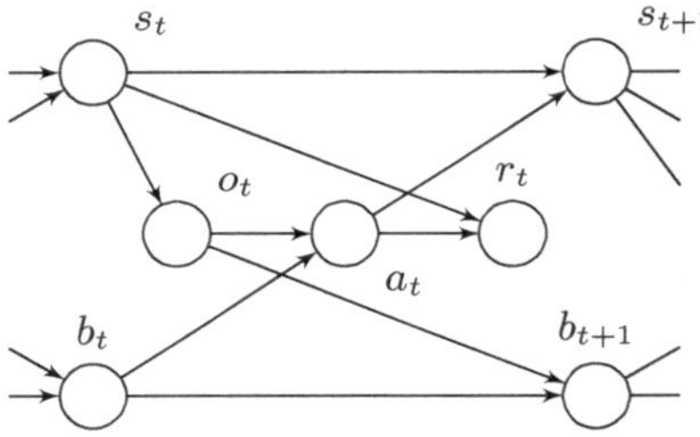

(a)

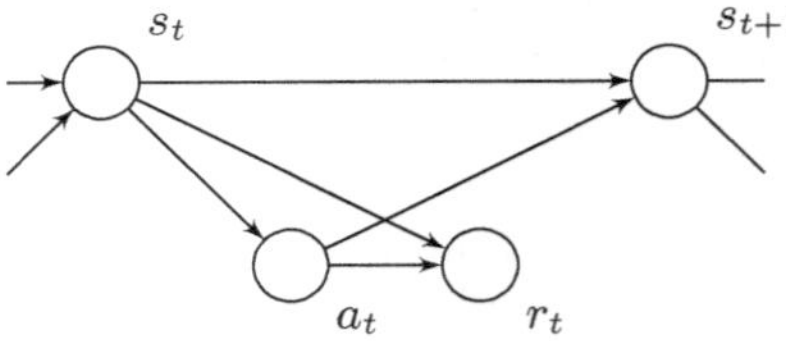

(b)

Fig. 2.1. The variables of a POMDP and an MDP involved in a single state transition from state s_t to state s_{t+1} after the agent performs action a_t and receives reward r_t. Each node represents a random variable, and each arrow indicates a dependency between two variables. (a) shows the transition in a POMDP, where the state s_t is hidden from the agent which observes o_t instead. The agent's action depends on the agent's belief b_t about the real state of the environment and the currently observed state o_t. Based on this action and the environment's hidden state, a reward r_t is received and the environment performs a transition to the next state s_{t+1}. Additionally, the agent update its belief b_{t+1}, based on the observed state o_t. (b) shows the same transition in an MDP where the agent can directly observe the environment's state s_t, and performs action a_t based on that. This causes the agent to receive reward r_t and the environment to perform a state transition to s_{t+1}.

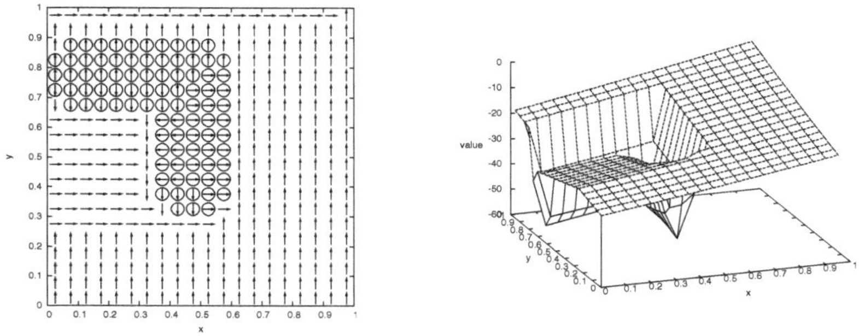

(a) (b)

Fig. 2.2. Optimal policy and value function for a discretised version of the "puddleworld" task [208]. The agent is located on a 1x1 square and can perform steps of size 0.05 into either of four directions. Each step that the agent performs results in a reward of -1, expect for actions that cause the agent to end up in a puddle, resulting in a reward of -20. The zero-reward absorbing goal state is in the upper right corner of the square. Thus, the task is to reach this state in the smallest number of steps while avoiding the puddle. The circles in (a) show the location of the puddle. (b) illustrates the optimal value function for this task, which gives the maximum expected sum of rewards for each state, and clearly shows the impact of the high negative reward of the puddle. Knowing this value function allows constructing the optimal policy, as given by the arrows in (a), by choosing the action in each state that maximises the immediate reward and the value of the next state.

by the previous hidden state and the performed action. For each such state transitions the agent receives a scalar *reward* or *payoff* that can depend on the previous hidden and observable state and the chosen action. The aim of the agent is to learn which actions to perform in each observed state (called the *policy*) such that the received reward is maximised in the long run.

Such a task definition is known as a Partially Observable Markov Decision Process (POMDP) [122]. Its variables and their interaction is illustrated in Fig. 2.1(a). It is able to describe a large number of seemingly different problems types. Consider, for example, a rat that needs to find the location of food in a maze: in this case the rat is the agent and the maze is the environment, and a reward of -1 is given for each movement that the rat performs until the food is found, which leads the rat to minimise the number of required movements to reach the food. A game of chess can also be described by a POMDP, where the white player becomes the agent, and the black player and the chess board define the environment. Further examples include path planning, robot control, stock market prediction, and network routing.

While the POMDP framework allows the specification of complex tasks, finding their solution is equally complicated. Its difficulty arises mostly due to the agent not having access to the true state of the environment. Thus, most of the recent work in LCS has focused on a special case of POMDP problems that treat the hidden and observable states of the environment as equivalent. Such problems are known as Markov Decision Processes (MDPs), as illustrated in Fig. 2.1(b), and are dealt with in more detail in Chap. 9. They are approached by LCS by the use of reinforcement learning which is centred on learning the expected sum of rewards for each state when following the optimal policy. Thus, the intermediate aim is to learn a *value function* that maps the states into their respective expected sum of rewards, which is a univariate regression problem. An example of such a value function and the policy derived from it is shown in Fig. 2.2.

Even though the ultimate aim of LCS is to handle MDPs and POMDPs, they firstly need to be able to master univariate regression problems. With that in mind, this work focuses on LCS models and approaches to handle such problems, and how the same approach can equally well be applied to multivariate regression and classification problems. In addition, a separate chapter describes how the same approach can be potentially extended to handle MDPs, and which additional considerations need to be made. Nonetheless, it needs to be emphasised that the theoretical basis of applying LCS to MDPs and POMDPs is still in its infancy, and further work on this topic is urgently required. Still, due to their initial focus on POMDPs, these are the tasks that will be considered when introducing LCS.

2.2 Early Learning Classifier Systems

The primary problems that LCS were designed to handle are sequential decision tasks that can be defined by POMDPs. In LCS it is assumed that each observed state is a composite element that is identified by the collection of its features, such that the agent is able to associate the choice of action with certain features of the state. This allows the agent to generalise over certain features and possibly also over certain states when defining its choice of action for each of the states. The aim of LCS is not only so find the optimal policy for a given POMDP, but also to exploit the possible generalisations to find the minimal solution representation.

At the time of their initial introduction the link between the tasks that LCS aim at solving and POMDPs was not yet established. As a consequence, there was neither a clear understanding that the regression task underlying value function learning is an intermediate step that needs to be achieved in order to efficiently learn optimal policies for given POMDPs, nor were objective functions available that captured all facets of their aim. Rather, their design was approached by the definition of sub-problems that each LCS has to solve, and a description of the various LCS subsystems. Only over the last 15 years the relation between LCS, MDPs and regression tasks became clearer, which resulted in exciting developments of new LCS and a more transparent understanding of their structure. The chronological introduction to LCS aims at capturing this paradigm shift.

2.2.1 Initial Idea

Although some of Holland's earlier work [109, 110, 111] had already introduces some ideas for LCS, a more specific framework was finally defined in [114].

The motivation was to escape the brittleness of popular expert systems of that time by evolving a set of cooperative and competing rules in a market-inspired economy. In particular, Holland addressed the following three problems [115]:

Parallelism and coordination. Complex situations are to be decomposed into simpler building blocks, called *rules*, that handle this situation cooperatively. The problem is to provide for the interaction and coordination of a large number of rules that are active simultaneously.

Credit assignment. To decide which rules in a rule-based system are responsible for its success, one needs to have a mechanism which accredits each rule with its responsibility to that success. Such mechanism become particularly complex when rules act collectively, simultaneously and sequentially. Furthermore, complex problems do not allow for exhaustive search over all possible rule combinations, and so this mechanism has to operate locally rather than globally.

Rule discovery. Only in toy problems can one evaluate all possible rules exhaustively. Real-world problems require the search for better rules based on current knowledge to generate plausible hypotheses about situations that are currently poorly understood.

Holland addressed these questions by proposing a rule-based system that can be viewed as a message processing system acting on a current set of messages, either internal or generated by a set of detectors to the environment and thus representing the environment's observable state. Credit assignment is handled by a market-like situation with bidders, suppliers and brokers. Rule discovery facilitates an evolutionary computation-based process that discovers and recombines building blocks of previously successful rules.

While the original framework is not replicate in full detail, the following section gives an overview of the most common features among some of the LCS implementations derived from this framework. A detailed overview and comparison of different early LCS is given in Chap. 2 of Barry's Ph.D. thesis [10].

2.2.2 The General Framework

In LCS the agent's behaviour is determined by a set of classifiers (Holland's rules), each consisting of at least one condition and an action. On sensing the state of the environment though a detector, the sensor reading of the agent is injected as a message into an internal message list, containing both internal and external messages. Classifier conditions are then tested for matching any of the messages on the message list. The matching classifiers are activated, promoting their actions by putting their message on the message list. The message on the list can be either interpreted to perform actions or to be kept on the list to act as an input for the next cycle. If several actions are promoted at the same time, a *conflict resolution subsystem* decides which action to perform. Once this is completed, the cycle starts again by sensing the new state of the environment. Figure 2.3 provides a schematic illustration of the message flow in LCS with a single message list.

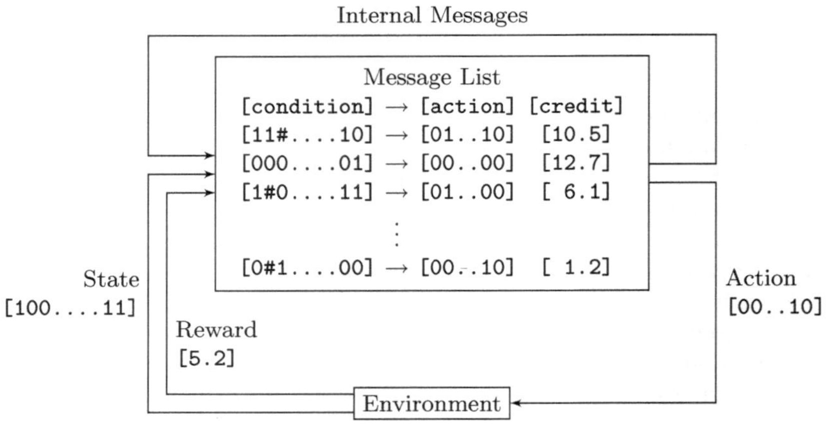

Fig. 2.3. Schematic illustration of an LCS with a single message list. Its operation is described in the main text.

All of the messages are usually encoded using binary strings. Hence, to allow matching of messages by classifier conditions, we are required to encode conditions and actions of classifiers as binary strings as well. A classifier can generalise over several different input messages by introducing *don't care* symbols "#" into its condition that match both both 1's and 0's in the corresponding position of the input message. The condition "0#1", for example, matches inputs "001" and "011" equally. Similarly, actions of the same length as classifier conditions can also contain the "#" symbol (in this case called *pass-through*), which implies that specific bits of the matching message are passed though to the actions, allowing a single classifier to perform different actions depending on the input message. The latter feature of generalisation in the classifier actions is much less frequently used than generalisation in the classifier condition.

The description above covers how the agent decides which actions to perform (called the *performance subsystem*) but does not explain how such an agent can react to external reward to optimise its behaviour in a given environment. Generally, the behaviour is determined by the population of classifiers and the conflict resolution subsystem. Hence, considering that the functionality of the conflict resolution subsystem is determined by properties of the classifiers, learning can be achieved by evaluating the quality of each classifier and aiming at a population that only contains classifiers of high quality. This is achieved by a combination of the *credit allocation subsystem* and the *rule induction subsystem*. The role of the former is to distribute externally received reward to classifiers that promoted the actions responsible for receiving this reward. The latter system creates new rules based on classifiers with high credit to promote the ones that are assumed to be of good quality.

2.2.3 Interacting Subsystems

To summarise, LCS aim at maximising external reward by an interaction of the following subsystems:

Performance Subsystem. This subsystem is responsible for reading the input message, activating the classifiers based on their condition matching any message in the message list, and performing actions that are promoted by messages that are posted by the active classifiers.

Conflict Resolution Subsystem. If the classifiers promote several conflicting actions, this subsystem decides for one action, based upon the quality rating of the classifiers that promote these actions.

Credit Allocation Subsystem. On receiving external reward, this subsystem decides how this reward is credited to the classifiers that promoted the actions causing the reward to be given.

Rule Induction Subsystem. This subsystem creates new classifiers based on current high-quality classifiers in the population. As the population size is usually limited, introducing new classifiers into the population requires the deletion of other classifiers from the population, which is an additional task of this subsystem.

Although the exact functionality for each of the systems was given in the original paper [114], further developments introduce changes to the operation of some subsystems, which is why only a general description is given here. Section 2.2.5 discusses some properties of these LCS, and point out the major problems that led the way to a new class of LCS that feature major performance improvements.

2.2.4 The Genetic Algorithm in LCS

Holland initially introduced Learning Classifier Systems as an extension of Genetic Algorithms to Machine Learning. GA's are a class of algorithms that are based on the principles of evolutionary biology, driven by mutation, selection and recombination. In principle, a population of candidate solutions is evolved and, by allowing more reproductive opportunities to fitter solutions, the whole population is pushed towards higher fitness. Although GA's were initially applied as function optimisers (for example [95]), Holland's idea was to adapt them to act as the search process in Machine Learning, giving rise to LCS.

In an LCS, the GA operates as the core of the rule induction subsystem, aiming at replicating classifiers of higher fitness to increase the quality of the whole population. New classifiers are created by selecting classifiers of high quality from the population, performing cross-over of their conditions and actions and mutating their offspring. The offspring is then reintroduced into the population, eventually causing deletion of lower quality classifiers due to bounded population size. Together with the credit allocation subsystem, which is responsible for rating the quality of the classifiers, this process was intended to generate a set of classifiers that promote optimal behaviour in a given environment.

2.2.5 The Problems of Early LCS

In most earlier classifier systems[1] each classifier in the population had an associated scalar strength. This strength was assigned by the credit allocation subsystem and acted as the fitness and hence quality rating of the classifier.

On receiving external reward, this reward contributed to the strength of all classifiers that promoted the action leading to that reward. Learning immediate reward alone is not sufficient, as sequential decision tasks might require a sequence of actions before any reward is received. Thus, reward needs to be propagated back to all classifiers in the action sequence that caused this reward to be received. The most popular scheme to perform this credit allocation was the *Implicit Bucket Brigade* [112, 186, 187].

Even though this schema worked fairly well, performance in more complicated tasks was still not satisfactory. According to Kovacs [133, 132], the main problem was the use of classifier strength as its reproductive fitness. This causes only high-reward classifiers to be maintained, and thus the information about low-rewarding areas of the environment is lost, and with it the knowledge about if the performed actions are indeed optimal. A related problem is that if the credit assignment is discounted, that is, if classifiers that are far away from the rewarding states receive less credit for causing this reward, then such classifiers have a lower fitness and are more likely to be removed, causing sub-optimal action selection in areas distant to rewarding states. Most fundamentally, however, is the problem that if the classifier strength is not shared between the classifiers, then environments with layered payoff will lead to the emergence of classifiers that match a large number of states, despite them not promoting the best action in all of those states. Examples for such environments are the ones that describe sequential decision tasks. It needs to be pointed out that Kovacs does not consider fitness sharing in his investigations, and that according to Bull and Hurst [34] optimal performance can be achieved even with strength-based fitness as long as fitness sharing is used, but "[...] suitable system parameters must be identified for a given problem", and how to do this remains open to further investigation.

It has also been shown by Forrest and Miller [88] that the stochastic selection of matching classifiers can lead to instabilities in any LCS that after each performed action reduces the strength of all classifiers by a *life tax* and has a small message list such that not all active classifiers can post their messages at once. In addition to these problems, Smith [196] investigated the emergence of parasitic classifiers that do not directly contribute to action selection but gain from the successful performance of other classifiers in certain LCS types with internal message lists.

Even though various taxation techniques, fitness sharing [34], and other methods have been developed to overcome the problems of overly general and parasitic classifiers, LCS still did not feature satisfactory performance in more complex tasks. A more drastic change was required.

---

[1] See [10, Chap. 2] for a description and discussion of earlier LCS.

2.3 The LCS Renaissance

Before introducing XCS, Wilson developed ZCS [236] as a minimalist classifier systems that aimed through its reductionist approach to provide a better understanding of the underlying mechanisms. ZCS still uses classifier fitness based on strength by using a version of the implicit bucket brigade for credit assignment, but utilises fitness sharing to penalise overly general classifiers.

Only a year after having published ZCS, Wilson introduced his XCS [237] that significantly influenced future LCS research. Its distinguishing feature is that the fitness of a classifier is not its strength anymore, but its accuracy in predicting the expected reward[2]. Consequently, XCS does maintain information about low-rewarding areas of the environment and penalises classifiers that match overly large areas, as their reward prediction becomes inaccurate. By using a niche GA that restricts the reproduction of classifiers to the currently observed state and promote the performed action, and removing classifiers independent of their matching, XCS prefers classifiers that match more states as long as they are still accurate, thus aiming towards optimally general classifiers[3]. More information about Wilson's motivation for the development, and an in-depth description of its functionality can be found in Kovacs' Ph.D. thesis [133]. A short introduction to XCS from the model-based perspective is given in App. B.

After its introduction, XCS was frequently modified and extended, and its theoretical properties and exact working analysed. This makes it, at the time of this writing, the most used and best analysed LCS available. These modifications also enhanced the intuitive understanding of the role of the classifiers within the system, and as the proposed LCS model borrows much of its design and intuition from XCS, the following sections give further background on the role of a classifier in XCS and its extensions. In the following, only single-step tasks, where a reward is received after each action, are considered. The detailed description of multi-step tasks is postponed to Chap. 9.

2.3.1 Computing the Prediction

Initially, each classifier in XCS only provided a single prediction for all states that it matches, independent of the nature of these states [237, 238, 239]. In XCSF [240, 241], this was extended such that each classifier represents a straight line and thus is able to vary its prediction over the states that it matches, based on the numerical value of the state. This concept was soon picked up by other researchers and was quickly extended to higher-order polynomials [141, 142, 143],

[2] Using measures different than strength for fitness was already suggested before but was never implemented in the form of pure accuracy. Even in the first LCS paper, Holland suggested that fitness should be based not only on the reward but also on the consistency of the prediction [111], which was also implemented [116]. Later, however, Holland focused purely on strength-based fitness [237]. A further LCS that uses some accuracy-like fitness measure is Booker's GOFER-1 [21].

[3] Wilson and others calls *optimally general* classifiers *maximally general* [237], which could lead to the misinterpretation that these classifiers match all states.

to the use of neural networks to compute the prediction [35, 175, 176, 156], and even Support Vector Machines (SVMs) [157].

What became clear was that each classifier approximates the function that is formed by a mapping from the value of the states to their associated payoffs, over the states that it matches [241]. In other words, each classifier provides a localised model of that function, where the localisation is determined by the condition and action of the classifier — even in the initial XCS, where the model is provided by a simple averaging over the payoff of all matched states [78]. This concept is illustrated in Fig. 2.4.

2.3.2 Localisation and Representation

Similar progress was made in how the condition of a classifier can be represented: while XCS initially used ternary strings for that task [237, 238], the representational repertoire was soon increased to real-numbered interval representations to handle real-valued states [239], as a prerequisite to function approximation with computed predictions [240, 241]. Amongst other representations used with XCS(F) to determine the matching of a classifier are now hyper-ellipsoids [41, 41], neural networks [38], S-expressions [144], and convex hulls [147]. Fuzzy classifier representations [60] additionally introduce matching by degree which — despite a different approach to their design – makes them very similar to the model that is presented here.

The possibility of using arbitrary representations in XCS(F) to determine matching of a classifier was highlighted in [241]. In fact, classifiers that model the payoff for a particular set of states and a single action can conceptually be seen as perform matching in the space of states *and* actions, as they only model the payoff if their condition matches the state, *and* their action is the one that is performed. Similarly, classifiers without actions, such as the ones used for function approximation [240, 241], perform matching in the space of states alone.

2.3.3 Classifiers as Localised Maps from Input to Output

To summarise, classifiers in XCS are localised models of the function that maps the value of the states to their associated payoffs. The localisation is determined by the condition/action pair that specifies which states and which actions of the environment are matched.

When LCS are applied to regression tasks, the standard machine learning terminology is to call the state/action pair the *input* and the associated payoff the *output*. Thus, the localised model of a classifier provides a mapping from the input to the output, and its localisation is determined by the input alone, as shown in Fig. 2.4.

Sequential decision tasks can be mapped onto the same concept by specifying an input by the state/action pair, and its associated output by the payoff. Similarly, in classification tasks the input is given by the attributes, and the output is the class label, as used in UCS [161], which is a variant of XCS specialised

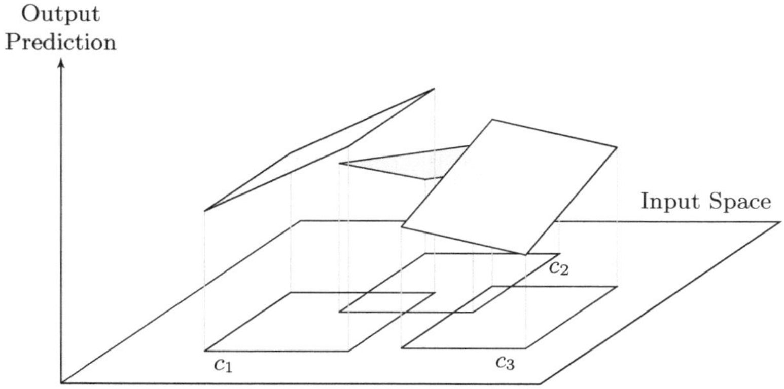

Fig. 2.4. Classifiers as localised maps from the input space into the output space. The illustration shows three classifiers c_1, c_2, and c_3 that match different areas of the input space. Their location in the input space is determined by the classifier's condition, which, in this example, is given by intervals on the coordinates of the input space. Each classifier provides an input-dependent prediction of the output. In this illustration, the classifiers form their prediction through a linear combination of the input space coordinates, thus forming planes in the input/output space.

for classification tasks. Therefore, the concept of classifiers providing a localised model that maps inputs to outputs generalises over all LCS tasks, which will be exploited when developing the LCS model.

In the light of the above, calling the localised models "classifiers" is a misnomer, as they are not necessarily classification models. In fact, their use for classification has only emerged recently, and before that they have been mostly represented by regression models. However, to make this work easily accessible, the LCS jargon of calling these models "classifiers" will be maintained. The reader, however, is urged to keep in mind that this term is not related to classification in the sense discussed in this book.

2.3.4 Recovering the Global Prediction

Several classifiers can match the same input but each might provide a different predictions for its output. To get a single output prediction for each input, the classifiers' output predictions need to be combined, and in XCS and all its variants this is done by a weighted average of these predictions, with weights proportional to the fitness of the associated classifiers [237, 238].

The component responsible for combining the classifier predictions in XCS and LCS has mostly been ignored, until is was shown that combining the classifier predictions in proportion to the inverse variance of the classifier models gives a lower prediction error than when using the inverse fitness [83]. At the same time, Brown, Kovacs and Marshall have demonstrated that the same component can be improved in UCS by borrowing concepts from ensemble learning [29].

Even though rarely discussed, the necessity of combining the classifier predictions is an important component of the developed model, as will become apparent in later chapters. This is a particular property of XCS-like models that treat their classifiers in some sense independently and thus require their combination at a later stage. For other LCS model types (for example ones that resemble ZCS), this might not be the case, as will be discussed in the following chapter.

2.3.5 Michigan-Style vs. Pittsburgh-Style LCS

What has been ignored so far is that there are in fact two distinct types of LCS: Michigan-style and Pittsburgh-style LCS. In Michigan-style LCS all classifiers within a population cooperate to collectively provide a solution. Examples are the first LCS, Cognitive System 1 (CS-1) [116], SCS [95], ZCS [236] and XCS [237]. In the less common Pittsburgh-style LCS several sets of classifiers compete against each other to provide a solution with a single fitness value for the set, with examples for such systems given by LS-1 [198, 199, 200], GALE [151] and CCS [153, 154].

Even though "Michigan and Pittsburgh systems are really quite different approaches to learning [...]" [133], they share the common goal of finding sets of classifiers that provide a solution to the task at hand. Consequently, it is asserted that their classifier populations can be represented by the same LCS model, but their way of improving that model is different.

In developing the LCS model we do not distinguish between the two styles, not even when defining the optimal set of classifiers in Chap. 7, in order to emphasise that they are just two different implementations that have the same goal. The point at which this distinction has to be made is as soon as implementation details will be discussed in Chap. 8.

2.4 Existing Theory

As with the creation of a model for LCS the aim is to also advance the theoretical understanding of LCS in general, let us review some previous theoretical work in LCS. Starting with theoretical approaches that consider all LCS subsystems at once, the focus subsequently shifts to work that concentrates on the GA in LCS, followed by discussing approaches that have analysed the function approximation and RL side of LCS.

2.4.1 The Holistic View

The first and currently only LCS model that allows studying the interaction with the environment and generalisation in the same model was developed by Holland just after the introduction of the LCS framework [113].

He describes the set of states that the system can take by combining all possible environmental states and internal states of the LCS, and defines a transition matrix that describes the Markov chain probabilities of transiting from one system state to another. Thus, changes in the environment and the LCS are tracked simultaneously.

Environmental similarities are exploited in the model by partitioning the Markov matrix into equivalence classes to get a sub-Markov matrix that collapses similar states into one. From this, reset times, upper bounds on expected experiment repetition times and other properties can be derived.

The model was created before the emergence of modern RL[4] and so cannot refer to its theoretical advances, and was not updated to reflect those. Additionally, the inclusion of the LCS state into the model causes the number of states to be uncountable due to the real-valued parametrisation of LCS. Thus, it is unclear if the model will provide significant advances in the understanding of LCS. Rather, one should rely on RL theory to study the performance of LCS in sequential decision tasks, as discussed in Chap. 9.

2.4.2 Approaches from the Genetic Algorithm Side

As many researchers consider LCS as Genetic-based Machine Learners (GBML), they are most frequently analysed from the GA perspective. Particularly when considering single-step problems, when each action is immediately mediated by a reward, the task is a regression task and does not require an RL component. Due to its similarity to the LCS model that will be introduced, we will mainly consider the analyses performed on XCS. Note, however, that none of these analyses is of direct importance to the work presented here, as they study a single algorithm that performs a task which is here only define by its aim, rather than by how it is performed. Nonetheless, the analysis of XCS has given valuable insights into the set of classifiers that XCS aims at evolving – a topic that is reconsidered in Sect. 7.1.1.

Single-Step Tasks

Single-step problems are essentially regression tasks where XCS aims at learning a complete mapping from the input space to the output space. In XCS, such problems are handled by an RL method that for these tasks reduces to a gradient-based supervised learning approach, as will be shown in Sects. 5.3.3 and 5.3.4.

Most of the analysis of XCS in single-step tasks has been performed by Butz et al. in an ongoing effort [51, 53, 44, 56, 49, 46, 50, 58] restricted to binary string representations, and using a what they call *facet-wise approach*. Their approach is to look at single genetic operators, analyse their functionality and then assemble a bigger picture from the operators' interaction, sometimes taking simplifying assumptions to make the analysis tractable.

They analyse the various evolutionary pressures in XCS, showing that the *set pressure* pushes towards less specific classifiers [53], as already conjectured in

---

[4] "Emergence of modern RL" refers to Sutton's development of TD [207] and Watkin's Q-Learning [228].

Wilson's *Generalization Hypothesis* [237]. Mutation is shown to push towards 50% or 66% specificity, and no quantitative values are derived for the fitness and subsumption pressure. Overall, it is qualitatively shown that XCS pushes towards optimally general classifiers, but the quantitative results should be treated with care due to their reliance of several significant assumptions.

In a subsequent series of work [51, 44, 46, 58], Butz et al. derive various time and population bounds to analyse how XCS scales with the size of the input and the problem complexity, where the latter expresses how strongly the values of various input bits depend on each other. Combining these bounds, they show that the computational complexity of XCS grows linearly with respect to the input space size and exponentially with the problem complexity. Thus they state that XCS is a Probably Approximately Correct (PAC)[5] learner [58]. While this claim might be correct, the work that is presented is certainly not sufficient to support it – in particular due to the simplifying assumptions made to derive these bounds. More work is required to formally support this claim.

In addition to analysing the genetic pressures and deriving various bounds, a wide range of further work has been performed, like the empirical and theoretical analysis of various selection policies in XCS (for example [56, 49, 85, 181]), or improving the XCS and UCS performance of classification problems with strong class imbalance [178, 179, 180]. None of these studies is directly related to the work presented here and therefore will not be discussed in detail.

Multi-Step Tasks

Very little work been has performed to analyse the GA in multi-step problems, where a sequence of action rather than a single action lead to the reward that is to be maximised. The only relevant study might be the one by Bull [31], where he has firstly shown in single-step tasks that overly general classifiers are supported in strength-based LCS but not in accuracy-based LCS. The model is then extended to a 2-step task, showing that "effective selection pressure can vary over time, possibly dramatically, until an equilibrium is reached and the constituency of the co-evolving match sets stop changing" [31]. The model even shows a pressure towards lower payoff rules in some cases, although this might be an artifact of the model.

2.4.3 Approaches from the Function Approximation Side

XCS was, for the first time, used for function approximation in XCSF [240] by allowing classifiers to compute their predictions from the values of the inputs. It has been shown that due to the use of gradient descent, such classifiers might only converge slowly to the correct model [142, 143], and a training algorithms based on Recursive Least Squares (RLS) [105], and the Kalman filter [78] were proposed to improve their speed of convergence.

[5] A PAC learner is guaranteed to have a low generalisation error with a high probability. Thus, it is probably approximately correct. See [127] for more information.

How classifiers are combined to form the global prediction is essential to function approximation but has been mostly ignored since the initial introduction of XCS. Only recently, new light has been shed on this component [83, 29], but there is certainly still room for advancing its understanding.

2.4.4 Approaches from the Reinforcement Learning Side

Again concentrating on XCS, its exact approach to performing reinforcement learning has been discussed by Lanzi [138] and Butz, Goldberg and Lanzi [45]. In the latter study, Butz et al. show the parallels between XCS and Q-Learning and aim at adding gradient descent to XCS's update equations. This modification is additionally published in [47], and was later analysed many times [223, 224, 142, 79, 140, 139], but with mixed results. Due to the current controversy about this topic, its detailed discussion to Sect. 9.3.6.

Another study that is directly relevant to RL is the limits of XCS in learning long sequences of actions [11, 12]. As this limitation emerges from the type of classifier set model that XCS aims at, it is also relevant to this work, and thus will be discussed in more detail in Sect. 9.5.1.

There has been no work on the stability of XCS when used for sequential decision tasks, even though such stability is not guaranteed (for example, [25]). Wada et al. claim in [223, 224] that XCS does not perform Q-Learning correctly – a claim that is question in Sect. 9.3.6 – and consequently introduce a modification of ZCS in [224] that makes it equivalent to Q-Learning with linear function approximation. They demonstrate its instability in [222], and present a stable variant in [224]. As described in Sect. 4.6, their LCS model is not compatible with XCS, as they do not train their classifiers independently. For an XCS-like model structure, stability considerations are discussed in Sect. 9.4.

2.5 Discussion and Conclusion

LCS have come a long way since their initial introduction, and still continue to be improved. From this historical overview of LCS and in particular XCS we can see that LCS are traditionally approached algorithmically and also analysed as such. Even in the first LCS, CS-1, most of the emphasis is put on how to approach the problem, and little on the problem itself. Given that many non-LCS approaches handle the same problem class (for example, [17, 209]), an algorithmic description of LCS emphasises the features that distinguishes LCS from non-LCS methods. But even with such statements one needs to be careful: considering the series of 11 short essays under the title "What is a Learning Classifier System?" [115] it becomes clear that there is no common agreement about what defines an LCS.

Based to these essays, Kovacs discusses in [134] if LCS should be seen as GA's or algorithms that perform RL. He concludes that while strength-based LCS are more similar to GA's, accuracy-based LCS shift their focus more towards RL. Thus, there is no universal concept that applies to all LCS, particularly when considering that there exist LCS that cannot handle sequential decision tasks (for example, UCS [161]), and others that do not have a GA (for example, MACS [92, 89]).

The extensive GA-oriented analysis in recent years has shed some light into which problems XCS can handle and where it might fail, and how to set some of its extensive set of system parameters. Nonetheless, questions still emerge if accuracy-based fitness is indeed better than strength-based fitness in all situations, or if we even need some definition of fitness at all [22]? Furthermore, the correct approach to reinforcement learning in LCS is still not completely clear (see Sect. 9.3.6). In any case, what should be emphasised is that both the GA and RL in LCS are just methods to reach some goal, and without a clear definition of this goal it is impossible to determine if any method is ever able to reach it.

This is why the promoted approach for the analysis of LCS differs from looking further at existing algorithms and figuring out what they actually do and how they might be improved. Rather, as already alluded to in the previous chapter, it might be better to take a step back and concentrate firstly on the problem itself before considering an approach to finding its solution. This requires a clear definition of the problems that are to be solved, followed by the formulation a model that determines the assumptions that are made about the problem structure. To ensure that the resulting method can be considered as an LCS, the design of this model is strongly inspired by the structure of LCS, and in particular XCS.

Having a problem and an explicit model definition allows for the application of standard machine learning methods to train this model. The model in combination with its training defines the method, and as we will see, the resulting algorithms are indeed close to the ones of XCS, but with all the advantages that were already described in the previous chapter. Additionally, we do not need to explicitly handle questions about possible fitness definitions or the correctness of the reinforcement learning method used, as they emerge naturally through deriving training methods for the model. From that perspective, the proposed approach handles many of the current issues in LCS more gracefully and holistically than previous attempts.

3 A Learning Classifier Systems Model

Specifying the model that is formed by a set of classifiers is central to the model-based approach. On one hand it explicitly defines the assumptions that are made about the problem that we want to solve, and on the other hand it determines the training methods that can be used to provide a solution. This chapter gives a conceptual overview over the LCS model, which is turned into a probabilistic formulation in the next chapter.

As specified in Chap. 1, the tasks that LCS are commonly applied to are regression tasks, classification tasks, and sequential decision tasks. The underlying theme of providing solutions to these tasks is to build a model that maps a set of observed inputs to their associated outputs. Taking the generative view, we assume that the observed input/output pairs are the result of a possibly stochastic process that generates an output for each associated input. Thus, the role of the model is to provide a good representation of the data-generating process.

As the data-generating process is not directly accessible, the number of available observations is generally finite, and the observations themselves possibly noisy, the process properties need to be induced from these finite observations. Therefore, we are required to make assumptions about the nature of this process which are expressed through the model that is assumed.

Staying close to the LCS philosophy, this model is given by a set of localised models that are combined to a global model. In LCS terms the localised models are the classifiers with their localisation being determined by which inputs they match, and the global model is determined by how the classifier predictions are combined to provide a global prediction. Acquiring such a model structure has several consequences on how it is trained, the most significant being that it is conceptually separable into a two-step procedure: firstly, we want to find a good number of classifiers and their localisation, and secondly we want to train this set of classifiers to be a seemingly good representation of the data-generation process. Both steps are closely interlinked and need to be dealt with in combination.

A more detailed definition of the tasks and the general concept of modelling the data-generating process is given in Sect. 3.1, after which Sect. 3.2 introduces

J. Drugowitsch: Des. & Anal. of Learn. Class. Sys.: A Prob. Approach, SCI 139, pp. 29–44, 2008.
springerlink.com

the model that describes a set of classifiers as a member of the class of parametric models. This includes an introduction to parametric models in Sect. 3.2.1, together with a more detailed definition of the localised classifier models and the global classifier set model in Sect. 3.2.3 and 3.2.4. After discussing how the model structure influences its training and how the model itself relates to Holland's initial LCS idea in Sects. 3.2.6 and 3.2.7, a brief overview is given of how the concepts introduced in this chapter propagate through the chapters to follow.

3.1 Task Definitions

In previous sections the different problem classes that LCS are applied to have already been described informally. Here, they are formalised to serve as the basis for further development. We differentiate between regression tasks, classification tasks, and sequential decision tasks.

Let us assume that we have a finite set of observations generated by noisy measurements of a stochastic process. All tasks have at their core the formation of a model that describes a hypothesis for the data-generating process. The process maps an input space $\mathcal{X}$ into an output space $\mathcal{Y}$, and so each observation (x, y) of that process is formed by an input $x \in \mathcal{X}$ that occurred and the associated measured output $y \in \mathcal{Y}$ of the process in reaction to the input. The set of all inputs $\mathbf{X} = \{x_1, x_2, \dots\}$ and associated outputs $\mathbf{Y} = \{y_1, y_2, \dots\}$ is called the *training set* or *data* $\mathcal{D} = \{\mathbf{X}, \mathbf{Y}\}$.

A model of that process provides a hypothesis for the mapping $\mathcal{X} \rightarrow \mathcal{Y}$, induced by the available data. Hence, given a new input x, the model can be used to predict the corresponding output y that the process is expected to generate. Additionally, an inspection of the hypothesis structure can reveal regularities within the data. In sequential decision tasks the model represents the structure of the task and is employed as the basis of decision-making.

Before going into the the similarities and differences between the regression, classification and sequential decision tasks, let us firstly consider the difficulty of forming good hypotheses about the nature of the data-generating process from only a finite number of observations. For this purpose we assume *batch learning*, that is, the whole training set with N observations of the form (x_n, y_n) is available at once. In a later section, this approach is contrasted with *incremental learning*, where the model is updated incrementally with each observation.

3.1.1 Expected Risk vs. Empirical Risk

In order to model a data-generating process, one needs to be able to express this process by a smooth stationary function $f : \mathcal{X} \rightarrow \mathcal{Y}$ that generates the observation (x, y) by $y = f(x) + \epsilon$, where ϵ is a zero-mean random variable. Thus, it needs to be given by a function such that the same expected output is generated for the same input. That is, given two inputs x, x' such that $x = x'$, the expected output of the process needs to be the same for both inputs. Were this not the case, then one would be unable to detect any regularities within the process and so it cannot be modelled in any meaningful way.

Smoothness of the function is required to express that the process generates similar outputs for similar inputs. That is, given two inputs x, x' that are close in $\mathcal{X}$, their associated outputs y, y' on average need to be close in $\mathcal{Y}$. This property is required in order to make predictions: if it did not hold, then we could not generalise over the training data, as relations between inputs do not transfer to relations between outputs, and thus we would be unable to predict the output for an input that is not in the training set. There are several ways of ensuring the smoothness of a function, such as by limiting its energy of high frequencies in the frequency domain [94]. Here, smoothness is dealt with from an intuitive perspective rather than in any formal way.

As discussed before, the process may be stochastic and the measurements of the output may be noisy. This stochasticity is modelled by the random variable ϵ, which has zero mean, such that for an observation (x, y) we have $\mathbb{E}(y) = f(x)$. The distribution of ϵ is determined by the process stochasticity and the measurement noise.

With this formulation, a model with structure $\mathcal{M}$ has to provide a hypothesis of the form $\hat{f}_{\mathcal{M}} : \mathcal{X} \rightarrow \mathcal{Y}$. In order to be a good model, $\hat{f}_{\mathcal{M}}$ has to be close to f. To be more specific, let $\mathrm{L} : \mathcal{Y} \times \mathcal{Y} \rightarrow \mathbb{R}^+$ be a loss function that describes a distance metric in $\mathcal{Y}$, that is $\mathrm{L}(y, y') > 0$ for all $y \neq y'$, and $\mathrm{L}(y, y') = 0$ otherwise. To get a hypothesis $\hat{f}_{\mathcal{M}}$ close to f we want to minimise the *expected risk*

$$\int^{\mathcal{X}} \mathrm{L}(f(x), \hat{f}_{\mathcal{M}}(x)) \mathrm{d}p(x), \tag{3.1}$$

where $p(x)$ is the probability density of having input x. In other words, our aim is to minimise the distance between the output of the data-generating process and our model of it, for each input x weighted by the probability of observing it.

The expected risk cannot be minimised directly, as f is only accessible by a finite set of observations. Thus, when constructing the model one needs to rely on an approximation of the expected risk, called the *empirical risk* and defined as

$$\frac{1}{N} \sum_{n=1}^{N} \mathrm{L}(y_n, \hat{f}_{\mathcal{M}}(x_n)), \tag{3.2}$$

which is the average loss of the model over all available observations. Depending on the definition of the loss function, minimising the empirical risk can result in least squares learning or the principle of maximum likelihood [218]. By the law of large numbers, the empirical risk converges to the expected risk *almost surely* with the number of observations tending to infinity, but for a small set of observations the two measures might be quite different. How to minimise the expected risk based on the empirical risk forms the basis of statistical learning theory, for which Vapnik has written a good introduction with slightly different definitions [218].

We could simply proceed by minimising the empirical risk. That this approach will not lead to an adequate result is shown by the following observation: the model that minimises the empirical risk is the training set itself. However, assuming noisy measurements, the data is almost certainly not completely correct. Hence, we want to find a model that represents the general pattern in the training data but does not model its noise. The field that deals with this issue is known as *model selection*. Learning a model such that it perfectly fits the training set but does not provide a good representation of f is known as *overfitting*. The opposite, that is, learning a model where the structural bias of the model dominates over the information included from the training set, is called *underfitting*.

While in LCS several heuristics have been applied to deal with this issue, it has never been characterised explicitly. In this and the following chapters the aim is considered to be the minimisation of the empirical risk. In Chap. 7, we return to the topic of model selection, and show how it can be handled with respect to LCS it in a principled manner.

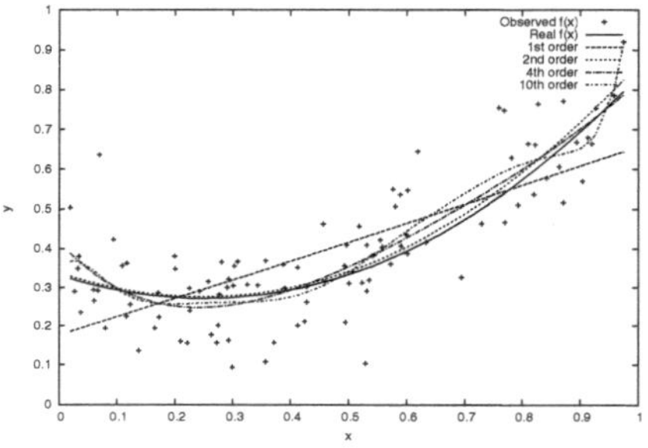

(a)

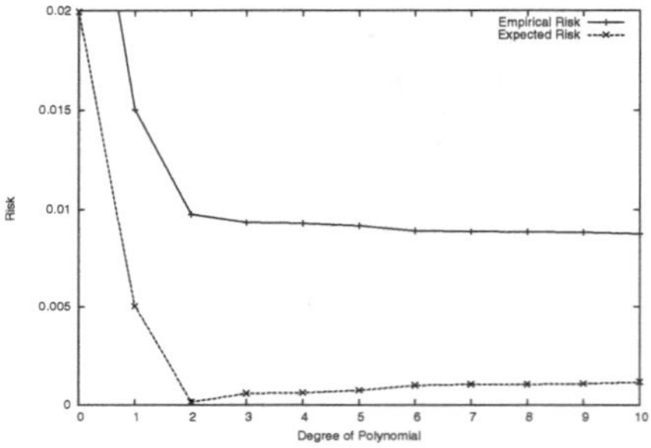

(b)

Fig. 3.1. Comparing the fit of polynomials of various degrees to 100 noisy observations of a 2nd-order polynomial. (a) shows the data-generating function, the available observations, and the least-squares fit of polynomials of degree 1, 2, 4, and 10. (b) shows how the expected and empirical risk changes with the degree of the polynomial. More information is given in Example 3.1.

Example 3.1 (Expected and Empirical Risk of Fitting Polynomials of Various Degree). Consider the data-generating function $f(x) = 1/3 - x/2 + x^2$, whose observations, taken over the range $x \in [0, 1]$, are perturbed by Gaussian noise with a standard deviation of 0.1. Assuming no knowledge of $f(x)$, and given only its observations, let us hypothesise that the data was indeed generated by a polynomial of some degree d, as described by the model

$$\hat{f}_d(x; \boldsymbol{\theta}) = \sum_{n=0}^{d} \theta_n x^n, \tag{3.3}$$

where $\boldsymbol{\theta} \in \mathbb{R}^{d+1}$ is the parameter vector of that model. The aim is to find the degree d that best describes the given observations.

The true function $f(x)$ and the given observations are shown in Fig. 3.1(a), together with fitted polynomials of degree 1, 2, 4, and 10, using the loss function $\mathrm{L}(y, y') = (y' - y)^2$. The 1st-degree polynomial $\hat{f}_1$ (that is, the straight line) clearly underfits the data. This is confirmed by its high expected and empirical risk when compared to other models, as shown in Fig. 3.1(b). On the other hand, the 2nd-degree polynomial $\hat{f}_2$, that conforms to the true data-generating model, represents the data well and is close to $f(x)$ (but not equivalent, due to the finite number of observations). Still, having no knowledge of $f(x)$ one has no reason to stop at $d = 2$, particularly when observing in Fig. 3.1(b) that increasing d reduces the empirical risk further. The expected risk, however, rises, which indicates that the models start to overfit the data by modelling its noise. This is clearly visible for the fit of $\hat{f}_{10}$ to the data in Fig. 3.1(a), which is closer to the observations than $\hat{f}_2$, but further away from f.

The trend of the expected and the empirical risk in Fig. 3.1(b) is a common one: an increase of the model complexity (which is in our case represented by d) generally causes a decrease in the empirical risk. The expected risk, however, only decreases up to a certain model complexity, from which on it starts to increase due to the model overfitting the data. Thus, the aim is to identify the model that minimises the expected risk, which is complicated by the fact that this risk measure is usually not directly accessible. One needs to resort to using the empirical risk in combination with some measure of the complexity of the model, and finding such a measure makes finding the best model a non-trivial problem.

3.1.2 Regression

Both regression and classification tasks aim at finding a hypothesis for the data-generating process such that some risk measure is minimised, but differ in the nature of the input and output space. A regression task is characterised by a multidimensional real-valued input space $\mathcal{X} = \mathbb{R}^{D_\mathcal{X}}$ with $D_\mathcal{X}$ dimensions and a multidimensional real-valued output space $\mathcal{Y} = \mathbb{R}^{D_\mathcal{Y}}$ with $D_\mathcal{Y}$ dimensions. Thus, the inputs are column vectors $\mathbf{x} = (x_1, \dots, x_{D_\mathcal{X}})^T$ and the corresponding outputs are column vectors $\mathbf{y} = (y_1, \dots, y_{D_\mathcal{Y}})^T$. In the case of batch learning it is assumed that N observations $(\mathbf{x}_n, \mathbf{y}_n)$ are available in the form of the input matrix $\mathbf{X}$ and output matrix $\mathbf{Y}$,

$$\mathbf{X} \equiv \begin{pmatrix} -\mathbf{x}_1^T- \\ \vdots \\ -\mathbf{x}_N^T- \end{pmatrix}, \qquad \mathbf{Y} \equiv \begin{pmatrix} -\mathbf{y}_1^T- \\ \vdots \\ -\mathbf{y}_N^T- \end{pmatrix}. \tag{3.4}$$

The loss function is commonly the L_2 norm, also known as the *Euclidean distance*, and is defined by $\mathrm{L}_2(\mathbf{y}, \mathbf{y}') \equiv \|\mathbf{y}, \mathbf{y}'\|_2 = \left(\sum_i (y_i' - y_i)^2\right)^{1/2}$. Hence, the loss increases quadratically in all dimensions with the distance from the desired value. Alternatively, the L_1 norm, also known as the absolute distance, and defined as $L_1(\mathbf{y}, \mathbf{y}') \equiv \|\mathbf{y}, \mathbf{y}'\|_1 = \sum_i |y'_i - y_i|$, can be used. The L_1 norm has the advantage that it only increases linearly with distance and is therefore more resilient to outliers. Using the L_2 norm, on the other hand, makes analytical solutions easier.

All LCS developed so far only handle univariate regression, which is characterised by a 1-dimensional output space, that is $\mathcal{Y} = \mathbb{R}$. Consequently, the output vectors $\mathbf{y}$ collapse to scalars $y \in \mathbb{R}$ and the output matrix $\mathbf{Y}$ becomes a column vector $\mathbf{y} \in \mathbb{R}^N$. For now we will also follow this convention, but will return to multivariate regression with $D_\mathcal{Y} > 1$ in Chap. 7.

3.1.3 Classification

The task of classification is characterised by an input space that is mapped into a subset of a multidimensional real-valued space $\mathcal{X} \subseteq \mathbb{R}^{D_\mathcal{X}}$ of $D_\mathcal{X}$ dimensions, and an output space $\mathcal{Y}$ that is a finite set of labels, mapped into a subset of the natural numbers $\mathcal{Y} \subset \mathbb{N}$. Hence, the inputs are again real-valued column vectors $\mathbf{x} = (x_1, \dots, x_{D_\mathcal{X}})^T$, and the outputs are natural numbers y. The elements of the input vectors are commonly referred to as *attributes*, and the outputs are called the *class labels*. An alternative formulation is for the output space to be $\mathcal{Y} = \{0, 1\}^{D_\mathcal{Y}}$, where $D_\mathcal{Y}$ is the number of classes. Rather than using natural numbers to represent the correct class label, the output is given by a vector $\mathbf{y}$ of 0s and a single 1. That 1 indicates which class the vector represents, with $\mathbf{y} = (1, 0, 0, \dots)^T$ standing for class 1, $\mathbf{y} = (0, 1, 0, \dots)^T$ representing class 2, and so on.

XCS approaches classification tasks by modelling them as regression tasks: each input vector $\mathbf{x}$ is augmented by its corresponding class label y, given by a natural number, to get the new input vector $\mathbf{x}' = (-\mathbf{x}^T-, y)^T$ that is mapped into some positive scalar that we can without loss of generality assume to be 1. Furthermore, each input vector in the training set is additionally augmented by any other valid class label except for the correct one (that is, as given by y) and maps into 0. Hence, the new input space becomes $\mathcal{X}' \subset \mathbb{R}^{D_\mathcal{X}} \times \mathbb{N}$, and the output space becomes $\mathcal{Y}' = [0, 1]$. Consequently, the correct class for a new input $\mathbf{x}$ can be predicted by augmenting the input by each possible class label and choosing the class for which the prediction of the model is closest to 1.

This procedure is not particularly efficient as it needlessly increases the size of the input space $\mathcal{X}'$ and subsequently also complicates the task of finding the best localisation of the classifiers in that space. UCS [161] is an XCS-derivative specialised on classification that handles this tasks more efficiently but still operates on the label-augmented input space $\mathcal{X}'$. A more efficient alternative formulation that does not require this augmentation is discussed in Sect. 4.2.2.

3.1.4 Sequential Decision

A sequential decision task, formulated as an MDP, requires an agent to maximise the long-term reward it receives through the interaction with an environment. At any time, the environment is in a certain state within the state space $\mathcal{X}$. A state transition occurs when the agent performs an action from the action set $\mathcal{A}$. Each of these state transitions is mediated by a scalar reward. The aim of the agent is to find a policy, which is a mapping $\mathcal{X} \rightarrow \mathcal{A}$ that determines the action in each state, that maximises the reward in the long run.

While it is possible to search the space of possible policies directly, a more efficient approach is to compute the *value function* $\mathcal{X} \times \mathcal{A} \rightarrow \mathbb{R}$ that determines for each state which long-term reward to expect when performing a certain action. If a model of the state transitions and rewards is known, *Dynamic Programming* (DP) can be used to compute this function. *Reinforcement Learning* (RL), on the other hand, deals with finding the value function if no such model is available. As the latter is commonly the case, Reinforcement Learning is also the approach employed by LCS.

There are two approaches to RL: either one learns a model of the transitions and rewards by observations and then uses dynamic programming to find the value function, called *model-based* RL, or one estimate the value function directly while interacting with the environment, called *model-free* RL.

In the model-based case, a model of the state transitions and rewards needs to be derived from the given observations, both of which are regression tasks. If the policy is to be computed while sampling the environment, the model needs to be updated incrementally, which requires an incremental learner.

In the model-free case, the function to model is the estimate of the value function, again leading to a regression task that needs to be handled incrementally. Additionally, the value function estimate is also updated incrementally, and as it is the data-generating process, this process is slowly changing. As a result, there is a dynamic interaction between the RL algorithm that updates the value function estimate and the incremental regression learner that models it, which is not in all cases stable and needs special consideration [25]. These are additional difficulties that need to be taken into account when performing model-free RL.

Clearly, although the sequential decision task was the prime motivator for LCS, it is also the most complex to tackle. Therefore, we deal with standard regression and classification tasks first, and come back to sequential decision tasks in Chap. 9. Even then it will be only dealt with from the theoretical perspective of stability, as it requires an incremental learning procedure that will not be developed here.

3.1.5 Batch vs. Incremental Learning

In batch learning it is assumed that the whole training set is available at once, and that the order of the observations in that set is irrelevant. Thus, the model can be trained with all data at once and in any order.

Incremental learning methods differ from batch learning in that the model is updated with each additional observation separately, and as such can handle observations that arrive sequentially as a stream. Revisiting the assumption of

Sect. 3.1.1, that the data-generating process f is expressible by a function, we can differentiate between two cases:

f is stationary. If the data-generating process does not change with time and the full training set is available at once, any incremental learning method is either only an incremental implementation of an equivalent batch learning algorithm, or an approximation to it.

f is non-stationary. Learning a model of a non-stationary generating process is only possible if the process is only slowly varying, that is, if it changes slowly with respect to the frequency that it is observed. Hence, it is reasonable to assume stationarity at least in a limited time-frame. It is modelled by putting more weight on later observations, as earlier observations give general information about the process but might reflect it in an outdated state. Such recency-weighting of the observations is very naturally achieved within incremental learning by assigning the current model a lower weight than new observations.

The advantage of incremental learning methods over batch learning methods are that the former can handle observations that arrive sequentially as a stream, and that they more naturally handle non-stationary processes, even though the second feature can also be simulated by batch learning methods by weighting the different observations according to their temporal sequence[1]. On the downside, when compared to batch learning, incremental learners are generally less transparent in what exactly they learn, and dynamically more complex.

With respect to the different tasks, incremental learners are particularly suited to model-free RL, where the value function estimate is learned incrementally and therefore changes slowly. Given that all data is available at once, regression and classification tasks are best handled by batch learners.

From the theoretical perspective, incremental learners can be derived from a batch learner that is applied to solve the same task. This has the advantage of preserving the transparency of the batch learning method and acquiring the flexibility of the incremental method. This principle is illustrated with the following example.

Example 3.2 (Relating Batch and Incremental Learning). We want to estimate the probability of a tossed coin showing head, without any initial bias about its fairness. We perform N experiments with no input $\mathcal{X} = \emptyset$ and outputs $\mathcal{Y} = \{0, 1\}$, where 0 and 1 stand for tail and head respectively. Adopting a frequentist approach, the probability of tossing a coin resulting in head can be estimated by

$$p_N(H) = \frac{1}{N} \sum_{n=1}^{N} y_n, \tag{3.5}$$

[1] Naturally, in the case of weighting observations according to their temporal sequence, the ordering of these observations *is* – in contrast to what was stated previously in the batch learning context – of significance.

where $p_N(H)$ stands for the estimated probability of head after N experiments. This batch learning approach can be easily turned into an incremental approach by

$$p_N(H) = \frac{1}{N} y_N + \frac{1}{N} \sum_{n=1}^{N-1} y_n = p_{N-1}(H) + \frac{1}{N}(y_N - p_{N-1}(H)), \tag{3.6}$$

starting with $p_1(H) = y_1$. Hence, to update the model $p_{N-1}(H)$ with the new observation y_N, one only needs to maintain the number N of experiments so far. Comparing (3.5) and (3.6) it is apparent that, whilst the incremental approach yields the same results as the batch approach, it is far less transparent in what it is actually calculating.

Let us now assume that the coin changes its properties slowly over time, and we therefore trust recent observations more. This is achieved by modifying the incremental update to

$$p_N(H) = p_{N-1}(H) + \gamma(y_N - p_{N-1}(H)), \tag{3.7}$$

where $0 < \gamma \leq 1$ is the recency factor that determines the influence of past observations to the current estimate. Recursive substitution of $p_n(H)$ results in the batch learning equation

$$p_N(H) = (1-\gamma)^N p_0(H) + \sum_{n=1}^{N} \gamma(1-\gamma)^{N-n} y_n. \tag{3.8}$$

Inspecting this equation reveals that observations n experiments back in time are weighted by $\gamma(1-\gamma)^n$. Additionally, it can be seen that an initial bias $p_0(H)$ is introduced that decays exponentially with the number of available observations. Again, the batch learning formulation has led to greater insight and transparency.

Are LCS Batch Learners or Incremental Learners?

LCS are often considered to be incremental learners. While they are usually implemented as such, there is no reason not to design them as batch learners when applying them to regression or classifications tasks, given that all data is available at once. Indeed, Pittsburgh-style LCS usually require an individual representing a set of classifiers to be trained on the full data, and hence can be interpreted as incrementally implemented batch learners when applied to regression and classification tasks.

Even Michigan-style LCS can acquire batch learning when the classifiers are trained independently: each classifier can be trained on the full data at once and is later only queried for its fitness evaluation and its prediction.

As the aim is to understand what LCS are learning, we – for now – will prefer transparency over performance. Hence, the LCS model is predominantly described from a batch learning perspective, although, throughout Chaps. 5, 6 and 7, incremental learning approaches that lead to similar results will also be discussed. Still, the prototype system that is developed is only fully described from the batch learning perspective. How to turn this system into an incremental learner is a topic of future research.

3.2 LCS as Parametric Models

While the term *model* may be used in many different ways, it is here defined as a collection of possible hypotheses about the data-generating process. Hence, the choice of model determines the available hypotheses and therefore biases the expressiveness about this process. Such a bias represents the assumptions that are made about the process and its stochasticity. Understanding the assumptions that are introduced with the model allows for making statements about its applicability and performance.

Example 3.3 (Different Linear Models and their Assumptions). A linear relation between inputs and outputs with constant-variance Gaussian noise ϵ leads to least squares (that is, using the L_2 loss function) linear regression. Alternatively, assuming the noise to have a Cauchy distribution results in linear regression using the L_1 loss function. As a Cauchy distribution has a longer tail than a Gaussian distribution, it is more resilient to outliers. Hence it is considered as being more robust, but the L_1 norm makes it harder to train [66]. This shows how an assumption of a model about the data-generating process can give us information about its expected performance.

Training a model means finding the hypothesis that is closest to what the data-generating process is assumed to be. For example, in a linear regression model the space of hypotheses is all hyper-planes in the input/output space, and performing linear regression means picking the hyper-plane that best explains the available observations.

The choice of model strongly determines how hard it is to train. While more complex models are usually able to express a larger range of possible hypotheses, this larger range also makes it harder for them to avoid overfitting and underfitting. Hence, very often, overfitting by minimising the empirical risk is counterbalanced by reducing the number of hypotheses that a model can express, thus making the assumptions that a model introduces more important.

Example 3.4 (Avoiding Overfitting in Artificial Neural Networks). Reducing the number of hidden neurons in a feed-forward neural network is a popular measure of avoiding overfitting the training data. This measure effectively reduces the number of possible hypothesis that the model is able to express and as such introduces a stronger structural bias. Another approach to avoiding overfitting in neural networks training is *weight decay* that exponentially decays the magnitude of the weight of the neural connections in the network. While not initially designed as such, weight decay is equivalent to assuming a zero mean Gaussian prior on the weights and hence biasing them towards smaller values. This prior is again equivalent to assuming smoothness of the target function [106].

Having underlined the importance of knowing the underlying model of a method, the family of parametric models is introduced, in order to identify LCS as a member of that family. The description is based on reflections on what classifiers actually are and do, and how they cooperate to form a model. While a general LCS model overview and its training is given, more details have to wait until after the a formal probabilistic LCS model is introduced in the following chapter.

3.2.1 Parametric Models

The chosen hypothesis during model training is usually determined by a set of adjustable parameters $\boldsymbol{\theta}$. Models for which the number of parameters is independent of the training set and remains unchanged during model training are commonly referred to as *parametric* models. In contrast, *non-parametric* models are models for which the number of adjustable parameters either depends on the training set, changes during training, or both.

Another property of a parametric model is its *structure* $\mathcal{M}$ (often also referred to as *scale*). Given a model family, the choice of structure determines which model to use from this family. For example, considering the family of feed-forward neural networks with a single hidden layer, the model structure is the number of hidden neurons and the model parameters are the weights of the neural connections. Hence, the model structure is the adjustable part of the model that remain unchanged during training but might determine the number of parameters.

With these definitions, our aims can be re-formulated: Firstly, and adequate model structure $\mathcal{M}$ is to be found that provides the model hypotheses $\hat{f}_{\mathcal{M}}(x; \boldsymbol{\theta})$. Secondly, the model parameter values $\boldsymbol{\theta}$ need to be found such that the expected risk for the chosen loss function is minimised.

3.2.2 An LCS Model

An LCS forms a *global model* by the combination of *local models*, represented by the classifiers. The number of classifiers can change during the training process, and so can the number of adjustable parameters by action of the GA. Hence, an LCS is not a parametric model per se.

An LCS can be turned into a parametric model by assuming that the number of classifiers is fixed, and that each classifier represents a parametric model. While this choice seems arbitrary at first, it becomes useful for later development. Its consequences are that both the number of classifiers and how they are located in the input space are part of the model structure $\mathcal{M}$ and are not modified while adjusting the model parameters. The model parameters $\boldsymbol{\theta}$ are the parameters of the classifiers and those required to combine their local models.

Consequently, training an LCS is conceptually split into two parts: Finding a good model structure $\mathcal{M}$, that is, the adequate number of classifiers and their location, and for that structure the values of the model parameters $\boldsymbol{\theta}$. This interpretation justifies calling LCS *adaptive models*.

Before providing more details on how to find a good model structure, let us first assume a fixed model structure with K classifiers and investigate in more detail the components of such a model.

3.2.3 Classifiers as Localised Models

In LCS, the combination of condition and action of a classifier determines the inputs that a classifier matches. Hence, given the training set, one classifier matches only a subset of the observations in that set. It can be said that a classifier is *localised* in the input space, where its location is determined by the inputs that it matches.

Matching

Let $\mathcal{X}_k \subseteq \mathcal{X}$ be the subset of the input space that classifier k matches. The classifier is trained by all observations that it matches, and hence its aim is to provide a local model $\hat{f}_k(x; \boldsymbol{\theta}_k)$ that maps $\mathcal{X}_k$ into $\mathcal{Y}$, where $\boldsymbol{\theta}_k$ is the set of parameters of the model of classifier k. More flexibly, matching can be defined by a matching function $m_k : \mathcal{X} \rightarrow [0, 1]$ specific to classifier k, and given by the indicator function for the set $\mathcal{X}_k$,

$$m_k(\mathbf{x}) = \begin{cases} 1 & \text{if } \mathbf{x} \in \mathcal{X}_k, \\ 0 & \text{otherwise.} \end{cases} \tag{3.9}$$

The advantage of using a matching function m_k rather than a set $\mathcal{X}_k$ is that the former allows for degrees of matching in-between 0 and 1 – a feature that we will be made use of in later chapters. Also note, that representing matching by $\mathcal{X}_k$ or the matching function m_k makes it independent of the choice of representation of the condition/action of a classifier. This is an important point, as it makes all future developments valid for *all* choices of representation.

Local Classifier Model

The local model of a classifier is usually a regression model with no particular restrictions. As discussed in Section 2.3.1, initially only simple averaging predictions were used, but more recently, classifiers have been extended to use linear regression models, neural networks, and SVM regression. While averagers are just a special case of linear models, neural networks might suffer from the problem of multiple local optima [104], and SVM regression has no clean approach to incremental implementations [157]. Hence, we will restrict ourselves to the well-studied class of linear models as a good trade-off between expressive power and complexity of training, and equally easily trainable classification models. Both are discussed in more depth in Chaps. 4 and 5.

Input to Matching and Local Models

Note that in LCS the input given to the matching function and that given to the classifier's model usually differ in that the input to the model is often formed by applying a transfer function to the input given to the matching mechanism. Nonetheless, to keep the notation uncluttered it is assumed that the given input $\mathbf{x}$ contains all available information and both matching and the local model selectively choose and modify the components that they require by an implicit transfer function.

Example 3.5 (Inputs to Matching and Local Model). Let us assume that both the input and the output space are 1-dimensional, that is, $\mathcal{X} = \mathbb{R}$ and $\mathcal{Y} = \mathbb{R}$, and that we perform interval matching over the interval $[l_k, u_k]$, such that $m_k(x) = 1$ if $l_k \leq x \leq u_k$, and $m_k(x) = 0$ otherwise. Applying the linear model $\hat{f}(x; w_k) = xw_k$ to the input, with w_k being the adjustable parameter of classifier k, one can only model straight lines through the origin. However, applying the transfer function $\phi(x) = (1, x)^T$ allows for the introduction of an additional bias to get $\hat{f}(x; \mathbf{w}_k) = \mathbf{w}_k^T \phi(x) = w_{k1} + xw_{k2}$, with $\mathbf{w}_k = (w_{k1}, w_{k2})^T \in \mathbb{R}^2$, which is an arbitrary straight line. In such a case, the input is assumed to be $\mathbf{x}' = (1, x)^T$, and the matching function to only operate on the second component of that input. Hence, both matching and the local model can be applied to the same input. A more detailed discussion about different transfer functions and their resulting models is given in Sect. 5.1.1.

3.2.4 Recovering the Global Model

To recover the global model from K local models, they need to be combined in some meaningful way. For inputs that only a single classifier matches, the best model is that of the matching classifier. However, there are no restrictions on how many classifiers can match a single input. Therefore, in some cases, it is necessary to mix the local models of several classifiers that match the same input.

There are several possible approaches to mixing classifier models, each corresponding to different assumptions about the data-generating process. A standard approach in introduced in Chap. 4 and alternatives are discussed in Chap. 6.

3.2.5 Finding a Good Model Structure

The model structure $\mathcal{M}$ is given by the number of classifiers and their localisation. As the localisation of a classifier k is determined by its matching function m_k, the model structure is completely specified by the number of classifiers K and their matching functions $\mathbf{M} = \{m_k\}$, that is, $\mathcal{M} = \{K, \mathbf{M}\}$.

To find a good model structure means to find a structure that allows for hypotheses about the data-generating process that are close to the process suggested by the available observations. Thus, finding a good model structure implies dealing with over and underfitting of the training set. A detailed treatment of this topic is postponed to Chap. 7, and for now its is assumed that a good model structure is known.

3.2.6 Considerations for Model Structure Search

The space of possible model structures is potentially huge, and hence to search this space, evaluating the suitability of a single model structure $\mathcal{M}$ to explain the data needs to be efficient to keep searching the model structure space computationally tractable. Additionally, one wants to guide the search by using all the available information about the quality of the classifiers within a certain model structure by fitting this model structure to the data.

Each classifier in the LCS model represents some information about the input/output mapping, limited to the subspace of the input space that it matches. Hence, while preserving classifiers that seem to provide a good model of the matched data, the model structure ought to be refined in areas of the input space for which none of the current classifiers provides an adequate model. This can be achieved by either modifying the localisation of current classifiers that do not provide an adequate fit, removing those classifiers, or adding new classifiers to compare their goodness-of-fit to the current ones. Intuitively, interpreting a classifier as a localised hypothesis for the data-generating process, we want to change or discard bad hypotheses, or add new hypotheses to see if they are favoured in comparison to already existing hypotheses.

In terms of the model structure search, the search space is better traversed by modifying the current model structure rather than discarding it at each search step. By only modifying part of the model, we have satisfied the aim of facilitating knowledge of the suitability of the current model structure to guide the structure search. Additionally, if only few classifiers are changed in their localisation in each step of the search, only modified or added classifiers need to be re-trained, given that the classifiers are trained independently. This is an important feature that makes the search more efficient, and that will be revisited in Sect. 4.4.

Such a search strategy clearly relates to how current LCS traverse the search space: In Michigan-style LCS, such as XCS, new classifiers are added either if no classifier is localised in a certain area of the input space, or to provide alternative hypotheses by merging and modifying the localisation structure of two other current classifiers with a high goodness-of-fit. Classifiers in XCS are removed with a likelihood that is proportional to on average how many other classifiers match the same area of the input space, causing the number of classifiers that match a particular input to be about the same for all inputs. Pittsburgh-style LCS also traverse the structure search space by merging and modifying sets of classifiers of two model structures that were evaluated to explain the data well. However, few current Pittsburgh-style LCS retain the evaluation of single classifiers to improve the efficiency of the search – a feature that is used in the prototype implementation described in Chap. 8.

3.2.7 Relation to the Initial LCS Idea

Recall that originally LCS addressed the problems of parallelism and coordination, credit assignment, and rule discovery, as described in Sect. 2.2.1. The following describes how these problems are addressed in the proposed model.

Parallelism is featured by allowing several classifiers to be overlapping, that is, to be localised partially in the same areas of the input space. Hence, they compete locally by providing different models for the same data, and cooperate globally by providing a global model only in combination. Coordination of the different classifiers is handled on one hand by the model component that combines the local models into a global model, and on the other hand by the model structure search that removes or changes classifiers based on their contribution to the full model.

Credit assignment is to assign external reward to different classifiers, and is mapped to regression and classification tasks that fit the model to the data, as the reward is represented by the output. In sequential decision tasks, credit assignment is additionally handled by the reinforcement learning algorithm, which will be discussed in detail in Chap. 9.

Lastly, the role of discovering new rules, that is, classifiers with a better localisation, is performed by the model structure search. How to use current knowledge to introduce new classifiers depends strongly on the choice of representation for the condition and action of a classifier. As the presented work does not make any assumptions about the representation, it does not deal with this issue in detail, but rather relies on the body of prior work (for example, [41, 38, 163, 144, 147, 203]) that is available on this topic.

3.3 Summary and Outlook

The task of LCS has been identified to find a good model that forms a hypothesis about the form of the data-generating process, based on a finite set of observations. The process maps an input space into an output space, and the model provides a possible hypothesis for this mapping. The task of finding a good model is made more complex as only a finite set of observations of the input/output mapping are available that are perturbed by measurement noise and the possible stochasticity of the process, and this task is dealt with by the field of model selection. The difference between minimising the expected risk, which is the difference between the real data-generating process and our model, and minimising the empirical risk, which is the difference between the observations available of that process and our model, has been emphasised.

Regression, classification and sequential decision tasks differ in the form of the input and output spaces and in the assumptions made about the data-generating process. For both regression and classification tasks it is assumed that the process to be representable by a smooth function with an additive zero-mean noise term. While sequential decision tasks as handled by RL also have a regression task at their core, they have special requirements on the stability of the learning method and therefore receive a separate treatment in Chap. 9.

A model was characterised as being a collection of possible hypotheses about the nature of the data-generating process, and training a model was defined as finding the hypothesis that is best supported by the available observations of that process. The class of parametric models was introduced, characterised by an unchanging number of model parameters while the model is trained, in contrast to the model structure of a parametric model, which is the part of the model that is adjusted before training it, and determines the number of adjustable parameters during model training.

The LCS model that was described in this chapter and forms the basis of further developments combines a set of local models (that is, the classifiers) to a global model. While LCS are not parametric models per se, they can be characterised as such by defining the model structure as the number of classifiers and their localisation, and the model parameters as the parameters of the classifiers and the ones required for combining the local models. As a result, the task of training LCS is conceptually split into finding a good model structure, that is, a good set of classifiers, and training these classifiers with the available training set.

Finding a good model structure requires us to deal with the topic of model selection and the trade-off between overfitting and underfitting. As this requires a good understanding of the LCS model itself, the problem of evaluating the quality of a model structure will not be handled before Chap. 7. Until then, the model structure $\mathcal{M}$ is assumed to be a constant.

The next chapter discusses how to train an LCS model given a certain model structure. In other words, it concerns how to adjust the model parameters in the light of the available data. The temporary aim at this stage is to minimise the empirical risk. Even though this might lead to overfitting, it still gives valuable insights into how to train the LCS model, and its underlying assumptions about the data-generating process. We proceed by formulating a probabilistic model of LCS in Chap. 4 based on a generalisation of the related Mixtures-of-Experts model. Furthermore, more details on training the classifiers are given in Chap. 5, and alternatives for combining the local classifier models to a global model are given in Chap. 6, assuming that the model structure remains unchanged. After that we return to developing a principled approach to finding a good set of classifiers, that is, a good model structure.

4 A Probabilistic Model for LCS

Having conceptually defined the LCS model, it will now be embedded into a formal setting. The formal model is initially designed for a fixed model structure $\mathcal{M}$; that is, the number of classifiers and where they are localised in the input space is constant during training of the model. Even though the LCS model could be characterised purely by its functional form [78], a probabilistic model will be developed instead. Its advantage is that rather than getting a point estimate $\hat{f}(\mathbf{x})$ for the output $\mathbf{y}$ given some input $\mathbf{x}$, the probabilistic model provides the probability distribution $p(\mathbf{y}|\mathbf{x}, \boldsymbol{\theta})$ that for some input $\mathbf{x}$ and model parameters $\boldsymbol{\theta}$ describes the probability density of the output being the vector $\mathbf{y}$. From this distribution its is possible to form a point estimate from its mean or its mode, and additionally to get information about the certainty of the prediction by the spread of the distribution.

This chapter concentrates on modelling the data by the principle of maximum likelihood: given a set of observations $\mathcal{D} = \{\mathbf{X}, \mathbf{Y}\}$, the best model parameters $\boldsymbol{\theta}$ are the ones that maximise the probability of the observations given the model parameters $p(\mathcal{D}|\boldsymbol{\theta})$. As described in the previous chapter this might lead to overfitting the data, but nonetheless it results in a first idea about how the model can be trained, and relates it closely to XCS, where overfitting is controlled on the model structure level rather than the model parameter level (see App. B). Chapter 7 generalises this model and introduces a training method that avoids overfitting.

The formulation of the probabilistic model is guided by a related machine learning model: the Mixtures-of-Expert (MoE) model [120, 121] fits the data by a fixed number of localised experts. Even though not identified by previous LCS research, there are strong similarities between LCS and MoE when relating the classifiers of LCS to the experts of MoE. However, they differ in that the localisation of the experts in MoE is changed by a gating network that assigns observations to experts, whereas in LCS the localisation of classifiers is defined by the matching functions and is fixed for a constant model structure. To relate these two approaches, the model is modified such that it acts as a generalisation to both the standard MoE model and LCS. Furthermore, difficulties in training the emerging model are solved by detaching expert training from training the gating network.

J. Drugowitsch: Des. & Anal. of Learn. Class. Sys.: A Prob. Approach, SCI 139, pp. 45–64, 2008.
springerlink.com

Firstly, the standard MoE model [121] is introduced, and its training and expert localisation is discussed. This is followed in Sect. 4.2 by a discussion of expert models for both regression and classification. To relate MoE to LCS, the MoE model is generalised in Sect. 4.3, together with how its training has to be modified to accommodate these generalisations. Identifying difficulties with the latter, a modified training scheme is introduced in Sect. 4.4, that makes the introduced model more similar to XCS.

4.1 The Mixtures-of-Experts Model

The MoE model is probably best explained from the generative point-of-view: given a set of K experts, each observation in the training set is assumed to be generated by one and only one of these experts. Let $\mathbf{z} = (z_1, \dots, z_K)^T$ be a random binary vector, where each of its elements z_k is associated with an expert and indicates whether that expert generated the given observation $(\mathbf{x}, \mathbf{y})$. Given that expert k generated the observation, then $z_j = 1$ for $j = k$, and $z_j = 0$ otherwise, resulting in a 1-of-K structure of $\mathbf{z}$. The introduced random vector is a *latent variable*, as its values cannot be directly observed. Each observation $(\mathbf{x}_n, \mathbf{y}_n)$ in the training set has such a random vector $\mathbf{z}_n$ associated with it, and $\mathbf{Z} = \{\mathbf{z}_n\}$ denotes the set of latent variables corresponding to each of the observations in the training set.

Each expert provides a probabilistic mapping $\mathcal{X} \to \mathcal{Y}$ that is given by the conditional probability density $p(\mathbf{y}|\mathbf{x}, \boldsymbol{\theta}_k)$, that is, the probability of the output being vector $\mathbf{y}$, given the input vector $\mathbf{x}$ and the model parameters $\boldsymbol{\theta}_k$ of expert k. Depending on whether we deal with regression or classification tasks, experts can represent different parametric models. Leaving the expert models unspecified for now, linear regression and classification models will be introduced in Sect. 4.2.

4.1.1 Likelihood for Known Gating

A common approach to training probabilistic models is to maximise the likelihood of the outputs given the inputs and the model parameters, a principle known as *maximum likelihood*. As will be shown later, maximum likelihood training is equivalent to minimising the empirical risk, with a loss function depending on the probabilistic formulation of the model.

Following the standard assumptions of independent observations, and additionally assuming knowledge of the values of the latent variables $\mathbf{Z}$, the likelihood of the training set is given by

$$p(\mathbf{Y}|\mathbf{X}, \mathbf{Z}, \boldsymbol{\theta}) = \prod_{n=1}^{N} p(\mathbf{y}_n|\mathbf{x}_n, \mathbf{z}_n, \boldsymbol{\theta}), \tag{4.1}$$

where $\boldsymbol{\theta}$ stands for the model parameters. Due to the 1-of-K structure of each $\mathbf{z}_n$, the likelihood for the nth observation is given by

$$p(\mathbf{y}_n|\mathbf{x}_n, \mathbf{z}_n, \boldsymbol{\theta}) = \prod_{k=1}^{K} p(\mathbf{y}_n|\mathbf{x}_n, \boldsymbol{\theta}_k)^{z_{nk}}, \tag{4.2}$$

where z_{nk} is the kth element of $\mathbf{z}_n$. As only one element of $\mathbf{z}_n$ can be 1, the above expression is equivalent to the jth expert model such that $z_{nj} = 1$.

As the logarithm function is monotonically increasing, maximising the logarithm of the likelihood is equivalent to maximising the likelihood. Combining (4.1) and (4.2), the log-likelihood $\ln p(\mathbf{Y}|\mathbf{X}, \mathbf{Z}, \boldsymbol{\theta})$ results in

$$\ln p(\mathbf{Y}|\mathbf{X}, \mathbf{Z}, \boldsymbol{\theta}) = \sum_{n=1}^{N} \sum_{k=1}^{K} z_{nk} \ln p(\mathbf{y}_n|\mathbf{x}_n, \boldsymbol{\theta}_k). \tag{4.3}$$

Inspecting (4.3) we can see that each observation n is assigned to the single expert for which $z_{nk} = 1$. Hence, it is maximised by maximising the likelihood of the expert models separately, for each expert based on its assigned set of observations.

4.1.2 Parametric Gating Network

As the latent variables $\mathbf{Z}$ are not directly observable, we do not know the values that they take and therefore cannot maximise the likelihood introduced in the previous section directly. Rather, a parametric model for $\mathbf{Z}$, known as the *gating network*, is used instead and trained in combination with the experts.

The gating network used in the standard MoE model is based on the assumption that the probability of an expert having generated the observation $(\mathbf{x}, \mathbf{y})$ is log-linearly related to the input $\mathbf{x}$. This is formulated by

$$g_k(\mathbf{x}) \equiv p(z_k = 1|\mathbf{x}, \mathbf{v}_k) \propto \exp(\mathbf{v}_k^T \mathbf{x}), \tag{4.4}$$

stating that the probability of expert k having generated observation $(\mathbf{x}, \mathbf{y})$ is proportional to the exponential of the inner product of the input $\mathbf{x}$ and the gating vector $\mathbf{v}_k$ of the same size as $\mathbf{x}$. Normalising $p(z_k = 1|\mathbf{x}, \mathbf{v}_k)$, we get

$$g_k(\mathbf{x}) \equiv p(z_k = 1|\mathbf{x}, \mathbf{v}_k) = \frac{\exp(\mathbf{v}_k^T \mathbf{x})}{\sum_{j=1}^{K} \exp(\mathbf{v}_j^T \mathbf{x})}, \tag{4.5}$$

which is the well-known *softmax* function, and corresponds to the multinomial logit model in Statistics that is often used to model consumer choice [165]. It is parametrised by one gating vector $\mathbf{v}_k$ per expert, in combination forming the set $\mathbf{V} = \{\mathbf{v}_k\}$. Fig. 4.1 shows the directed graphical model that illustrates the structure and variable dependencies of the Mixtures-of-Experts model.

To get the log-likelihood $l(\boldsymbol{\theta}; \mathcal{D}) \equiv \ln p(\mathbf{Y}|\mathbf{X}, \boldsymbol{\theta})$, we use the 1-of-$K$ structure of $\mathbf{z}$ to express the probability of having a latent random vector $\mathbf{z}$ for a given input $\mathbf{x}$ and a set of gating parameters $\mathbf{V}$ by

$$p(\mathbf{z}|\mathbf{x}, \mathbf{V}) = \prod_{k=1}^{K} p(z_k = 1|\mathbf{x}, \mathbf{v}_k)^{z_k} = \prod_{k=1}^{K} g_k(\mathbf{x})^{z_k}. \tag{4.6}$$

Thus, by combining (4.2) and (4.6), the joint density over $\mathbf{y}$ and $\mathbf{z}$ is given by

$$p(\mathbf{y}, \mathbf{z}|\mathbf{x}, \boldsymbol{\theta}) = \prod_{k=1}^{K} g_k(\mathbf{x})^{z_k} p(\mathbf{y}|\mathbf{x}, \boldsymbol{\theta}_k)^{z_k}. \tag{4.7}$$

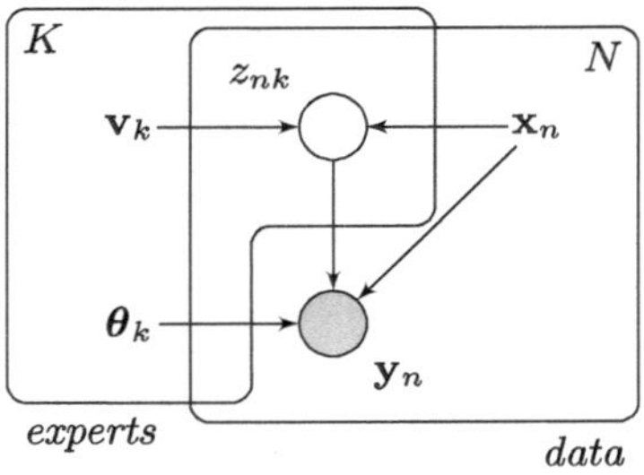

Fig. 4.1. Directed graphical model of the Mixtures-of-Experts model. The circular nodes are random variables (z_{nk}), which are observed when shaded ($\mathbf{y}_n$). Labels without nodes are either constants ($\mathbf{x}_n$) or adjustable parameters ($\boldsymbol{\theta}_k$, $\mathbf{v}_k$). The boxes are "plates", comprising replicas of the entities inside them. Note that z_{nk} is shared by both boxes, indicating that there is one z for each expert for each observation.

By marginalising[1] over $\mathbf{z}$, the output density results in

$$p(\mathbf{y}|\mathbf{x},\boldsymbol{\theta}) = \sum_{z}\prod_{k=1}^{K} g_k(\mathbf{x})^{z_k} p(\mathbf{y}|\mathbf{x},\boldsymbol{\theta}_k)^{z_k} = \sum_{k=1}^{K} g_k(\mathbf{x}) p(\mathbf{y}|\mathbf{x},\boldsymbol{\theta}_k), \tag{4.8}$$

and subsequently, the log-likelihood $l(\boldsymbol{\theta};\mathcal{D})$ is

$$l(\boldsymbol{\theta};\mathcal{D}) = \ln \prod_{n=1}^{N} p(\mathbf{y}_n|\mathbf{x}_n|\boldsymbol{\theta}) = \sum_{n=1}^{N} \ln \sum_{k=1}^{K} g_k(\mathbf{x}_n) p(\mathbf{y}_n|\mathbf{x}_n,\boldsymbol{\theta}_k). \tag{4.9}$$

Example 4.1 (Gating Network for 2 Experts). Let us consider the input space $D_{\mathcal{X}} = 3$, where an input is given by $\mathbf{x} = (1, x_1, x_2)^T$. Assume two experts with gating parameters $\mathbf{v}_1 = (0,0,1)^T$ and $\mathbf{v}_2 = (0,1,0)^T$. Then, Fig. 4.2 shows the gating values $g_1(\mathbf{x})$ for Expert 1 over the range $-5 \leq x_1 \leq 5$, $-5 \leq x_2 \leq 5$. As can be seen, we have $g_1(\mathbf{x}) > 0.5$ in the input subspace $x_1 - x_2 < 0$. Thus, with the given gating parameters, Expert 1 mainly models observations in this subspace. Overall, the gating network causes a soft linear partitioning of the input space along the line $x_1 - x_2 = 0$ that separates the two experts.

4.1.3 Training by Expectation-Maximisation

Rather than using gradient descent to find the experts and gating network parameters $\boldsymbol{\theta}$ that maximise the log-likelihood (4.9) [120], we can make use of the latent variable structure and apply the expectation-maximisation (EM) algorithm

---

[1] Given a joint density $p(x,y)$, one can get $p(y)$ by *marginalising* over x by

$$p(y) = \int p(x,y)\mathrm{d}x.$$

The same principle applies to getting $p(y|z)$ from the conditional density $p(x,y|z)$.

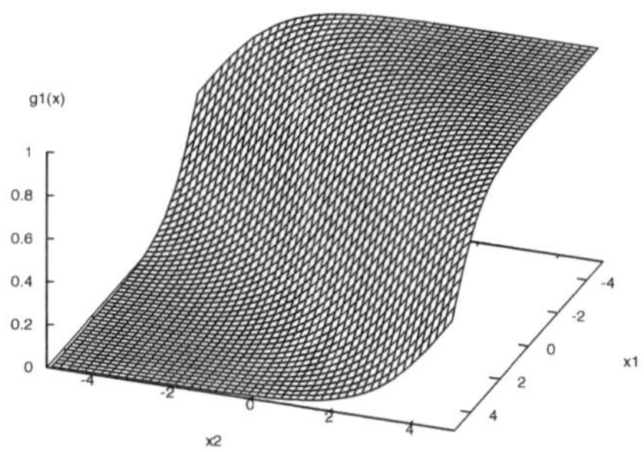

Fig. 4.2. Plot of the softmax function $g_1(\mathbf{x})$ by (4.5) with inputs $\mathbf{x} = (1, x_1, x_2)^T$, and gating parameters $\mathbf{v}_1 = (0, 0, 1)$, $\mathbf{v}_2 = (0, 1, 0)$

[71, 121]. It begins with the observation that maximisation of the likelihood is simplified if the values of the latent variables were known, as in (4.3). Hence, assuming that $\mathbf{Z}$ is part of the data, $\mathcal{D} = \{\mathbf{X}, \mathbf{Y}\}$ is referred to as the *incomplete data*, and $\mathcal{D} \cup \{\mathbf{Z}\} = \{\mathbf{X}, \mathbf{Y}, \mathbf{Z}\}$ is known as the *complete data*. The EM-algorithm proceeds with the expectation step, by finding the expectation of the complete data log-likelihood $\mathbb{E}_Z(l(\boldsymbol{\theta}; \mathcal{D} \cup \{\mathbf{Z}\}))$ with the current model parameters $\boldsymbol{\theta}$ fixed, where $l(\boldsymbol{\theta}; \mathcal{D} \cup \{\mathbf{Z}\}) \equiv \ln p(\mathbf{Y}, \mathbf{Z}|\mathbf{X}, \boldsymbol{\theta})$ is the logarithm of the joint density of the outputs and the values of the latent variables. In the maximisation step the above expectation is maximised with respect to the model parameters. When iterating this procedure, the incomplete data log-likelihood $l(\boldsymbol{\theta}; \mathcal{D})$ is guaranteed to increase monotonically until a maximum is reached [174]. More details on the application of the EM-algorithm to train the MoE model are given by Jordan and Jacobs [121]. We will now consider each step in turn.

The Expectation Step

Using (4.7), the complete-data log-likelihood is given by

$$\begin{aligned} l(\boldsymbol{\theta}; \mathcal{D} \cup \{\mathbf{Z}\}) &\equiv \ln p(\mathbf{Y}, \mathbf{Z}|\mathbf{X}, \boldsymbol{\theta}) \\ &= \ln \prod_{n=1}^{N} p(\mathbf{y}_n, \mathbf{z}_n|\mathbf{x}_n, \boldsymbol{\theta}) \\ &= \sum_{n=1}^{N} \sum_{k=1}^{K} z_{nk} \left(\ln g_k(\mathbf{x}_n) + \ln p(\mathbf{y}_n|\mathbf{x}_n, \boldsymbol{\theta}_k)\right) \end{aligned} \tag{4.10}$$

where $\boldsymbol{\theta}$ is the set of expert parameters $\{\boldsymbol{\theta}_1, \dots, \boldsymbol{\theta}_K\}$ and gating parameters $\mathbf{V}$. When fixing these parameters, the latent variables are the only random variables in the likelihood, and hence its expectation is

$$\mathbb{E}_Z\left(l(\boldsymbol{\theta}; \mathcal{D} \cup \{\mathbf{Z}\})\right) = \sum_{n=1}^{N} \sum_{k=1}^{K} r_{nk} \left(\ln g_k(\mathbf{x}_n) + \ln p(\mathbf{y}_n|\mathbf{x}_n, \boldsymbol{\theta}_k)\right), \tag{4.11}$$

where $r_{nk} \equiv \mathbb{E}(z_{nk})$ is commonly referred to as the *responsibility* of expert k for observation n [19] and by the use of Bayes' rule and (4.8) evaluates to

$$\begin{aligned} r_{nk} \equiv \mathbb{E}(z_{nk}) &= p(z_{nk} = 1|\mathbf{x}_n, \mathbf{y}_n, \boldsymbol{\theta}) \\ &= \frac{p(z_{nk} = 1|\mathbf{x}_n, \mathbf{v}_k)p(\mathbf{y}_n|\mathbf{x}_n, \boldsymbol{\theta}_k)}{p(\mathbf{y}_n|\mathbf{x}_n, \boldsymbol{\theta})} \\ &= \frac{g_k(\mathbf{x}_n)p(\mathbf{y}_n|\mathbf{x}_n, \boldsymbol{\theta}_k)}{\sum_{j=1}^K g_j(\mathbf{x_n})p(\mathbf{y}_n|\mathbf{x}_n, \boldsymbol{\theta}_j)}. \end{aligned} \tag{4.12}$$

Hence, the responsibilities are distributed according to the current gating and goodness-of-fit of an expert in relation to the gating and goodness-of-fit of the other experts.

The Maximisation Step

In the maximisation step we aim at adjusting the model parameters to maximise the expected complete data log-likelihood. $g_k(\mathbf{x}_n)$ and $p(\mathbf{y}_n|\mathbf{x}_n, \boldsymbol{\theta}_k)$ do not share any parameters, and so maximising (4.11) results in the two independent maximisation problems

$$\max_{\mathbf{V}} \sum_{n=1}^N \sum_{k=1}^K r_{nk} \ln g_k(\mathbf{x}_n), \tag{4.13}$$

$$\max_{\boldsymbol{\theta}} \sum_{n=1}^N \sum_{k=1}^K r_{nk} \ln p(\mathbf{y}_n|\mathbf{x}_n, \boldsymbol{\theta}_k). \tag{4.14}$$

Note that the responsibilities are evaluated with the previous model parameters and are not considered as being functions of these parameters. The function concerning the gating parameters $\mathbf{V}$ can be maximised by the Iteratively Re-weighted Least Squares (IRLS) algorithm as described in Chap. 6 (see also [121, 19]). The expert parameters can be modified independently, and the method depends on the expert model. Their training is described when introducing their models in Sect. 4.2.

To summarise, $l(\boldsymbol{\theta}; \mathcal{D})$ is maximised by iterating over the expectation and the maximisation steps. In the expectation step, the responsibilities are computed for the current model parameters. In the maximisation step, the model parameters are updated with the computed responsibilities. Convergence of the algorithm can be determined by monitoring the result of (4.9).

4.1.4 Localisation by Interaction

The experts in the standard MoE model are localised in the input space through the interaction of expert and gating network training: after the gating is randomly initialised, the responsibilities are calculated by (4.12) according to how well the experts fit the data in the areas of the input space that they are assigned to. In the maximisation step, performing (4.13) tunes the gating parameters such that the gating network fits best the previously calculated responsibilities. Equation (4.14) causes the experts to be only trained on the areas that they are assigned to by the responsibilities. The next expectation step re-evaluates the responsibilities according to the new fit of the experts, and the maximisation step adapts the gating network and the experts again. Hence, iterating the expectation and the maximisation step causes the experts to be distributed according to their best fit to the data.

The pattern of localisation is determined by the form of the gating model. As previously demonstrated, the softmax function causes a soft linear partition of the input space. Thus, the underlying assumption of the model is that the data was generated by some processes that are linearly separated in the input space. The model structure becomes richer by adding hierarchies to the gating network [121]. That would move MoE to far away from LCS, which is why it will not be discussed any further.

4.1.5 Training Issues

The likelihood function of MoE is neither convex nor unimodal [20]. Hence, training it by using a hill-climbing procedure such as the EM-algorithm will not guarantee that we find the global maximum. Several approaches have been developed to deal with this problem (for example, [20, 4]), all of which are either based on random restart or stochastic global optimisers. Hence, they require several training epochs and/or a long training time. While this is not an issue for MoE where the global optimum only needs to be found once, it is not an option for LCS where the model needs to be (at least partially) re-trained for each change in the model structure. A potential LCS-related solution will be presented in Sect. 4.4.

4.2 Expert Models

So far, $p(\mathbf{y}|\mathbf{x}, \boldsymbol{\theta}_k)$ has been left unspecified. Its form depends on the task that is to be solved, and differs for regression and classification tasks. Here, we only deal with the LCS-related univariate regression task and the multiclass classification tasks, for which the expert models are introduced in the following sections.

4.2.1 Experts for Linear Regression

For each expert k, the linear univariate regression model (that is, $D_{\mathcal{Y}} = 1$) is characterised by a linear relation of the input $\mathbf{x}$ and the adjustable parameter $\mathbf{w}_k$, which is a vector of the same size as the input. Hence, the relation between the input $\mathbf{x}$ and the output y is modelled by a hyper-plane $\mathbf{w}_k^T\mathbf{x} - y = 0$. Additionally, the stochasticity and measurement noise are modelled by a Gaussian. Overall, the probabilistic model for expert k is given by

$$p(y|\mathbf{x}, \mathbf{w}_k, \tau_k) = \mathcal{N}(y|\mathbf{w}_k^T\mathbf{x}, \tau_k^{-1}) = \left(\frac{\tau_k}{2\pi}\right)^{1/2} \exp\left(-\frac{\tau_k}{2}(\mathbf{w}_k^T\mathbf{x} - y)^2\right), \quad (4.15)$$

where $\mathcal{N}$ stands for a Gaussian, and the model parameters $\boldsymbol{\theta}_k = \{\mathbf{w}_k, \tau_k\}$ are the $D_{\mathcal{X}}$-dimensional weight vector $\mathbf{w}_k$ and the noise precision (that is, inverse variance) τ_k. The distribution is centred on the inner product $\mathbf{w}_k^T \mathbf{x}$, and its spread is inversely proportional to τ_k and independent of the input.

As we give a detailed discussion about the implications of assuming this expert model and various forms of its incremental training in Chap. 5, let us here only consider how it specifies the maximisation step of the EM-algorithm for training the MoE model, in particular with respect to the weight vector $\mathbf{w}_k$: Combining (4.14) and (4.15), the term to maximise becomes

$$\begin{aligned}\sum_{n=1}^{N}\sum_{k=1}^{K} r_{nk} \ln p(y_n|\mathbf{x}_n, \mathbf{w}_k, \tau_k) &= \sum_{n=1}^{N}\sum_{k=1}^{K} r_{nk}\left(\frac{1}{2}\ln\frac{\tau_k}{2\pi} - \frac{\tau_k}{2}(\mathbf{w}_k^T\mathbf{x}_n - y_n)^2\right)\\ &= -\sum_{k=1}^{K}\frac{\tau_k}{2}\sum_{n=1}^{N} r_{nk}(\mathbf{w}_k^T\mathbf{x}_n - y_n)^2 + \text{const.},\end{aligned}$$

where the constant terms absorbs all terms that are independent of the weight vectors. Considering the experts separately, the aim for expert k is to find

$$\min_{\mathbf{w}_k} \sum_{n=1}^{N} r_{nk}(\mathbf{w}_k^T\mathbf{x}_n - y_n)^2, \tag{4.16}$$

which is a weighted linear least squares problem. This shows how the assumption of a Gaussian noise locally leads to minimising the empirical risk with the L_2 loss function.

4.2.2 Experts for Classification

For classification, assume that the number of classes is $D_{\mathcal{Y}}$, and the outputs are the vectors $\mathbf{y} = (y_1, \dots, y_{D_{\mathcal{Y}}})^T$ with all elements being $y_j = 0$, except for the element $y_{\bar{j}} = 1$, where $\bar{j}$ is the class associated with this output vector. Thus, similarly to the latent vector $\mathbf{z}$, the different $\mathbf{y}$'s obey a 1-of-$D_{\mathcal{Y}}$ structure.

The expert model $p(\mathbf{y}|\mathbf{x}, \boldsymbol{\theta}_k)$ gives the probability of the expert having generated an observation of the class specified by $\mathbf{y}$. Analogous to the gating network (4.4), this model could assume a log-linear relationship between this probability and the input $\mathbf{x}$, which implies that $p(\mathbf{y}|\mathbf{x}, \boldsymbol{\theta}_k)$ is assumed to vary with $\mathbf{x}$. However, to simplify interpretation of the expert model, it will be assumed that this probability remains constant over all inputs that the expert is responsible for, that is

$$p(\mathbf{y}|\mathbf{x}, \mathbf{w}_k) = \prod_{j=1}^{D_{\mathcal{Y}}} w_{kj}^{y_j}, \qquad \text{with } \sum_{j=1}^{D_{\mathcal{Y}}} w_{kj} = 1. \tag{4.17}$$

Thus, $p(\mathbf{y}|\mathbf{x}, \mathbf{w}_k)$ is independent of the input $\mathbf{x}$ and parametrised by $\boldsymbol{\theta}_k = \mathbf{w}_k$, and for any given $\mathbf{y}$ representing class $\bar{j}$, the model's probability is given by $w_{\bar{j}}$, the $\bar{j}$th element of $\mathbf{w}_k$.

By combining (4.14) and (4.17), the term to maximise in the M-step of the EM algorithm becomes

$$\sum_{n=1}^{N}\sum_{k=1}^{K} r_{nk} \ln p(\mathbf{y}_n|\mathbf{x}_n, \mathbf{w}_k) = \sum_{n=1}^{N}\sum_{k=1}^{K} r_{nk} \sum_{j=1}^{D_{\mathcal{Y}}} y_{nj} \ln w_{kj},$$

under the constraint $\sum_j w_{kj} = 1$ for all k. Considering each expert separately, expert k has to solve the constraint optimisation problem

$$\max_{\mathbf{w}_k} \sum_{n=1}^{N} r_{nk} \sum_{j=1}^{D_{\mathcal{Y}}} y_{nj} \ln w_{nj}, \tag{4.18}$$

$$\text{subject to } \sum_{j=1}^{D_{\mathcal{Y}}} w_{kj} = 1.$$

While the concepts introduced in the following sections are valid for any form of expert models, a detailed description of how to train the above expert models to find its parameters is given in Chap. 5.

4.3 Generalising the MoE Model

The standard MoE model assumes that each observation was generated by one and only one expert. In this section, the model will be made more LCS-like by replacing the term "expert" with "classifier", and by introducing the additional assumption that a classifier can only have produced the observation if it matches the corresponding input. The following sections implement this assumption and discuss its implications.

4.3.1 An Additional Layer of Forced Localisation

Let us recall that for a certain observation $(\mathbf{x}, \mathbf{y})$, the latent variable $\mathbf{z}$ determines which classifier generated this observation. The generalisation that is introduced assumes that a classifier k can have only generated this observation, that is, $z_k = 1$, if it matches the corresponding input.

Let us introduce an additional binary random vector $\mathbf{m} = (m_1, \dots, m_K)^T$, each element being associated with one classifier [2]. The elements of $\mathbf{m}$ are 1 if and only if the associated classifier matches the current input x, and 0 otherwise. Unlike $\mathbf{z}$, $\mathbf{m}$ does not comply to the 1-of-K structure, as more than one classifier can match the same input. The elements of the random vector are linked to the matching function by

$$p(m_k = 1|\mathbf{x}) = m_k(\mathbf{x}), \tag{4.19}$$

[2] While the symbol m also refers to the matching function, its use as either the matching function or the random variable that determines matching is apparent from its context.

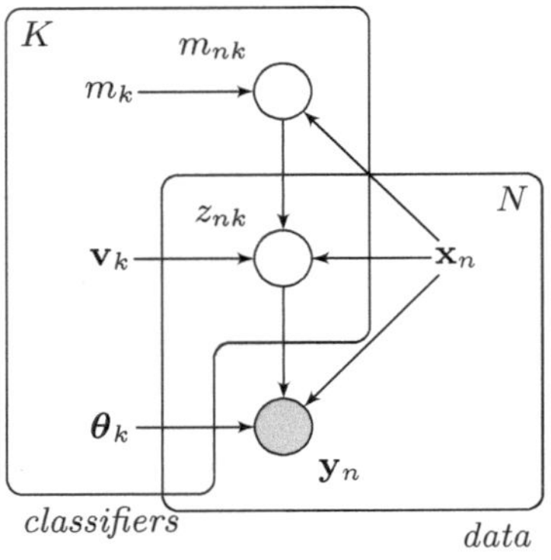

Fig. 4.3. Directed graphical model of the generalised Mixtures-of-Experts model. See the caption of Fig. 4.1 for instructions on how to read this graph. When compared to the Mixtures-of-Expert model in Fig. 4.1, the latent variables z_{nk} depends additionally on the matching random variables m_{nk}, whose values are determined by the mixing functions m_k and the inputs $\mathbf{x}_n$

that is, the value of a classifier's matching function determines the probability of that classifier matching a certain input.

To enforce matching, the probability for classifier k having generated observation $(\mathbf{x}, \mathbf{y})$, given by (4.4), is redefined to be

$$p(z_k = 1|\mathbf{x}, \mathbf{v}_k, m_k) \propto \begin{cases} \exp(\mathbf{v}_k^T \phi(\mathbf{x})) & \text{if } m_k = 1 \text{ for } \mathbf{x}, \\ 0 & \text{otherwise,} \end{cases} \tag{4.20}$$

where ϕ is a transfer function, whose purpose will be explained later and which can for now be assumed to be the identity function, $\phi(\mathbf{x}) = \mathbf{x}$. Thus, the differences from the previous definition (4.4) are the additional transfer function and the condition on m_k that locks the generation probability to 0 if the classifier does not match the input. Removing the condition on m_k by marginalising it out results in

$$\begin{aligned} g_k(\mathbf{x}) \equiv p(z_k = 1|\mathbf{x}, \mathbf{v}_k) &\propto \sum_{m \in \{0,1\}} p(z_k = 1|\mathbf{x}, \mathbf{v}_k, m_k) p(m_k = m|\mathbf{x}) \\ &= 0 + p(z_k = 1|\mathbf{x}, \mathbf{v}_k, m_k) p(m_k = 1|\mathbf{x}) \\ &= m_k(\mathbf{x}) \exp(\mathbf{v}_k^T \phi(\mathbf{x})). \end{aligned} \tag{4.21}$$

Adding the normalisation term, the gating network is now defined by

$$g_k(\mathbf{x}) \equiv p(z_k = 1|\mathbf{x}, \mathbf{v}_k) = \frac{m_k(\mathbf{x}) \exp(\mathbf{v}_k^T \phi(\mathbf{x}))}{\sum_{j=1}^K m_j(\mathbf{x}) \exp(\mathbf{v}_j^T \phi(\mathbf{x}))}. \tag{4.22}$$

As can be seen when comparing it to (4.5), the additional layer of localisation is specified by the matching function, which reduces the gating to $g_k(\mathbf{x}) = 0$ if the classifier does not match $\mathbf{x}$, that is, if $m_k(\mathbf{x}) = 0$.

As classifiers can only generate observations if they match the corresponding input, the classifier model itself does not require any modification. Additionally, (4.9) is still valid, as $z_k = 1$ only if $m_k = 1$ by (4.20). Figure 4.3 shows the graphical model that, when compared to Fig. 4.1, illustrates the changes that are introduces by generalising the MoE model.

4.3.2 Updated Expectation-Maximisation Training

The only modifications to the standard MoE are changes to the gating network, expressed by g_k. As (4.12), (4.13) and (4.14) are independent of the functional form of g_k, they are still valid for the generalised MoE. Therefore, the expectation step of the EM-algorithm is again performed by evaluating the responsibilities by (4.12), and the gating and classifier models are updated by (4.13) and (4.14). Convergence of the algorithm is again monitored by (4.9).

4.3.3 Implications on Localisation

Localisation of the classifiers is achieved on one hand by the matching function of the classifiers, and on the other hand by the combined training of gating networks and classifiers.

Let us first consider the case when the nth observation $(\mathbf{x}_n, \mathbf{y}_n)$ is matched by one and only one classifier k, that is $m_j(\mathbf{x}_n) = 1$ only if $j = k$, and $m_j(\mathbf{x}_n) = 0$ otherwise. Hence, by (4.22), $g_j(\mathbf{x}_n) = 1$ only if $j = k$, and $g_j(\mathbf{x}_n) = 0$ otherwise, and consequently by (4.12), $r_{nj} = 1$ only if $j = k$, and $r_{nj} = 0$ otherwise. Therefore, full responsibility for the observation is given to the one and only matching classifier, independent of its goodness-of-fit.

On the other hand, assume that the same observation $(\mathbf{x}_n, \mathbf{y}_n)$ is matched by all classifiers, that is $m_j(\mathbf{x}_n) = 1$ for all $j \in \{1, \dots, K\}$, and assume the identity transfer function $\phi(\mathbf{x}) = \mathbf{x}$. In that case, (4.22) reduces to the standard MoE gating network (4.5) and we perform a soft linear partitioning as described in Sect. 4.1.4.

In summary, localisation by matching determines for which areas of the input space the classifiers attempt to model the observations. In areas where they match, they are distributed by soft linear partitions as in the standard MoE model. Hence, we can acquire a two-layer intuition on how localisation is performed: Matching determines the rough areas where classifiers are responsible to model the observations, and the softmax function then performs the fine-tuning in areas of overlap between classifiers.

4.3.4 Relation to Standard MoE Model

The only difference between the generalised MoE model and the standard MoE model is the definition of the gating model g_k. Comparing the standard model (4.5) with its generalisation (4.22), the standard model is recovered from the generalisation by having $m_k(\mathbf{x}) = 1$ for all k and $\mathbf{x}$, and the identity transfer function $\phi(\mathbf{x}) = \mathbf{x}$ for all $\mathbf{x}$. Defining the matching functions in such a way is equivalent to having each classifier match all inputs. This results in a set of classifiers that all match the whole input space, and localisation is performed by soft linear partitioning of the gating network.

4.3.5 Relation to LCS

This generalised MoE model satisfies all characteristics of LCS outlined in Sect. 3.2: Each classifier describes a localised model with its localisation determined by the model structure, and the local models are combined to form a global model. So given that the model can be trained efficiently, and that there exists a good mechanism for searching the space of model structures, do we already have an LCS? While some LCS researchers might disagree — partially because there is no universal definition of what an LCS is and LCS appear to be mostly thought of in algorithmic terms rather than in terms of the model that they describe — the author believes that this is the case.

However, the generalised MoE model has a feature that no LCS has ever used: beyond the localisation of classifiers by their matching function, the responsibilities of classifiers that share matching inputs is further distributed by the softmax function. While this feature might lead to a better fit of the model to the data, it blurs the observation/classifier association by extending it beyond the matching function. Nonetheless, the introduced transfer function ϕ can be used to level this effect: when defined as the identity function $\phi(\mathbf{x}) = \mathbf{x}$, then by (4.21) the probability of a certain classifier generating an observation for a matching input is log-linearly related to the input $\mathbf{x}$. However, by setting $\phi(\mathbf{x}) = 1$ for all $\mathbf{x}$, the relation is reduced to $g_k(\mathbf{x}) \propto m_k(\mathbf{x}) \exp(v_k)$, where the gating parameter $\mathbf{v}_k$ reduces to the scalar v_k. Hence, the gating weight becomes independent of the input (besides the matching) and only relies on the constant v_k through $\exp(v_k)$. In areas of the input space that several classifiers match, classifiers with a larger v_k have a stronger influence when forming a prediction of the global model, as they have a higher gating weight. To summarise, setting $\phi(\mathbf{x}) = 1$ makes gating independent of the input (besides the matching) and the gating weight for each classifier is determined by a single scalar that is independent of the current input $\mathbf{x}$ that it matches. Further details and alternative models for the gating network are discussed in Chap. 6.

Note that $\phi(\mathbf{x}) = 1$ is not applicable in the standard MoE model, that is, when all classifiers match the full input space. In this case, we have neither localisation by matching nor by the softmax function. Hence, the global model is not better at modelling the data than a single local model applied to the whole data.

Example 4.2 (Localisation by Matching and the Softmax Function). Consider the same setting as in Example 4.1, and additionally $\phi(\mathbf{x}) = \mathbf{x}$ for all $\mathbf{x}$ and the matching functions

$$m_1(\mathbf{x}) = \begin{cases} 1 & \text{if } \sqrt{x_1^2 + x_2^2} \leq 3, \\ 0 & \text{otherwise,} \end{cases} \tag{4.23}$$

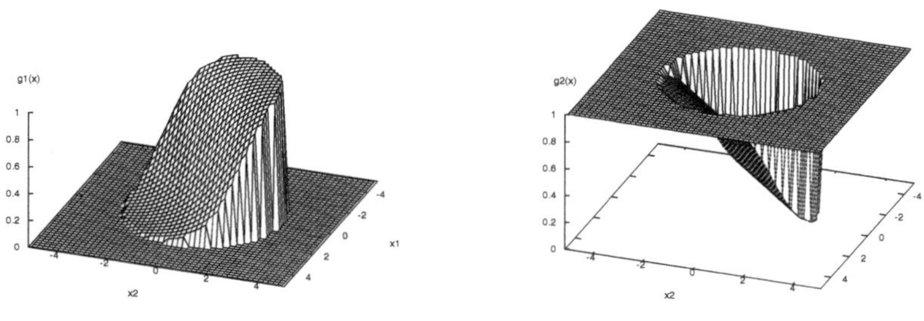

(a) (b)

Fig. 4.4. Plots showing the generalised softmax function (4.22) for 2 classifiers with inputs $\mathbf{x} = (1, x_1, x_2)^T$ and $\phi(\mathbf{x}) = \mathbf{x}$, where Classifier 1 in plot (a) has gating parameters $\mathbf{v}_1 = (0, 0, 1)^T$ and matches a circle of radius 3 around the origin, and Classifier 2 in plot (b) has gating parameters $\mathbf{v}_2 = (0, 1, 0)^T$ and matches all inputs

and $m_2(\mathbf{x}) = 1$ for all $\mathbf{x}$. Therefore, Classifier 1 matches a circle of radius 3 around the origin, and Classifier 2 matches the whole input space. The values for $g_1(\mathbf{x})$ and $g_2(\mathbf{x})$ are shown in Figs. 4.4(a) and 4.4(b), respectively. As can be seen, the whole part of the input space that is not matched by Classifier 1 is fully assigned to Classifier 2 by $g_2(\mathbf{x}) = 1$. In the circular area where both classifiers match, the softmax function performs a soft linear partitioning of the input space, just as in Fig. 4.2.

The effect of changing the transfer function to $\phi(\mathbf{x}) = 1$ is visualised in Fig. 4.5, and shows that in such a case no linear partitioning takes place. Rather, in areas of the input space that both classifiers match, (4.22) assigns the generation probabilities input-independently in proportion the exponential of the gating parameters $v_1 = 0.7$ and $v_2 = 0.3$.

Besides localisation beyond matching, the generalised MoE model has another feature that distinguishes it from any previous LCS [3]: it allows for matching by a degree of the range $[0, 1]$ rather than by just specifying where a classifier matches and where it does not (as, for example, specified by set $\mathcal{X}_k$ and (3.9)). Additionally, by (4.19), this degree has the well-defined meaning of the probability $p(m_k = 1|\mathbf{x})$ of classifier k matching input $\mathbf{x}$. Alternatively, by observing that $\mathbb{E}(m_k|\mathbf{x}) = p(m_k = 1|\mathbf{x})$, this degree can also be interpreted as the

[3] While Butz seems to have experimented with matching by a degree in [41], he does not describe how it is implemented and only states that "Preliminary experiments in that respect [...] did not yield any further improvement in performance". Furthermore, his hyper-ellipsoidal conditions [41, 52] might look like matching by degree on initial inspection, but as he determines matching by a threshold on the basis function, matching is still binary. Fuzzy LCS (for example, [60]), on the other hand, provide matching by degree but are usually not developed from the bottom up which makes modifying the parameter update equations difficult.

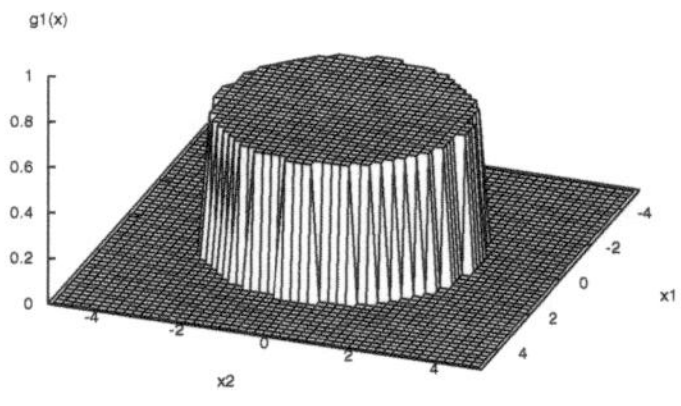

(a)

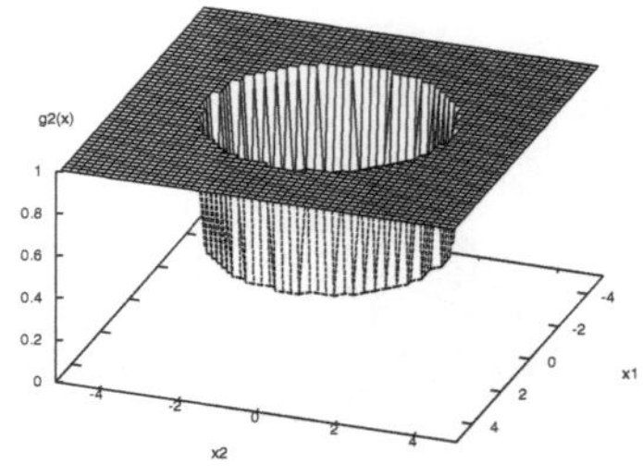

(b)

Fig. 4.5. Plots showing the generalised softmax function (4.22) for 2 classifiers with inputs $\mathbf{x} = (1, x_1, x_2)^T$ and $\phi(\mathbf{x}) = 1$, where Classifier 1 in plot (a) has gating parameters $v_1 = 0.7$ and matches a circle of radius 3 around the origin, and Classifier 2 in plot (b) has gating parameters $v_2 = 0.3$ and matches all inputs

expectation of the classifier matching the corresponding input. Overall, matching by a degree allows the specification of soft boundaries of the matched areas which can be interpreted as the uncertainty about the exact area to match[4], justified by the limited number of data available. This might solve issues with hard classifier matching boundaries when searching for good model structures, which can occur when the input space $\mathcal{X}$ is very large or even infinite, leading to a possibly infinite number of possible model structures. In that case, smoothing the classifier matching boundaries – as employed in Chap. 8 – makes fully covering the input space with classifiers easier.

4.3.6 Training Issues

If each input is only matched by a single classifier, each classifier model is trained separately, and the problem of getting stuck in local maxima does not occur, analogous to the discussion that follows in Sect. 4.4.3. Classifiers with overlapping matching areas, on the other hand, cause the same training issues as already discussed for the standard MoE model in Sect. 4.1.5, which causes the model training to be time-consuming.

In the presented approach, LCS training is conceptually split into two parts: training the model for a fixed model structure, and searching the space of possible model structures. To do the latter, evaluation of a single model structure by training the model needs to be efficient. Hence, the current training strategy is hardly a viable option. However, identifying the cause for local maxima allows for modifying the model to avoid those and therefore make model training more efficient, as shown in the next section.

[4] Thanks to Dr. Dan Richardson, University of Bath, for this interpretation.

4.4 Independent Classifier Training

The assumption of the standard MoE model is that any observation is generated by one and only one classifier. This was generalised by adding the restriction that any classifier can only have generated an observation if it matches the input associated with this observation, thereby adding an additional layer of forced localisation of the classifiers in the input space.

Here, a change rather than a generalisation is introduced to the model assumptions: as before it is assumed that the data is generated by a combination of localised processes, but the role of the classifiers is changed from cooperating with other classifiers in order to locally model the observations that it matches to modelling *all* observations that it matches, independent of the other classifiers that match the same inputs. This distinction becomes clearer once the resulting formal differences have been discussed in Sects. 4.4.2 and 4.4.3.

The motivation behind this change is twofold: firstly, it removes local maxima and thus simplifies classifier training, and secondly, it simplifies the intuition behind what a classifier models. Firstly, these motivations are discussed in more details, followed by their implication on training the model and the assumptions about the data-generating process.

4.4.1 The Origin of Local Maxima

Following the discussion in Sect. 4.1.5, local maxima of the likelihood function are the result of the simultaneous training of the classifiers and the gating network. In the standard MoE model, this simultaneous training is necessary to provide the localisation of the classifiers in the input space. In the introduced generalisation, on the other hand, a preliminary layer of localisation is provided by the matching function, and the interaction between classifiers and the gating network is only required for inputs that are matched by more than one classifier. This was already demonstrated in Sect. 4.3.3, where it was shown that classifiers acquire full responsibility for inputs that they match alone. Hence, in the generalised MoE, local maxima only arise when classifiers overlap in the input space.

4.4.2 What Does a Classifier Model?

By (4.14), a classifier aims at maximising the sum of log-likelihoods of all observations, weighted by the responsibilities. By (4.12) and (4.22), these responsibilities can only be non-zero if the classifier matches the corresponding inputs, that is, $r_{nk} > 0$ only if $m_k(\mathbf{x}_n) > 0$. Hence, by maximising (4.14), a classifier only considers observations that it matches.

Given that an observation $(\mathbf{x}_n, \mathbf{y}_n)$ is matched by a single classifier k, it was established in Sect. 4.3.3 that $r_{nk} = 1$ and $r_{nj} = 0$ for all $j \neq k$. Hence, (4.14) assigns full weight to classifier k when maximising the likelihood of this observation. Consequently, given that all observations that a classifier matches are matched by only this classifier, the classifier models these observations in full, independent of the other classifiers[5].

Let us consider how observations are modelled that are matched by more than one classifier: as a consequence of (4.12), the non-negative responsibilities of all matching classifiers sum up to 1, and are therefore between 0 and 1. Hence, by (4.14), each matching classifier assigns less weight to modelling the observation than if it would be the only classifier matching it. Intuitively, overlapping classifiers "share" the observation when modelling it.

To summarise, i) a classifier only models observations that it matches, ii) it assigns full weight to observations that no other classifier matches, and iii) it assigns partial weight to observations that other classifiers match. Expressed differently, a classifier fully models all observations that it matches alone, and partially models observations that itself and other classifiers match. Consequently, the local model provided by a classifier cannot be interpreted by their matching function alone, but also requires knowledge of the gating network parameters. Additionally, when changing the model structure as discussed in Sect. 3.2.6 by adding, removing, or changing the localisation of classifiers, all other overlapping classifiers need to be re-trained as their model is now incorrect due to changing responsibilities. These problems can be avoided by training the classifiers independently of each other, making the classifier model more transparent.

4.4.3 Introducing Independent Classifier Training

Classifiers are trained independently if we replace the responsibilities r_{nk} in (4.14) by the matching functions $m_k(\mathbf{x}_n)$ to get

$$\max_{\theta} \sum_{n=1}^{K} \sum_{k=1}^{K} m_k(\mathbf{x}_n) \ln p(\mathbf{y}_n|\mathbf{x}_n, \boldsymbol{\theta}_k). \tag{4.24}$$

Hence, a classifier models all observations that it matches, independent of the other classifiers. Thus, the first goal of simplifying the intuition about what a single classifier models is reached. While this does not cause any change for observations that are matched by a single classifier, observations that are matched by several classifiers are modelled by each of these classifiers independently rather than shared between them. This independence is shown by the graphical model in Fig. 4.6, which illustrates the model of a single classifier k.

With this change, the classifiers are independent of the responsibilities and subsequently also of the gating network. Thus, they can be trained completely independently, and the model structure can be modified by adding, removing, or changing classifier locations without re-training the other classifiers that are currently in the model, and thereby make searching the space of possible model structures more efficient.

[5] XCS has the tendency to evolve sets of classifiers with little overlap in the areas that they match. In such cases, all classifiers model their assigned observations in full, independent of if they are trained independently or in combination.

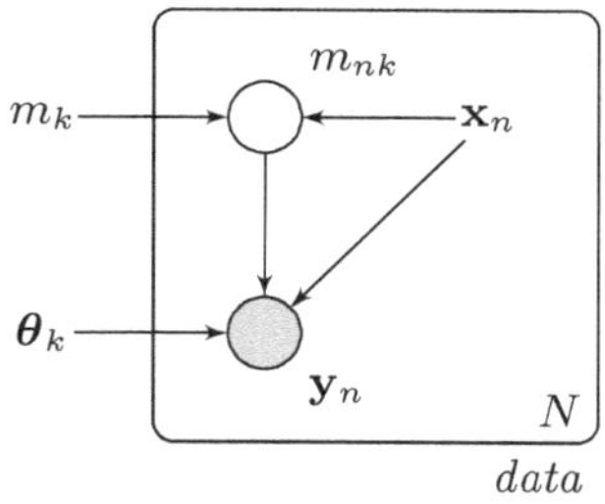

Fig. 4.6. Directed graphical model for training classifier k independently. See the caption of Fig. 4.1 for instructions on how to read this graph. Note that the values of the matching random variables m_{nk} are determined by the matching function m_k and the inputs $\mathbf{x}_n$.

An additional consequence of classifiers being trained independently of the responsibilities is that for standard choices of the local models (see, for example [121]), the log-likelihood (4.24) is concave for each classifier. Therefore, it has a unique maximum and consequently we cannot get stuck in local maxima when training individual classifiers.

4.4.4 Training the Gating Network

Training the gating network remains unchanged, and therefore is described by (4.12) and (4.13). Given a set of trained classifiers, the responsibilities are fully specified by evaluating (4.12). Hence, the log-likelihood of the gating network (4.13) is a concave function (for example, [20]), and therefore has a unique maximum.

Thus, the classifier models have unique optima and can be trained independently of the gating network by maximising a concave log-likelihood function. Furthermore, the gating network depends on the goodness-of-fit of the classifiers, but as they are trained independently, the log-likelihood function of the gating network is also concave. Therefore, the complete model has a unique maximum likelihood, and as a consequence, the second goal of removing local maxima to ease training of the model is reached.

4.4.5 Implications on Likelihood and Assumptions about the Data

Letting a classifier model match each observation with equal weight violates the assumption that each observation was generated by one and only one classifier for observations that are matched by more than one classifier. Rather, the model of each classifier can be interpreted as a hypothesis for a data-generating process that generated all observations of the matched area of the input space.

The gating network, on the other hand, was previously responsible for modelling the probabilities of some classifier having produced some observation, and the classifiers were trained according to this probability. While the gating network still has the same purpose when the classifiers are trained independently, the estimated probability is not fed back to the classifiers anymore. The cost of this lack of feedback is a worse fit of the model to the data, which results in a lower likelihood of the data given the model structure.

Note, however, that independent classifier training only causes a change in the likelihood in areas where more than one classifier matches the same input. Hence, we only get a lower likelihood if classifiers have large areas of overlap, and it is doubtful that such a solution is ever desired. Nonetheless, the potentially worse fit needs to be offset by the model structure search to find solutions with sufficiently minimal overlap between different classifiers.

As the gating network is not gating the observations to the different classifiers anymore, but rather mixes the independently trained classifier models to best explain the available observations, it will in the remaining chapters be referred to as the *mixing model* rather than the gating network.

4.5 A Brief Comparison to Linear LCS Models

The LCS model introduced in this chapter is a non-linear model as both the classifiers and the mixing model have tunable parameters. It is in its structure very similar to XCS and its derivatives, as well as to other LCS that train is classifiers independently (for example, CCS [153, 154]).

Another popular structure for LCS models is a linear one, which is characterised by the output estimate $\hat{f}_{\mathcal{M}}(\mathbf{x};\boldsymbol{\theta})$ being a linear function of the model parameters $\boldsymbol{\theta}$. Assuming a linear classifier model (4.15) and a output estimate $\hat{f}_{\mathcal{M}}(\mathbf{x};\boldsymbol{\theta})$ formed by the mean of $p(\mathbf{y}|\mathbf{x},\boldsymbol{\theta})$ by (4.8), this estimate is given by

$$\hat{f}_{\mathcal{M}}(\mathbf{x};\boldsymbol{\theta}) = \mathbb{E}(\mathbf{y}|\mathbf{x},\boldsymbol{\theta}) = \sum_{k=1}^{K} g_k(\mathbf{x})\mathbf{w}_k^T\mathbf{x}. \tag{4.25}$$

In order for $\hat{f}_{\mathcal{M}}$ to be a linear in $\boldsymbol{\theta}$, the g_k's have to be independent of the parameters, unlike for the generalised MoE where they are parametrised by $\mathbf{V} \in \boldsymbol{\theta}$. This causes the log-likelihood $l(\mathcal{D};\boldsymbol{\theta})$ to be convex with respect to the w_k's, with a unique maximum that is easy to find.

The linear LCS model can have different instantiations by specifying the g_k's differently. Due to their current use in LCS, two of these instantiations are of particular interest. The first is given by $g_k(\mathbf{x}) = m_k(\mathbf{x})$, such that (4.25) becomes

$$\hat{f}_{\mathcal{M}}(\mathbf{x};\boldsymbol{\theta}) = \sum_{k=1}^{K} m_k(\mathbf{x})\mathbf{w}_k^T\mathbf{x}. \tag{4.26}$$

Therefore, for each input $\mathbf{x}$, the models of all matching classifiers are effectively agglomerated. This clearly shows that the classifiers do not form their predictions independently. As a consequence, classifiers cannot be interpreted as localised models, but are rather localised components of the model that is formed by the combination of all classifiers. While this by itself is not necessarily a drawback, the need to re-train overlapping classifiers when adding or removing classifiers to search the model structure space is clearly a disadvantage of the linear structure, and generally of all structures that do not train classifiers independently. Also, due to the interaction of overlapping classifiers, there is no clear indicator of the quality of a single classifier. LCS instances that use this agglomerating structure are ZCS [236], as identified by Wada et al. [224], and an LCS developed by Booker [23]. In both cases, the quality measure of classifier k is a measure of the magnitude of its parameters $\mathbf{w}_k$ – a method called "fitness sharing" in the case of ZCS[6].

An alternative to agglomerating classifiers in linear models is to average over them by using $g_k(\mathbf{x}) = m_k(\mathbf{x}) / \sum_{\bar{k}} m_{\bar{k}}(\mathbf{x})$, such that (4.25) becomes

$$\hat{f}_{\mathcal{M}}(\mathbf{x}; \boldsymbol{\theta}) = \sum_{k=1}^{K} \frac{m_k(\mathbf{x})}{\sum_{\bar{k}} m_{\bar{k}}(\mathbf{x})} \mathbf{w}_k^T \mathbf{x}. \tag{4.27}$$

Note that this form is very similar to the gating network (4.22) of the generalised MoE, with the difference that the average is not weighted by the quality of the prediction of the classifiers. Thus, the fit of this model will be certainly worse than the weighted averaging of the generalised MoE. Also, even though now the predictions of overlapping classifiers do not directly depend one each other, the value of $g_k(\mathbf{x})$ still depends on other classifiers matching the same input $\mathbf{x}$. Thus, classifiers are not trained independently, and they needs to be re-trained in case of the removal or addition of overlapping classifiers. An instance of this form of linear LCS was introduced by Wada et al. as a linearised version of XCS [223].

It needs to be emphasised that this section is not supposed to demonstrate the superiority of the introduced LCS model and its currently used instances over LCS based on linear models. Rather, it attempts to point out significant differences between these two model types and its consequences. Having a linear model structure removes the need of an explicit mixing model and simplifies finding the model parameters for a fixed model structure, but this comes at the price of having to re-train the model once this structure changes. Using non-linear models, on the other hand, requires a mixing model and the introduction of independent classifier training (as a rather unsatisfying solution) to simplify the training of a single model structure, but simplifies changing this structure and provides a clearer interpretation of the model formed by a single classifier.

---

[6] It is not clear if such a quality measure is indeed useful in all occasions. Booker proposed to consider classifiers with low parameter values as bad classifiers, as "The ones with large weights are the most important terms in the approximation" [24], but would that also work in cases where low parameter values are actually good parameter values? One can easily imaging a part of a function that is constantly 0 and thus requires 0 parameter values to model it.

4.6 Discussion and Summary

In this chapter, a probabilistic LCS model was introduced as a generalisation of the MoE model, by adding matching as a form of forced localisation of the experts. Additionally, training was simplified by handling the classifiers independently of the gating network. The resulting probabilistic LCS model acts as the basis for further development in this book. In fact, solving (4.24) to train the classifiers forms the basis of the next chapter. The chapter thereafter deals with the mixing model by describing how the solution to (4.13) can be found exactly and by approximation. Thus, in combination, the following two chapters describe in detail how the model can be trained by maximum likelihood, both by batch learning and incrementally.

Even though we have approached the LCS model from a different perspective, the resulting structure is very similar to a currently existing LCS: XCS and its derivatives follow the same path of independently training the classifier models and combining them by a mixing model. While in XCS it is not explicitly identified that the classifiers are indeed trained independently, this fact becomes apparent in the next chapter, where it is shown that the classifier parameter update equations that result from independent classifier training resemble those of XCS. The mixing model used by XCS does not conform to the generalised softmax function but rather relies on heuristics, as is demonstrated in Chap. 6.

Independent classifier training moves LCS closer to ensemble learning. This similarity has been exploited recently by Brown, Marshall and Kovacs [29, 162], who have used knowledge from ensemble learning and other machine learning methods to improve the performance of UCS [161]. Even though this direction is very promising, the direct link between LCS and ensemble learning will not be considered further in this book.

In summary, amongst currently popular LCS, the presented model is most similar to XCS(F). It combines independently trained classifiers by a mixing model to provide a global model that aims at explaining the given observations. This particular model type was chosen not to represent the "best" LCS model, but as an example to demonstrate the model-based approach. Other LCS model are equally amendable to this approach, but for the beginning, only a single model type is fully considered. As in this model type the classifiers are trained independently of each other, it is possible to concentrate on the training of a single classifier, as is done in the following chapter.

5 Training the Classifiers

The model of a set of classifiers consists of the classifiers themselves and the mixing model. The classifiers are localised linear regression or classification models that are trained independently of each other, and their localisation is determined by the matching function m_k. This chapter is entirely devoted to the training of a single classifier and mainly focuses on the linear regression models, but also briefly discusses classification at the end of the chapter.

The linear classifier model was already introduced in Sec. 4.2.1, but here more details are provided about its underlying assumptions, and how it can be trained in both a batch learning and an incremental learning way. Most of the concepts and methods in this chapter are well known in statistics (for example, [97]) and adaptive filter theory (for example, [105]), but have not been put into the context of LCS before.

In training a classifier we focus on solving (4.24), which emerges from applying the principle of maximum likelihood to the LCS model. Maximising the likelihood minimises the empirical rather than the expected risk, which might lead to overfitting. Nonetheless, it provides a first approach to training the classifiers, and results in parameter update equations that are for regression models mostly equivalent to the ones used in XCS(F), which confirms that the LCS model is in its structure similar to XCS(F). Chapter 7 returns to dealing with over- and underfitting, with methods that are closely related to the methods derived in this chapter.

The classifier model parameters to estimate are the weight vector and its noise variance for the linear regression model, and the weight vector alone for the classification model. The noise variance is a good indicator of the goodness-of-fit of the linear model and is also used in a modified form to estimate the accuracy of a classifier in XCS and its variants. In general, it is useful to guide the model structure search as we have already discussed in Sect. 3.2.6, and thus having a good estimate of the noise variance is advantageous. Thus, we put special emphasis on how to estimate it efficiently and accurately. For the classification model, a classifier quality measure emerges naturally from the estimated weight vector and does not need to be estimated separately, as shown in Sect. 5.5.

J. Drugowitsch: Des. & Anal. of Learn. Class. Sys.: A Prob. Approach, SCI 139, pp. 65–99, 2008.
springerlink.com

Since each classifier is trained independently (see Sect. 4.4), this chapter focuses exclusively on the training of a single classifier k. To keep the notation uncluttered, the subscript k is dropped; that is, the classifier's matching function m_k is denoted m, the model parameters $\boldsymbol{\theta}_k = \{\mathbf{w}_k, \tau_k\}$ become $\mathbf{w}$ and τ, and the estimate $\hat{f}_k$ provided by classifier k is denoted $\hat{f}$. For any further variables introduced throughout this chapter it will be explicitly stated whether they are local to a classifier.

Firstly, the linear regression classifier model and its underlying assumptions are introduced, followed in Sect 5.2 by how to estimate its parameters if all training data is available at once. Incremental learning approaches are discussed in Sect. 5.3, where gradient-based and exact methods of tracking the optimal weight vector estimate are described. Estimating the noise variance simultaneously is discussed for both methods in Sect. 5.3.7. In Sect. 5.4, slow convergence of gradient-based methods is demonstrated empirically. Turning to classification, the training of these models is discussed in Sect. 5.5, after which the chapter is summarised by putting its content into the context of current LCS.

5.1 Linear Classifier Models and Their Underlying Assumptions

Linear regression models were chosen as a good balance between the expressiveness of the model and the ease of training the model (see Sect. 3.2.3). The univariate linear model has already been introduced Sect. 4.2.1, but here, its underlying assumptions and implications are considered in more detail.

5.1.1 Linear Models

A linear model assumes a linear relation between the inputs and the output, parametrised by a set of model parameters. Given an input vector $\mathbf{x}$ with $D_{\mathcal{X}}$ elements, the model is parametrised by the equally-sized random vector $\boldsymbol{\omega}$ with realisation $\mathbf{w}$, and assumes that the scalar output random variable υ with realisation y follows the relation

$$\upsilon = \boldsymbol{\omega}^T \mathbf{x} + \epsilon, \tag{5.1}$$

where ϵ is a zero-mean Gaussian random variable that models the stochasticity of the process and the measurement noise. Hence, ignoring for now the noise term ϵ, its is assumed that the process generates the output by a weighted sum of the components of the input, as becomes very clear when considering a realisation $\mathbf{w}$ of $\boldsymbol{\omega}$, and rewriting the inner product

$$\mathbf{w}^T \mathbf{x} \equiv \sum_i w_i x_i, \tag{5.2}$$

where w_i and x_i are the ith element of $\mathbf{w}$ and $\mathbf{x}$ respectively.

While linear models are usually augmented by a bias term to offset them from the origin, it will be assumed that the input vector always contains a single constant element (which is usually fixed to 1), which has the equal effect. For example, consider the input space to be the set of reals; that is $\mathcal{X} = \mathbb{R}$, $D_{\mathcal{X}} = 1$ and both x and w are scalars. In such a case, the assumption of a linear model implies that the observed output follows xw, which is a straight line through the origin with slope w. To add the bias term, we can instead assume an augmented input space $\mathcal{X}' = \{1\} \times \mathbb{R}$, with input vectors $\mathbf{x}' = (1, \mathbf{x})^T$, resulting in the linear model $\mathbf{w}^T\mathbf{x}' = w_1 + w_2 x$ – a straight line with slope w_2 and bias w_1. Equally, the input vector can be augmented by other elements to extend the expressiveness of the linear model, as shown in the following example:

Example 5.1 (Common Classifier Models used in XCS(F)). Initially, classifiers in XCS [237, 238] only provided a single prediction, independent of the input. Such behaviour is equivalent to having the scalar input $x_n = 1$ for all n, as the weight w then models the output as an average over all matched outputs, as will be demonstrated in Example 5.2. Hence, such classifiers will be called *averaging classifiers*.

Later, Wilson introduced XCSF (the F standing for "function"), that initially used straight lines as the local models [241]. Hence, in the one-dimensional case, the inputs are given by $\mathbf{x}_n = (1, i_n)$ to model the output by $w_1 + w_2 i_n$, where i_n is the variable part of the input. This concept was taken further by Lanzi et al. [141] by applying 2nd and 3rd order polynomials, using the input vectors $\mathbf{x}_n = (1, i_n, i_n^2)^T$ and $\mathbf{x}_n = (1, i_n, i_n^2, i_n^3)^T$ respectively. Naturally, the input vector does not need to be restricted to taking i_n to some power, but allows for the use of arbitrary functions. These functions are known as *basis functions*, as they construct the base of the input space. Nonetheless, increasing the complexity of the input space makes it harder to interpret the local models. Hence, if it is the aim to understand the localised model, these models should be kept simple – such as straight lines.

5.1.2 Gaussian Noise

The noise term ϵ captures the stochasticity of the data-generating process and the measurement noise. In the case of linear models, the inputs and outputs are assumed to stand in a linear relation. Every deviation from this relation is captured by ϵ and is interpreted as noise. Hence, assuming the absence of measurement noise, the fluctuation of ϵ gives information about the adequacy of assuming a linear model. In other words, if the variance of ϵ is small, then inputs and outputs do indeed follow a linear relation. Hence, the variance of ϵ can be used as a measure of how well the local model fits the data. For that reason, the aim is not only to find a weight vector that maximises the likelihood, but also to simultaneously estimate the variance of ϵ.

For linear models it is common to assume that the random variable ϵ representing the noise has zero mean, constant variance, and follows a normal distribution [97], that is $\epsilon \sim \mathcal{N}(0, \tau^{-1})$, where τ is the noise precision (inverse noise variance). Hence, in combination with (5.1), and for some realisation $\mathbf{w}$ of $\boldsymbol{\omega}$ and input $\mathbf{x}$, the output is modelled by

$$\upsilon \sim p(y|\mathbf{x}, \mathbf{w}, \tau^{-1}) = \mathcal{N}(y|\mathbf{w}^T\mathbf{x}, \tau^{-1}) = \left(\frac{\tau}{2\pi}\right)^{1/2} \exp\left(-\frac{\tau}{2}(\mathbf{w}^T\mathbf{x} - y)^2\right), \quad (5.3)$$

which defines the probabilistic model of a linear regression and forms the core of its investigation.

That the assumption of Gaussian noise is sensible is discussed at length by Maybeck [164, Chap. 1].

5.1.3 Maximum Likelihood and Least Squares

To model the matched observations, a classifier aims at maximising the probability of these observations given its model, as formally described by (4.24). Combined with the linear model (5.3), the term to maximise by a single classifier k is given by

$$\sum_{n=1}^{N} m(\mathbf{x}_n) \ln p(y_n|\mathbf{x}_n, \mathbf{w}, \tau^{-1}) =$$

$$\sum_{n=1}^{N} m(\mathbf{x}_n) \left(-\frac{1}{2} \ln(2\pi) + \frac{1}{2} \ln \tau - \frac{\tau}{2} (\mathbf{w}^T \mathbf{x}_n - y_n)^2 \right). \tag{5.4}$$

As already shown in Sect. 4.2.1, maximising (5.4) with respect to the weight vector $\mathbf{w}$ results in the weighted least squares problem,

$$\min_{w} \sum_{n=1}^{N} m(\mathbf{x}_n) \left(\mathbf{w}^T \mathbf{x}_n - y_n\right)^2, \tag{5.5}$$

where the weights are given by the classifier's matching function. Thus, to determine $\mathbf{w}$ by maximum likelihood, we only consider observations for which $m(\mathbf{x}_n) > 0$, that is, which are matched.

To determine the noise precision of the fitted model, we maximise (5.4) with respect to τ, resulting in the problem

$$\max_{\tau} \left(\ln(\tau) \sum_{n=1}^{N} m(\mathbf{x}_n) + \tau \sum_{n=1}^{N} m(\mathbf{x}_n) \left(\mathbf{w}^T \mathbf{x}_n - y_n\right)^2 \right), \tag{5.6}$$

where $\mathbf{w}$ is the weight vector determined by (5.5).

The rest of this chapter is devoted to discussing batch and incremental learning solutions to (5.5) and (5.6), starting with batch learning.

5.2 Batch Learning Approaches to Regression

When performing batch learning, all data $\mathcal{D}$ is assumed to be available at once (see Sect. 3.1.5). Hence, we have full knowledge of $\{\mathbf{x}_n, y_n\}$, N and, knowing the current model structure $\mathcal{M}$, also of the classifier's matching function m.

Let us now apply this approach to find the classifier's model parameters by solving (5.5) and (5.6).

Notation

The following notation is used in this and the remaining chapters: let $\mathbf{x}, \mathbf{y} \in \mathbb{R}^M$ be vectors, and $\mathbf{A} \in \mathbb{R}^M \times \mathbb{R}^M$ a diagonal matrix. Let $\langle \mathbf{x}, \mathbf{y} \rangle \equiv \mathbf{x}^T \mathbf{y}$ be the inner product of $\mathbf{x}$ and $\mathbf{y}$, at let $\langle \mathbf{x}, \mathbf{y} \rangle_A \equiv \mathbf{x}^T \mathbf{A} \mathbf{y}$ be the inner product weighted by $\mathbf{A}$, forming the inner product space $\langle \cdot, \cdot \rangle_A$. Then, $\|\mathbf{x}\|_A \equiv \sqrt{\langle \mathbf{x}, \mathbf{x} \rangle_A}$ is the norm associated with the inner produce space $\langle \cdot, \cdot \rangle_A$. Any two vectors $\mathbf{x}, \bar{\mathbf{x}}$ are said to be $\mathbf{A}$-orthogonal, if $\langle \mathbf{x}, \bar{\mathbf{x}} \rangle_A = 0$. Note that $\|\mathbf{x}\| \equiv \|\mathbf{x}\|_I$ is the Euclidean norm, where $\mathbf{I}$ is the identity matrix.

5.2.1 The Weight Vector

Using the matrix notation introduced in (3.4), and defining the diagonal $N \times N$ matching matrix $\mathbf{M}_k$ of classifier k by $\mathbf{M}_k = \mathrm{diag}(m(\mathbf{x}_1), \dots, m(\mathbf{x}_N))$, in this chapter simply denoted $\mathbf{M}$, (5.5) can be rewritten to

$$\min_w \left((\mathbf{X}\mathbf{w} - \mathbf{y})^T \mathbf{M} (\mathbf{X}\mathbf{w} - \mathbf{y}) \right) = \min_w \|\mathbf{X}\mathbf{w} - \mathbf{y}\|_M^2. \tag{5.7}$$

Thus, the aim is to find the $\mathbf{w}$ that minimises the weighted distance between the estimated outputs $\mathbf{X}\mathbf{w}$ and the observed outputs $\mathbf{y}$ in the inner product space $\langle \cdot, \cdot \rangle_M$. This distance is convex with respect to $\mathbf{w}$ and therefore has a unique minimum [26]. Note that as the output space is single-dimensional, the set of observed outputs is given by the vector $\mathbf{y}$ rather than the matrix $\mathbf{Y}$.

The solution to (5.7) is found by setting its first derivative to zero, resulting in

$$\hat{\mathbf{w}} = \left(\mathbf{X}^T \mathbf{M} \mathbf{X} \right)^{-1} \mathbf{X}^T \mathbf{M} \mathbf{y}. \tag{5.8}$$

Alternatively, a numerically more stable solution that can also be computed if $\mathbf{X}^T \mathbf{M} \mathbf{X}$ is singular and therefore cannot be inverted, is

$$\hat{\mathbf{w}} = \left(\sqrt{\mathbf{M}} \mathbf{X} \right)^+ \sqrt{\mathbf{M}} \mathbf{y}, \tag{5.9}$$

where $\mathbf{X}^+ \equiv (\mathbf{X}^T \mathbf{X})^{-1} \mathbf{X}^T$ denotes the pseudo-inverse of matrix $\mathbf{X}$ [19].

Using the weight vector according to (5.8), the matching-weighted vector of estimated outputs $\mathbf{X}\hat{\mathbf{w}}$ evaluates to

$$\mathbf{X}\hat{\mathbf{w}} = \mathbf{X} \left(\mathbf{X}^T \mathbf{M} \mathbf{X} \right)^{-1} \mathbf{X}^T \mathbf{M} \mathbf{y}. \tag{5.10}$$

Observe that $\mathbf{X}(\mathbf{X}^T \mathbf{M} \mathbf{X})^{-1} \mathbf{X}^T \mathbf{M}$ is a projection matrix that projects the vector of observed outputs $\mathbf{y}$ onto the hyperplane $\{\mathbf{X}\mathbf{w} | \mathbf{w} \in \mathbb{R}^{D_\mathcal{X}}\}$ with respect to $\langle \cdot, \cdot \rangle_M$. This result is intuitively plausible, as the $\mathbf{w}$ that minimises the weighted distance $\|\mathbf{X}\mathbf{w} - \mathbf{y}\|_M$ between the observed and the estimated outputs is the closest point on this hyperplane to $\mathbf{y}$ with respect to $\langle \cdot, \cdot \rangle_M$, which is the orthogonal projection of $\mathbf{y}$ in $\langle \cdot, \cdot \rangle_M$ onto this plane. This concept will be used extensively in Chap. 9.

5.2.2 The Noise Precision

Equation (5.6) needs to be solved in order to get the maximum likelihood noise precision. As before, we evaluate the maximum of (5.6) by setting its first derivative with respect to τ to zero, to get

$$\hat{\tau}^{-1} = c^{-1} \|\mathbf{X}\hat{\mathbf{w}} - \mathbf{y}\|_{\mathbf{M}}^2, \tag{5.11}$$

where

$$c_k = \sum_{n=1}^{N} m_k(\mathbf{x}_n) = \mathrm{Tr}(\mathbf{M}_k), \tag{5.12}$$

is the *match count* of classifier k, and is in this chapter simply denoted c. $\mathrm{Tr}(\mathbf{M})$ denotes the trace of the matrix $\mathbf{M}$, which is the sum of its diagonal elements. Hence, the inverse noise precision, that is, the noise variance, is given by the average squared error of the model output estimates over all matched observations.

Note, however, that the precision estimate is biased, as it is based on another estimate $\hat{\mathbf{w}}$ [97, Chap. 5]. This can be accounted for by instead using

$$\hat{\tau}^{-1} = (c - D_{\mathcal{X}})^{-1} \|\mathbf{X}\hat{\mathbf{w}} - \mathbf{y}\|^2_{\mathbf{M}}, \tag{5.13}$$

which is the unbiased estimate of the noise precision.

To summarise, the maximum likelihood model parameters of a classifier using batch learning are found by first evaluating (5.8) to get $\hat{\mathbf{w}}$ and then (5.13) to get $\hat{\tau}$.

Example 5.2 (Batch Learning with Averaging Classifiers). Averaging classifiers are characterised by using $x_n = 1$ for all n for their linear model. Hence, we have $\mathbf{X} = (1, \dots, 1)^T$, and evaluating (5.8) results in the scalar weight estimate

$$\hat{w} = c^{-1} \sum_{n=1}^{N} m(\mathbf{x}_n) y_n, \tag{5.14}$$

which is the outputs y_n averaged over all matched inputs. Note that, as discussed in Sect. 3.2.3, the inputs to the matching function as appearing in $m(\mathbf{x}_n)$ are not necessarily the same as the ones used to build the local model. In the case of averaging classifiers this differentiation is essential, as the inputs $x_n = 1$ used for building the local models do not carry any information that can be used for localisation of the classifiers.

The noise precision is determined by evaluating (5.13) and results in

$$\hat{\tau}^{-1} = (c - 1)^{-1} \sum_{n=1}^{N} m(\mathbf{x}_n)(\hat{w} - y_n)^2, \tag{5.15}$$

which is the unbiased average over the squared deviation of the outputs from their average, and hence gives an indication of which prediction error can be expected from the linear model.

5.3 Incremental Learning Approaches to Regression

Having derived the batch learning solution, let us now consider the case where we want to update our model with each additional observation. In particular, assume that the model parameters $\hat{\mathbf{w}}_N$ and $\hat{\tau}_N$ are based on N observations, and the new observations $(\mathbf{x}_{N+1}, y_{N+1})$ are to be incorporated, to get the updated parameters $\hat{\mathbf{w}}_{N+1}$ and $\hat{\tau}_{N+1}$. The following notation will be used: $\mathbf{X}_N, \mathbf{y}_N, \mathbf{M}_N$, and c_N denote the input, output, matching matrix, and match count respectively, after N observations. Similarly, $\mathbf{X}_{N+1}, \mathbf{y}_{N+1}, \mathbf{M}_{N+1}, c_{N+1}$ stand for the same objects after knowing the additional observation $(\mathbf{x}_{N+1}, y_{N+1})$.

Several methods can be used to perform the model parameter update, starting with computationally simple gradient-based approaches, to more complex, but also more stable methods. Since quickly obtaining a good idea of the quality of the model of a classifier is important, and as the noise precision quality measure after (5.6) relies on the weight estimate, the speed of convergence with respect to estimating both $\mathbf{w}$ and τ needs to be considered in addition to the computational costs of the methods.

Firstly, a well-known adaptive filter theory principle concerning the optimality of incremental linear models will be derived. Then we consider some gradient-based approaches, followed by approaches that recursively track the least-squares solution. All this only concerns the weight vector update $\mathbf{w}$. Similar methods will be applied to the noise precision τ in Sect. 5.3.7.

5.3.1 The Principle of Orthogonality

The Principle of Orthogonality determines when the weight vector estimate $\hat{\mathbf{w}}_N$ is optimal in the weighted least squares sense of (5.5):

Theorem 5.3 (Principle of Orthogonality (for example, [105])). *The weight vector estimate $\hat{\mathbf{w}}_N$ after N observations is optimal in the sense of (5.5) if the sequence of inputs $\{\mathbf{x}_1, \dots, \mathbf{x}_N\}$ is $\mathbf{M}_N$-orthogonal to the sequence of estimation errors $\{(\hat{\mathbf{w}}_N^T\mathbf{x}_1 - y_1), \dots, (\hat{\mathbf{w}}_N^T\mathbf{x}_N - y_N)\}$, that is*

$$\langle \mathbf{X}_N, \mathbf{X}_N\hat{\mathbf{w}}_N - \mathbf{y}_N\rangle_{\mathbf{M}_N} = \sum_{n=1}^{N} m(\mathbf{x}_n)\mathbf{x}_n\left(\hat{\mathbf{w}}_N^T\mathbf{x}_n - y_n\right) = 0. \tag{5.16}$$

Proof. The solution of (5.5) is found by setting the first derivative of (5.7) to zero to get

$$2\mathbf{X}_N^T\mathbf{M}_N\mathbf{X}_N\hat{\mathbf{w}}_N - 2\mathbf{X}_N^T\mathbf{M}_N\mathbf{y}_N = 0.$$

The result follows from rearranging the expression.

By multiplying (5.16) by $\hat{\mathbf{w}}_N$, a similar statement can be made about the output estimates:

Corollary 5.4 (Corollary to the Principle of Orthogonality (for example, [105])). *The weight vector estimate $\hat{\mathbf{w}}_N$ after N observations is optimal in the sense of (5.5) if the sequence of output estimates $\{\hat{\mathbf{w}}_N^T\mathbf{x}_1, \dots, \hat{\mathbf{w}}_N^T\mathbf{x}_N\}$ is $\mathbf{M}_N$-orthogonal to the sequence of estimation errors $\{(\hat{\mathbf{w}}_N^T\mathbf{x}_1 - y_1), \dots, (\hat{\mathbf{w}}_N^T\mathbf{x}_N - y_N)\}$, that is*

$$\langle \mathbf{X}_N\hat{\mathbf{w}}_N, \mathbf{X}_N\hat{\mathbf{w}}_N - \mathbf{y}_N\rangle_{\mathbf{M}_N} = \sum_{n=1}^{N} m(\mathbf{x}_n)\hat{\mathbf{w}}_N^T\mathbf{x}_n\left(\hat{\mathbf{w}}_N^T\mathbf{x}_n - y_n\right) = 0. \tag{5.17}$$

Hence, when having a $\hat{\mathbf{w}}_N$ that minimises $\|\mathbf{X}_N\hat{\mathbf{w}}_N - \mathbf{y}_N\|_{\mathbf{M}_N}$, both the sequence of inputs and output estimates are $\mathbf{M}_N$-orthogonal to the estimation errors. In other words, the hyperplane spun by the vectors $\mathbf{X}_N$ and $\mathbf{X}_N\hat{\mathbf{w}}_N$ is $\mathbf{M}_N$-orthogonal to the vector of estimation errors $(\mathbf{X}_N\hat{\mathbf{w}}_N - \mathbf{y}_N)$, and therefore, the output estimate is an orthogonal projection onto this hyperplane with respect to $\langle\cdot,\cdot\rangle_{M_N}$. This conforms to the batch learning solution introduced in Sect. 5.2.1.

5.3.2 Steepest Gradient Descent

Steepest gradient descent is a well-known method for function minimisation, based on following the gradient of that function. Applied to (5.5), it can be used to find the weight vector that minimises the squared error. However, it is only applicable if all observations are known at once, which is not the case when performing incremental learning. Nonetheless, it is discussed here as it gives valuable insights into the stability and speed of convergence of other gradient-based incremental learning methods that are described in later sections.

As for batch learning, let $\mathbf{X}, \mathbf{y}, \mathbf{M}$ and c be the output matrix, the input vector, the matching vector, and the match count respectively, given all N observations. Then, steepest gradient descent is defined by

$$\mathbf{w}_{n+1} = \mathbf{w}_n - \gamma_{n+1}\frac{1}{2}\nabla_{w_n}\left(\|\mathbf{X}\mathbf{w}_n - \mathbf{y}\|_M^2\right), \tag{5.18}$$

starting at some arbitrary $\mathbf{w}_0$, and hence generating a sequence of weight vectors $\{\mathbf{w}_0, \mathbf{w}_1, \dots\}$ by performing small steps along the gradient of the squared error. Note that n does in this case refer to the iteration number of the method rather than to the index of the observation, and $\gamma_n > 0$ is the step size in the nth iteration. Evaluating the gradient ∇_{w_n} with respect to $\mathbf{w}_n$ results in the algorithm

$$\mathbf{w}_{n+1} = \mathbf{w}_n - \gamma_{n+1}\mathbf{X}^T\mathbf{M}(\mathbf{X}\mathbf{w}_n - \mathbf{y}). \tag{5.19}$$

With each step along the gradient, steepest gradient descent reduces the squared error. As the error function is convex and hence has a unique minimum, following its gradient will lead to this minimum and hence, solves (5.5).

Stability Criteria

By definition, the step size γ_n can change at each iteration. When kept constant, that is $\gamma_n = \gamma$ for all $n > 0$, and the gradient is Lipschitz continuous[1], then the steepest gradient descent method is guaranteed to converge to the minimum (5.5), if that minimum exists [17, Prop. 3.4]. In our case the gradient as a function of $\mathbf{w}$ is Lipschitz continuous and hence, convergence for a constant step size is guaranteed.

[1] A function $f : A \to A$ is Lipschitz continuous if there exists a finite constant scalar K such that $\|f(a) - f(b)\| \leq K\|a - b\|$ for any $a, b \in A$. The magnitude K is a measure of the continuity of the function f.

Another condition for the stability of steepest gradient descent, which is easier to evaluate, is for the step size γ to hold

$$0 < \gamma < \frac{2}{\lambda_{max}}, \tag{5.20}$$

where λ_{max} is the largest eigenvalue of the input correlation matrix $c^{-1}\mathbf{X}^T\mathbf{M}\mathbf{X}$ [105, Chap. 4]. Hence, the step size that keeps the algorithm stable depends highly on the values of the input vectors.

Time Constant Bounds

Similar to the stability of the method, its rate of convergence is also dependent on the eigenvalues of the input correlation matrix. Let T be the *time constant*[2] of the weight vector update. This time constant is bounded by

$$\frac{1}{-\ln(1-\gamma\lambda_{max})} \leq T \leq \frac{1}{-\ln(1-\gamma\lambda_{min})}, \tag{5.21}$$

where λ_{max} and λ_{min} are the largest and the smallest eigenvalue of $c^{-1}\mathbf{X}^T\mathbf{M}\mathbf{X}$ respectively [105, Chap. 4]. As a low T implies a higher rate of convergence, we would prefer λ_{max} and λ_{min} to be close together for a tight bound, and large such that T is kept small. However, if the eigenvalues are widely spread, which is an indication of ill-conditioned inputs, then the settling time of the gradient descent algorithm is limited by λ_{min} [17, Chap. 3]. Therefore, the convergence rate is – as the stability criterion – dependent on the values of the input vectors.

Example 5.5 (Stability Criteria and Time Constant for Steepest Gradient Descent). Consider an averaging classifier that matches all inputs, that is $\mathbf{X} = (1, \dots, 1)^T$ and $\mathbf{M} = \mathbf{I}$, the identity matrix. The only eigenvalue of $c^{-1}\mathbf{X}^T\mathbf{M}\mathbf{X}$ is $\lambda = 1$, and therefore, according to (5.20), steepest gradient descent is stable for $0 \leq \gamma \leq 2$. Equation (5.21) results in the time constant $T = -\ln(1-\gamma)^{-1}$, and hence the method converges faster with a larger step size, as is intuitively expected.

The same analysis can be applied to classifiers with straight line models, with input vectors $\mathbf{x}_n = (1, i_n)^T$ with $i_n \in \mathbb{R}$ for all n. In that case, the input vector correlation matrix is given by

$$c^{-1}\mathbf{X}^T\mathbf{M}\mathbf{X} = \frac{1}{N}\sum_{n=1}^{N}\begin{pmatrix} 1 & i_n \\ i_n & i_n^2 \end{pmatrix}, \tag{5.22}$$

with eigenvalues $\lambda_1 = 0, \lambda_2 = 1 + N^{-1}\sum i_n^2$. Hence, the step size has to obey

$$0 \leq \gamma \leq \frac{2}{1 + N^{-1}\sum i_n^2}, \tag{5.23}$$

[2] The time constant is a measure of the *responsivity* of a dynamic system. A low time constant means that the systems response quickly to a changing input. Hence, it is inversely proportional to the rate of convergence.

which demonstrates that the larger the values of i_n, the smaller the step size has to be to still guarantee stability of the algorithm. The time constant is bounded by

$$\frac{-1}{\ln(1-\gamma(1+N^{-1}\sum i_n^2))} \leq T \leq \infty, \tag{5.24}$$

showing that a large eigenvalue spread $|\lambda_2 - \lambda_1|$ caused by on average high magnitudes of i_n pushes the time constant towards infinity, which results in very slow convergence. Therefore, the convergence rate of steepest gradient descent depends frequently on the range of the inputs[3]. This dependency is demonstrated empirically in Sect. 5.4.

5.3.3 Least Mean Squared

The Least Mean Squared (LMS) algorithm is an incremental approximation to steepest gradient descent. Rather than performing gradient descent on the error function given all observations, it follows the gradient of the error function given only the current observation. For this reason, it is also known as *Stochastic Incremental Steepest Gradient Descent*, *ADALINE*, or, after their developers Widrow and Hoff [234], the *Widrow-Hoff Update*.

By inspecting (5.5), the error function for the $(N+1)$th observation based on the model after N observations is $m(\mathbf{x}_{N+1})(\hat{\mathbf{w}}_N^T\mathbf{x}_{N+1} - y_{N+1})^2$, and its gradient with respect to $\mathbf{w}_N$ is therefore $2m(\mathbf{x}_{N+1})\mathbf{x}_{N+1}(\hat{\mathbf{w}}_N^T\mathbf{x}_{N+1} - y_{N+1})$. Using this local gradient estimate as a surrogate for the global gradient, the LMS update is given by

$$\hat{\mathbf{w}}_{N+1} = \hat{\mathbf{w}}_N + \gamma_{N+1} m(\mathbf{x}_{N+1})\mathbf{x}_{N+1}(y_{N+1} - \hat{\mathbf{w}}_N^T\mathbf{x}_{N+1}), \tag{5.25}$$

starting with an arbitrary $\mathbf{w}_0$.

As the gradient estimate is only based on the current input, the method suffers from *gradient noise*. Due to this noise, a constant step size γ will cause random motion close to the optimal approximation [105, Chap. 5].

Misadjustment Due to Local Gradient Estimate

Let $h_N(\mathbf{w}) = c_N^{-1}\|\mathbf{X}_N\mathbf{w} - \mathbf{y}_N\|^2$ be the mean squared error (MSE) after N observations as a function of the weight vector. The *excess mean square estimation error* is the difference between the MSE of the LMS algorithm and the minimal MSE after (5.16). The ratio between the excess MSE and the minimal MSE error is the *misadjustment*, which is a measure of how far away the convergence area of LMS is from the optimal estimate. The estimate error for some small constant step size can, according to [105, Chap. 5], be estimated by

$$h_N(\mathbf{w}_N^*) + \frac{\gamma h_N(\mathbf{w}_N^*)}{2}\sum_{j=1}^{J}\lambda_j, \tag{5.26}$$

[3] A similar LCS-related analysis was done by Lanzi et al. [142, 143], but there the stability criteria for steepest gradient descent were applied to the LMS algorithm.

where $\mathbf{w}_N^*$ is the weight vector that satisfies (5.16) and thus, $h_N(\mathbf{w}_N^*)$ is the minimal MSE, and λ_j is the jth of the J eigenvalues of $c^{-1}\mathbf{X}_N^T\mathbf{M}_N\mathbf{X}_N$. This shows that the excess MSE estimate is i) always positive, and ii) is proportional to the step size γ. Thus, reducing the step size also reduces the misadjustment. Indeed, under the standard stochastic approximation assumptions that $\sum_{n=1}^{\infty}\gamma_n = \infty$ and $\sum_{n=1}^{\infty}\gamma_t^2 < \infty$, the Lipschitz continuity of the gradient, and some Pseudo-gradient property of the gradient, convergence to the optimal estimate can be guaranteed [17, Prop. 4.1].

Stability Criteria and Average Time Constant

As the LMS filter is a traversal filter of length one, using only the current observation for its gradient estimate, no concrete bounds for the step size can be currently given [105, Chap. 6]. However, if the step size is small when compared to the inverse of the largest eigenvalue of the input vector correlation matrix, then the stability criteria are the same as for steepest gradient descent (5.20).

As the gradient changes with each step, we can only give an expression for the local time constant that varies with time (for more details see [78]). On average, however, the time constant can be bounded in the same way as for steepest gradient descent (5.21), with the same consequences.

This leaves us in a dilemma: it was already established that the misadjustment is proportional to the step size. On the other hand, the time constant is inversely proportional to it. Hence, we have conflicting requirements and can either aim for a low estimation error or a fast rate of convergence, but will not be able to satisfy both requirements with anything other than a compromise.

Relation to Batch Learning

To get a better intuitive understanding of how the LMS algorithm estimates the weight vector, let us reformulate it as a batch learning approach for the simplified case of an averaging classifier that matches all inputs, that is $x_n = 1, m(\mathbf{x}_n) = 1$ for all $n > 0$. In that case, (5.25) reduces to

$$\hat{w}_{N+1} = \hat{w}_N + \gamma_{N+1}(y_{N+1} - \hat{w}_N), \tag{5.27}$$

and by recursive substitution (as in Example 3.2) results in the batch learning formulation

$$\hat{w}_N = \sum_{n=1}^{N} y_n \gamma_n \prod_{m=n+1}^{N}(1-\gamma_m) + w_0 \prod_{n=1}^{N}(1-\gamma_n). \tag{5.28}$$

Hence, the nth observation y_n is weighted by $\gamma_n \prod_{m=n+1}^{N}(1-\gamma_m)$, which, for $0 < \gamma_{\bar{n}} < 1$ for all $0 < \bar{n} \leq n$, means that the lower n, the less y_n contributes to the weight estimate. Also, w_0 introduces a bias that decays exponentially with $\prod_{n=1}^{N}(1-\gamma_n)$. Comparing this insight to the results of Example 5.2, where it was shown that the optimal weight in the least squares sense for averaging classifiers is the average over all matched outputs, it becomes apparent that the LMS algorithm does not achieve this optimum for arbitrary step sizes. Nonetheless, it can be applied readily for recency-weighted applications, such as to handle non-stationary processes, as is required in reinforcement learning applications.

5.3.4 Normalised Least Mean Squared

As seen from (5.25), the magnitude of the weight update is directly proportional to the new input vector $\mathbf{x}_{N+1}$, causing *gradient noise amplification* [105, Chap. 6]. Thus, if some elements of the input vector are large, the correction based on a local error will be amplified and causes additional noise. This problem can be overcome by weighting the correction by the squared Euclidean norm of the input, resulting in the update

$$\hat{\mathbf{w}}_{N+1} = \hat{\mathbf{w}}_N + \gamma_t m(\mathbf{x}_{N+1}) \frac{\mathbf{x}_{N+1}}{\|\mathbf{x}_{N+1}\|^2} (y_{N+1} - \hat{\mathbf{w}}_N^T \mathbf{x}_{N+1}). \tag{5.29}$$

This update equation can also be derived by calculating the weight vector update that minimises the norm of the weight change $\|\hat{\mathbf{w}}_{N+1} - \hat{\mathbf{w}}_N\|^2$, subject to the constraint $m(\mathbf{x}_{N+1})\hat{\mathbf{w}}_{N+1}\mathbf{x}_{N+1} = y_{N+1}$. As such, the normalised LMS filter can be seen as a solution to a constrained optimisation problem.

Regarding stability, the step size parameter γ is now weighted by the inverted square norm of the input vector. Thus, stability in the MSE sense is dependent on the current input. The lower bound is still 0, and the upper bound will be generally larger than 2 if the input values are overestimated, and smaller than 2 otherwise. The optimal step size, located at the largest value of the mean squared deviation, is the centre of the two bounds [105, Chap. 6].

As expected, the normalised LMS algorithm features a rate of convergence that is higher than that of the standard LMS filter, as empirically demonstrated by Douglas [76]. One drawback of the modification is that one needs to check $\|\mathbf{x}_{N+1}\|^2$ for being zero, in which case no update needs to be performed to avoid division by zero.

To summarise, both variants of the LMS algorithm have low computational and space costs $\mathcal{O}(D_\mathcal{X})$, but only rely on the local gradient estimate and may hence feature slow convergence and misadjustment. The step size can be adjusted to either improve convergence speed or misadjustment, but one cannot improve both at the same time. Additionally, the speed of convergence is by (5.21) influenced by the value of the inputs and might be severely reduced by ill-conditioned inputs, as will be demonstrated in Sect. 5.4.

Let us recall that to quickly getting an idea of the goodness-of-fit of a classifier model, measured by the model variance (5.13), requires a good weight vector estimate. Despite their low computational cost, gradient-based methods are known to suffer from low speed of convergence and are therefore not necessarily the best choice for this task. The following sections describe incremental methods that are computationally more costly, but are able to recursively track the weight vector that satisfies (5.16) and are therefore optimal in the least squares sense.

5.3.5 Recursive Least Squares

The Principle of Orthogonality (5.16) is satisfied if the *normal equation*

$$\left(\mathbf{X}_N^T \mathbf{M}_N \mathbf{X}_N\right) \hat{\mathbf{w}_N} = \mathbf{X}_N^T \mathbf{M}_N \mathbf{y}_N, \tag{5.30}$$

holds. Using the $D_\mathcal{X} \times D_\mathcal{X}$ symmetric matrix $\mathbf{\Lambda}_N = \mathbf{X}_N^T \mathbf{M}_N \mathbf{X}_N$, $\mathbf{\Lambda}_N$ and $\mathbf{\Lambda}_{N+1}$ are related by

$$\mathbf{\Lambda}_{N+1} = \mathbf{\Lambda}_N + m(\mathbf{x}_{N+1}) \mathbf{x}_{N+1} \mathbf{x}_{N+1}^T, \tag{5.31}$$

with $\mathbf{\Lambda}_0 = \mathbf{0}$. Similarly, we have

$$\mathbf{X}_{N+1}^T \mathbf{M}_{N+1} \mathbf{y}_{N+1} = \mathbf{X}_N^T \mathbf{M}_N \mathbf{y}_N + m(\mathbf{x}_{N+1}) \mathbf{x}_{N+1} y_{N+1}, \tag{5.32}$$

which, in combination with (5.30) and (5.31), allows us to derive the relation

$$\mathbf{\Lambda}_{N+1} \hat{\mathbf{w}}_{N+1} = \mathbf{\Lambda}_{N+1} \mathbf{w}_N + m(\mathbf{x}_{N+1}) \mathbf{x}_{N+1} (y_{N+1} - \hat{\mathbf{w}}_N^T \mathbf{x}_{N+1}). \tag{5.33}$$

Pre-multiplying the above by $\mathbf{\Lambda}_{N+1}^{-1}$, we get the weight vector update

$$\hat{\mathbf{w}}_{N+1} = \hat{\mathbf{w}}_N + m(\mathbf{x}_{N+1}) \mathbf{\Lambda}_{N+1}^{-1} \mathbf{x}_{N+1} (y_{N+1} - \hat{\mathbf{w}}_N^T \mathbf{x}_{N+1}), \tag{5.34}$$

which, together with (5.31) and starting with $\mathbf{w}_0 = \mathbf{0}$, defines the recursive least squares (RLS) algorithm (for example, [105, Chap. 9] or [17, Chap. 3]).

Following this algorithm satisfies the Principle of Orthogonality with each additional observation, and as such provides an incremental approach of tracking the optimal weight vector in the least squares sense. This comes at the cost $\mathcal{O}(D_\mathcal{X}^3)$ of having to invert the matrix $\mathbf{\Lambda}$ with each additional observation that is to be included into the model. Alternatively, we can utilise the properties of $\mathbf{\Lambda}$ to derive the following modified update:

Operating on $\mathbf{\Lambda}^{-1}$

The Sherman-Morrison formula (also known as the Matrix Inversion Lemma, for example [105, Chap. 6]) provides a method of adding a dyadic product to an invertible matrix by operating directly on the inverse of this matrix. Hence, it is applicable to (5.31), and results in

$$\mathbf{\Lambda}_{N+1}^{-1} = \mathbf{\Lambda}_N^{-1} - m(\mathbf{x}_{N+1}) \frac{\mathbf{\Lambda}_N^{-1} \mathbf{x}_{N+1} \mathbf{x}_{N+1}^T \mathbf{\Lambda}_N^{-1}}{1 + m(\mathbf{x}_{N+1}) \mathbf{x}_{N+1}^T \mathbf{\Lambda}_N^{-1} \mathbf{x}_{N+1}}, \tag{5.35}$$

which is of cost $\mathcal{O}(D_\mathcal{X}^2)$ rather than $\mathcal{O}(D_\mathcal{X}^3)$ for inverting $\mathbf{\Lambda}$ in (5.34) at each update.

The drawback of this approach is that $\mathbf{\Lambda}$ cannot be initialised to $\mathbf{\Lambda}_0 = \mathbf{0}$, as the Sherman-Morrison formula is only valid for invertible matrices, and $\mathbf{\Lambda}_0 = \mathbf{0}$ is clearly not. This issue is usually handled by initialising $\mathbf{\Lambda}_0^{-1} = \delta \mathbf{I}$, where δ is a large positive scalar (to keep $\mathbf{\Lambda}_0$ close to $\mathbf{0}$), and $\mathbf{I}$ is the identity matrix. While this approach introduces an initial bias to the RLS algorithm, this bias decays exponentially, as will be shown in the next section.

Relation to Ridge Regression

It is easy to show that the solution $\hat{\mathbf{w}}_N$ to minimising

$$\|\mathbf{X}_N\mathbf{w} - \mathbf{y}_N\|^2_{M_N} + \lambda\|\mathbf{w}\|^2, \tag{5.36}$$

(λ is the positive scalar *ridge complexity*) with respect to $\mathbf{w}$ requires

$$(\mathbf{X}_N^T\mathbf{M}_N\mathbf{X}_N + \lambda\mathbf{I})\hat{\mathbf{w}}_N = \mathbf{X}_N^T\mathbf{M}_N\mathbf{y}_n \tag{5.37}$$

to hold. The above is similar to (5.30) with the additional term $\lambda\mathbf{I}$. Hence, (5.31) still holds when initialised with $\mathbf{\Lambda}_0 = \lambda\mathbf{I}$, and consequently so does (5.34). Therefore, initialising $\mathbf{\Lambda}_0^{-1} = \delta\mathbf{I}$ to apply (5.35) to operate on $\mathbf{\Lambda}^{-1}$ rather than $\mathbf{\Lambda}$ is equivalent to minimising (5.36) with $\lambda = \delta^{-1}$.

In addition to the matching-weighted squared error, (5.36) penalises the size of $\mathbf{w}$. This approach is known as *ridge regression* and was initially introduced to work around the problem of initially singular $\mathbf{X}_N^T\mathbf{M}_N\mathbf{X}_N$ for small N, that prohibited the solution of (5.30). However, minimising (5.36) rather than (5.7) is also advantageous if the input vectors suffer from a high noise variance, resulting in large $\mathbf{w}$ and a bad model for the real data-generating process. Essentially, ridge regression assumes that the size of $\mathbf{w}$ is small and hence computes better model parameters for noisy data, given that the inputs are normalised [102, Chap. 3].

To summarise, using the RLS algorithm (5.34) and (5.35) with $\mathbf{\Lambda}_0^{-1} = \delta\mathbf{I}$, a classifier performs ridge regression with ridge complexity $\lambda = \delta^{-1}$. As by (5.36), the contribution of $\|\mathbf{w}\|$ is independent of the number of observations N, its influence decreases exponentially with N.

A Recency-Weighted Variant

While the RLS algorithm provides a recursive solution such that (5.16) holds, it weights all observations equally. Nonetheless, we might sometimes require recency-weighting, such as when using LCS in combination with reinforcement learning. Hence, let us derive a variant of RLS that applies a scalar decay factor $0 \leq \lambda \leq 1$ to past observations.

More formally, after N observations, we aim at minimising

$$\sum_{n=1}^{N} m(\mathbf{x}_n)\lambda^{\sum_{j=n+1}^{N} m(\mathbf{x}_j)}(\mathbf{w}^T\mathbf{x}_n - y_n)^2 = \|\mathbf{X}_N\mathbf{w} - \mathbf{y}_N\|^2_{M_N^\lambda} \tag{5.38}$$

with respect to $\mathbf{w}$, where the λ-augmented diagonal matching matrix $\mathbf{M}_N^\lambda$ is given by

$$\mathbf{M}_N^\lambda = \begin{pmatrix} m(\mathbf{x}_1)\lambda^{\sum_{j=2}^{N} m(\mathbf{x}_j)} & & & \mathbf{0} \\ & m(\mathbf{x}_2)\lambda^{\sum_{j=3}^{N} m(\mathbf{x}_j)} & & \\ & & \ddots & \\ \mathbf{0} & & & m(\mathbf{x}_N) \end{pmatrix}. \tag{5.39}$$

Note that $\lambda^{\sum_{j=n+1}^{N} m(\mathbf{x}_j)}$ is used rather than simply λ^{N-n} to only decay past observations if the current observation is matched. As before, the solution $\hat{\mathbf{w}}_N$ that minimises (5.38) satisfies

$$(\mathbf{X}_N^T \mathbf{M}_N^\lambda \mathbf{X}_N)\hat{\mathbf{w}}_N = \mathbf{X}_N^T \mathbf{M}_N^\lambda \mathbf{y}_N. \tag{5.40}$$

Using $\mathbf{\Lambda}_N = \mathbf{X}_N^T \mathbf{M}_N^\lambda \mathbf{X}_N$ and the relations

$$\mathbf{\Lambda}_{N+1} = \lambda^{m(\mathbf{x}_{N+1})}\mathbf{\Lambda}_N + m(\mathbf{x}_{N+1})\mathbf{x}_{N+1}\mathbf{x}_{N+1}^T, \tag{5.41}$$

$$\mathbf{\Lambda}_{N+1}\hat{\mathbf{w}}_{N+1} = \lambda^{m(\mathbf{x}_{N+1})}\mathbf{\Lambda}_N\hat{\mathbf{w}}_N + m(\mathbf{x}_{N+1})\mathbf{x}_{N+1}\mathbf{y}_{N+1}, \tag{5.42}$$

the recency-weighted RLS weight vector update is given by

$$\hat{\mathbf{w}}_{N+1} = \lambda^{m(\mathbf{x}_{N+1})}\hat{\mathbf{w}}_N + m(\mathbf{x}_{N+1})\mathbf{\Lambda}_{N+1}^{-1}\mathbf{x}_{N+1}(y_{N+1} - \hat{\mathbf{w}}_N^T\mathbf{x}_{N+1}). \tag{5.43}$$

The matrix $\mathbf{\Lambda}$ can be updated by either using (5.41) or by applying the Sherman-Morrison formula to get

$$\begin{aligned}\mathbf{\Lambda}_{N+1}^{-1} &= \lambda^{-m(\mathbf{x}_{N+1})}\mathbf{\Lambda}_N^{-1} \\ &\quad - m(\mathbf{x}_{N+1})\lambda^{-m(\mathbf{x}_{N+1})}\frac{\mathbf{\Lambda}_N^{-1}\mathbf{x}_{N+1}\mathbf{x}_{N+1}^T\mathbf{\Lambda}_N^{-1}}{\lambda^{m(\mathbf{x}_{N+1})} + m(\mathbf{x}_{N+1})\mathbf{x}_{N+1}^T\mathbf{\Lambda}_N^{-1}\mathbf{x}_{N+1}}.\end{aligned} \tag{5.44}$$

All equations from this section reduce to the non-recency-weighted equivalents if $\lambda = 1$.

In summary, the RLS algorithm recursively tracks the solution according to the Principle of Orthogonality. As this solution is always optimal in the least squares sense, there is no need to discuss its convergence to the optimal solution, as was required for gradient-based algorithms. While the RLS can also be adjusted to perform recency-weighting, as developed in this section, its only drawback when compared to the LMS or normalised LMS algorithm is its higher computational cost. Nonetheless, if this additional cost is bearable, it should be always preferred to the gradient-based algorithm, as will be demonstrated in Sect. 5.4.

Example 5.6 (RLS Algorithm for Averaging Classifiers). Consider averaging classifiers, such that $\mathbf{x}_n = 1$ for all $n > 0$. Hence, (5.31) becomes

$$\Lambda_{N+1} = \Lambda_N + m(\mathbf{x}_{N+1}), \tag{5.45}$$

which, when starting with $\Lambda_0 = 0$ is equivalent to the match count $\Lambda_N = c_N$. The weight update after (5.34) reduces to

$$w_{N+1} = w_N + m(\mathbf{x}_{N+1})c_{N+1}^{-1}(y_{N+1} - w_N). \tag{5.46}$$

Note that this is equivalent to the LMS algorithm (5.25) for averaging classifiers when using the step size $\gamma_N = c_N^{-1}$. By recursive back-substitution of the above, and using $w_0 = 0$, we get

$$w_N = c_N^{-1}\sum_{n=1}^{N} m(\mathbf{x}_{N+1})y_n, \tag{5.47}$$

which is, as already derived for batch learning (5.14), the matching-weighted average over all observed outputs.

Interestingly, XCS applies the MAM update that is equivalent to averaging the input for the first γ^{-1} inputs, where γ is the step size, and then tracking the input using the LMS algorithm [237]. In other words, it bootstraps its weight estimate using the RLS algorithm, and then continues tracking of the output using the LMS algorithm. Note that this is only the case for XCS with averaging classifiers, and does not apply for XCS derivatives that use more complex models, such as XCSF. Even though not explicitly stated by Wilson [241] and others, it is assumed that the MAM update was not used for the weight update in those XCS derivatives, but is still used when updating its scalar parameters, such as the relative classifier accuracy and fitness.

5.3.6 The Kalman Filter

The RLS algorithm was introduced purely on the basis of the Principle of Orthogonality without consideration of the probabilistic structure of the random variables. Even though the Kalman filter results in the same update equations, it provides additional probabilistic information and hence supports better understanding of the method. Furthermore, its use is advantageous as "[...] the Kalman filter is optimal with respect to virtually any criterion that makes sense" [164, Chap. 1].

Firstly, the system model is introduced, from which the update equation in covariance form and inverse covariance form are derived. This is followed by considering how both the system state and the measurement noise can be estimated simultaneously by making use of the Minimum Model Error philosophy. The resulting algorithm is finally related to the RLS algorithm.

The System Model

The Kalman-Bucy system model [123, 124] describes how a noisy process modifies the state of a system, and how this affects the noisy observation of the system. Both the process and the relation between system state and observation is assumed to be linear, and all noise is zero-mean white (uncorrelated) Gaussian noise.

In our case, the process that generates the observations is assumed to be stationary, which is expressed by a constant system state. Additionally, the observations are in linear relation to the system state and all deviations from that linearity are covered by zero-mean white (uncorrelated) Gaussian noise. The resulting model is

$$\upsilon_n = \boldsymbol{\omega}^T \mathbf{x}_n + \epsilon_n, \tag{5.48}$$

where υ_n is the random variable that represents the observed nth scalar output of the system, $\boldsymbol{\omega}$ is the system state random variable, $\mathbf{x}_n$ is the known nth input vector to the system, and ϵ_n is the measurement noise associated with observing y_n.

The noise ϵ_n is modelled by a zero-mean Gaussian $\epsilon_n \sim \mathcal{N}(0, (m(\mathbf{x}_n)\tau_n)^{-1})$ with precision $m(\mathbf{x}_n)\tau_n$. Here, we utilise the matching function to blur observations that are not matched. Given, for example, that $\mathbf{x}_n$ is matched and so $m(\mathbf{x}_n) = 1$, the resulting measurement noise has variance τ_n^{-1}. However, if that state is not matched, that is if $m(\mathbf{x}_n) = 0$, then the measurement noise has infinite variance and the associated observation does not contain any information.

The system state $\boldsymbol{\omega}$ is modelled by the multivariate Gaussian model $\boldsymbol{\omega} \sim \mathcal{N}(\hat{\mathbf{w}}, \mathbf{\Lambda}^{-1})$ centred on $\hat{\mathbf{w}}$ and with precision matrix $\mathbf{\Lambda}$. Hence, the output υ_n is also Gaussian $\upsilon_n \sim \mathcal{N}(y_n, (m(\mathbf{x}_n)\tau_n)^{-1})$, and jointly Gaussian with the system state $\boldsymbol{\omega}$. More details on the random variables, their relations and distributions can be found in [164, Chap. 5] and [2, Chap. 1].

Comparing the model (5.48) to the previously introduced linear model (5.1), it can be seen that the system state corresponds to the weight vector, and that the only difference is the assumption that the measurement noise variance can change with each observation. Additionally, the Kalman-Bucy system model explicitly assumes a multivariate Gaussian model for the system state $\boldsymbol{\omega}$, resulting in the output υ also being modelled by a Gaussian.

The aim of the Kalman filter is to estimate the system state that can subsequently be used to predict the output given a new input. This is achieved by conditioning a prior $\boldsymbol{\omega}_0 \sim \mathcal{N}(\hat{\mathbf{w}}_0, \mathbf{\Lambda}_0^{-1})$ on the available observations. As before, we proceed by assuming that the current model $\boldsymbol{\omega}_N \sim \mathcal{N}(\hat{\mathbf{w}}_N, \mathbf{\Lambda}_N^{-1})$ results from incorporating the information of N observations, and we want to add the new observation $(\mathbf{x}_{N+1}, y_{N+1}, \tau_{N+1})$. Later it will be shown how to estimate the noise precision τ_{N+1}, but for now we assume that it is part of the observation.

Covariance Form

As the system state and the observation are jointly Gaussian, the Bayesian update of the model parameters is given by [2, Chap. 3]

$$\begin{aligned}
\hat{\mathbf{w}}_{N+1} &= \mathbb{E}\left(\boldsymbol{\omega}_N | \upsilon_{N+1} \sim \mathcal{N}(y_{N+1}, (m(\mathbf{x}_{N+1})\tau_{N+1})^{-1})\right) \\
&= \mathbb{E}(\boldsymbol{\omega}_N) + \mathrm{cov}(\boldsymbol{\omega}_N, \upsilon_{N+1})\mathrm{var}(\upsilon_{N+1})^{-1}(y_{N+1} - \mathbb{E}(\upsilon_{N+1})), \quad (5.49) \\
\mathbf{\Lambda}_{N+1}^{-1} &= \mathrm{cov}\left(\boldsymbol{\omega}_N, \boldsymbol{\omega}_N | \upsilon_{N+1} \sim \mathcal{N}(y_{N+1}, (m(\mathbf{x}_{N+1})\tau_{N+1})^{-1})\right) \\
&= \mathrm{cov}(\boldsymbol{\omega}_N, \boldsymbol{\omega}_N) - \mathrm{cov}(\boldsymbol{\omega}_N, \upsilon_{N+1})\mathrm{var}(\upsilon_{N+1})^{-1}\mathrm{cov}(\upsilon_{N+1}, \boldsymbol{\omega}_N). (5.50)
\end{aligned}$$

Evaluating the expectations, variances and covariances

$$\begin{aligned}
\mathbb{E}(\boldsymbol{\omega}_N) &= \hat{\mathbf{w}}_N, & \mathrm{cov}(\boldsymbol{\omega}_N, \boldsymbol{\omega}_N) &= \mathbf{\Lambda}_N^{-1}, \\
\mathbb{E}(\upsilon_{N+1}) &= \hat{\mathbf{w}}_N^T \mathbf{x}_{N+1}, & \mathrm{cov}(\upsilon_{N+1}, \boldsymbol{\omega}_N) &= \mathbf{x}_{N+1}^T \mathbf{\Lambda}_N^{-1}, \\
\mathrm{cov}(\boldsymbol{\omega}_N, \upsilon_{N+1}) &= \mathbf{\Lambda}_N^{-1} \mathbf{x}_{N+1}, &&
\end{aligned}$$

$$\mathrm{var}(\upsilon_{N+1}) = \mathbf{x}_{N+1}^T \mathbf{\Lambda}_N^{-1} \mathbf{x}_{N+1} + (m(\mathbf{x}_{N+1})\tau_{N+1})^{-1},$$

and substituting them into the Bayesian update results in

$$\zeta_{N+1} = m(\mathbf{x}_{N+1})\mathbf{\Lambda}_N^{-1}\mathbf{x}_{N+1}\left(m(\mathbf{x}_{N+1})\mathbf{x}_{N+1}^T\mathbf{\Lambda}_N^{-1}\mathbf{x}_{N+1} + \tau_{N+1}^{-1}\right)^{-1}, \quad (5.51)$$

$$\hat{\mathbf{w}}_{N+1} = \hat{\mathbf{w}}_N + \zeta_{N+1}\left(y_{N+1} - \hat{\mathbf{w}}_N^T\mathbf{x}_{N+1}\right), \quad (5.52)$$

$$\mathbf{\Lambda}_{N+1}^{-1} = \mathbf{\Lambda}_N^{-1} - \zeta_{N+1}\mathbf{x}_{N+1}^T\mathbf{\Lambda}_N^{-1}. \quad (5.53)$$

This form of the Kalman filter is known as the *covariance form* as it operates on the covariance matrix $\mathbf{\Lambda}^{-1}$ rather than the precision matrix $\mathbf{\Lambda}$.

The value ζ_{N+1} is the *Kalman gain* and is a temporary measure that depends on the current model ω_N and the new observation. It mediates how much ω_N is corrected, that is, how much the current input $\mathbf{x}_{N+1}$ influences $\mathbf{\Lambda}_{N+1}^{-1}$, and how the output residual $y_{N+1} - \hat{\mathbf{w}}_N^T\mathbf{x}_{N+1}$ contributes to computing $\hat{\mathbf{w}}_{N+1}$.

As the measurement noise variance τ_{N+1}^{-1} approaches zero, the gain ζ_{N+1} weights the output residual more heavily. On the other hand, as the weight covariance $\mathbf{\Lambda}_N^{-1}$ approaches zero, the gain ζ_{N+1} assigns less weight to the output residual [233]. This is the behaviour that would be intuitively excepted, as low-noise observations should influence the model parameters more strongly than high-noise observations. Also, the gain is mediated by the matching function and in the cases of non-matched inputs reduced to zero, which causes the model parameters to remain unchanged.

Inverse Covariance Form

Using the Kalman filter to estimate the system state requires the definition of a prior $\boldsymbol{\omega}_0$. In many cases, the correct prior is unknown and setting it arbitrarily might introduce an unnecessary bias. While complete lack of information can be theoretically induced as the limiting case of certain eigenvalues of $\mathbf{\Lambda}_0^{-1}$ going to infinity [164, Chap. 5.7], it cannot be used in practice due to large numerical errors when evaluating (5.51).

This problem can be dealt with by operating the Kalman filter in the *inverse covariance form* rather than the previously introduced covariance form. To update $\mathbf{\Lambda}$ rather than $\mathbf{\Lambda}^{-1}$ we substitute ζ_{N+1} from (5.51) into (5.53) and apply the Matrix Inversion Lemma (for example, [105, Chap. 9.2]) to get

$$\mathbf{\Lambda}_{N+1} = \mathbf{\Lambda}_N + m(\mathbf{x}_{N+1})\tau_{N+1}\mathbf{x}_{N+1}\mathbf{x}_{N+1}^T. \quad (5.54)$$

The weight update is derived by combining (5.51) and (5.53) to get

$$\zeta_{N+1} = m(\mathbf{x}_{N+1})\tau_{N+1}\mathbf{\Lambda}_{N+1}^{-1}\mathbf{x}_{N+1}, \quad (5.55)$$

which, when substituted into (5.52), gives

$$\hat{\mathbf{w}}_{N+1} = \hat{\mathbf{w}}_N + m(\mathbf{x}_{N+1})\tau_{N+1}\mathbf{\Lambda}_{N+1}^{-1}\mathbf{x}_{N+1}(y_{N+1} - \hat{\mathbf{w}}_N^T\mathbf{x}_{N+1}). \quad (5.56)$$

Pre-multiplying the above by $\mathbf{\Lambda}_{N+1}$ and substituting (5.54) for the first $\mathbf{\Lambda}_{N+1}$ of the resulting equation gives the final update equation

$$\mathbf{\Lambda}_{N+1}\hat{\mathbf{w}}_{N+1} = \mathbf{\Lambda}_N\hat{\mathbf{w}}_N + m(\mathbf{x}_{N+1})\tau_{N+1}\mathbf{x}_{N+1}y_{N+1}. \quad (5.57)$$

Thus, $\hat{\mathbf{w}}$ is updated indirectly through the vector $(\mathbf{\Lambda}\hat{\mathbf{w}}) \in \mathbb{R}^{D_\mathcal{X}}$ from which $\hat{\mathbf{w}}$ can be recovered by $\hat{\mathbf{w}} = \mathbf{\Lambda}^{-1}(\mathbf{\Lambda}\hat{\mathbf{w}})$. Even though the initial $\mathbf{\Lambda}$ might be singular and therefore cannot be inverted to calculate $\hat{\mathbf{w}}$, it can still be updated by (5.54) until it is non-singular and can be inverted. This allows for using the non-informative prior $\mathbf{\Lambda}_0 = \mathbf{0}$ that cannot be used when applying the covariance form of the Kalman filter.

Minimum Model Error Philosophy

For deriving the Kalman filter update equations we have assumed knowledge of the measurement noise variances $\{\tau_1^{-1}, \tau_2^{-1}, \dots\}$. In our application of the Kalman filter that is not the case, and so we have find a method that allows us to estimate the variances at the same time as the system state.

Assuming a different measurement noise variance for each observation makes estimating these prohibitive, as it would require estimating more parameters than there are observations. To reduce the degrees of freedom of the model it will be assumed that τ is constant for all observations, that is $\tau_1 = \tau_2 = \dots = \tau$. In addition, we adopt the *Minimum Model Error (MME)* philosophy [170] that aims at finding the model parameters that minimises the model error, which is determined by the noise variance τ. The MME is based on the *Covariance Constraint* condition, which states that the observation-minus-estimate error variance must match the observation-minus-truth error variance, that is

$$(y_n - \hat{\mathbf{w}}^T \mathbf{x}_n)^2 \approx (m(\mathbf{x}_n)\tau)^{-1}. \tag{5.58}$$

Given that constraint and the assumption of not having any process noise, the model error for the nth observation is given by weighting the left-hand side of (5.58) by the inverted right-hand side, which, for N observations results in

$$\tau \sum_{n=1}^{N} m(\mathbf{x}_n) \left(\hat{\mathbf{w}}^T \mathbf{x}_n - y_n \right)^2 . \tag{5.59}$$

Minimising the above is independent of τ and therefore equivalent to (5.5). Thus, assuming a constant measurement noise variance has led us back to minimising the error that we originally intended to minimise.

Relation to Recursive Least Squares

The Kalman filter update equations are very similar but not quite the same as the RLS update equations. Maybe the most obvious match is the inverse covariance update (5.54) of the Kalman filter, and (5.31) of the RLS algorithm, only differing by the additional term τ_{N+1} in (5.54). Similarly, (5.56) and (5.34) differ by the same term.

In fact, if all $\mathbf{\Lambda}$ in the RLS update equations are substituted by $\tau^{-1}\mathbf{\Lambda}$, in addition to assuming $\tau_1 = \tau_2 = \dots = \tau$ for the Kalman filter, these equations become equivalent. More specifically, the covariance form of the Kalman filter corresponds to the RLS algorithm that uses (5.35), and the inverse covariance form is equivalent to using (5.31). They also share the same characteristics: while (5.35) is computationally cheaper, it cannot be used with a non-informative prior, just like the covariance form. Conversely, using (5.31) allows the use of non-informative priors, but requires a matrix inversion with every additional update, as does the inverse covariance form to recover $\hat{\mathbf{w}}$ by $\hat{\mathbf{w}} = \mathbf{\Lambda}^{-1}(\mathbf{\Lambda}\hat{\mathbf{w}})$, making it computationally more expensive.

The information gain from this relation is manifold:

- The weight vector of the linear model corresponds to the system state of the Kalman filter. Hence, it can be modelled by a multivariate Gaussian, that, in the notation of the RLS algorithm, is given by $\boldsymbol{\omega}_N \sim \mathcal{N}(\hat{\mathbf{w}}_N, (\tau\mathbf{\Lambda}_N)^{-1})$. As τ is unknown, it needs to be substituted by its estimate $\hat{\tau}$.
- Acquiring this model for $\boldsymbol{\omega}$ causes the output random variable υ to become Gaussian as well. Hence, using the model for prediction, these predictions will be Gaussian. More specifically, given a new input $\mathbf{x}'$, the predictive density is
$$y' \sim \mathcal{N}\left(\hat{\mathbf{w}}^T\mathbf{x}', \hat{\tau}^{-1}(\mathbf{x}'^T\mathbf{\Lambda}^{-1}\mathbf{x}' + m(\mathbf{x}')^{-1})\right), \tag{5.60}$$
and is thus centred on $\hat{\mathbf{w}}^T\mathbf{x}'$. Its spread is determined on one hand by the estimated noise variance $(m(\mathbf{x}')\hat{\tau})^{-1}$ and the uncertainty of the weight vector estimate $\mathbf{x}'^T(\hat{\tau}\mathbf{\Lambda})^{-1}\mathbf{x}$. The $\mathbf{\Lambda}$ in the above equations refers to the one estimated by the RLS algorithm.
Following Hastie et al. [102, Chap. 8.2.1], the two-sided 95% confidence of the standard normal distribution is given by considering its 97.5% point (as $(100\% - 2 \times 2.5\%) = 95\%$), which is 1.96. Therefore, the 95% confidence interval of the classifier predictions is centred on the mean of (5.60) with 1.96 times the square root of the prediction's variance to either side of the mean.
- In deriving the Kalman filter update equations, matching was embedded as a modifier to the measurement noise variance, that is $\epsilon_n \sim \mathcal{N}(0, (m(\mathbf{x}_n)\tau)^{-1})$, which gives us a new interpretation for matching: A matching value between 0 and 1 for a certain input can be interpreted as reducing the amount of information that the model acquires about the associated observation by increasing the noise of the observation and hence reducing its certainty.
- A similar interpretation can be given for RLS with recency-weighting: the decay factor λ acts as a multiplier to the noise precision of past observations and hence reduces their certainty. This causes the model to put more emphasis on more recent observations due to their lower noise variance. Formally, modelling the noise for the nth observation after N observations by
$$\epsilon_n \sim \mathcal{N}\left(0, \left(m(\mathbf{x}_n)\tau\lambda^{\sum_{j=n+1}^{N} m(\mathbf{x}_j)}\right)^{-1}\right) \tag{5.61}$$
causes the Kalman filter to perform the same recency weighting as the recency-weighted RLS variant.

- The Gaussian prior on $\boldsymbol{\omega}$ provides a different interpretation of the ridge complexity λ in ridge regression: recalling that λ corresponds to initialising RLS with $\boldsymbol{\Lambda}_0^{-1} = \lambda^{-1}\mathbf{I}$, it is also equivalent to using the Kalman filter with the prior $\boldsymbol{\omega}_0 \sim \mathcal{N}(\mathbf{0}, (\lambda\tau)^{-1}\mathbf{I})$. Hence, ridge regression assumes the weight vector to be centred on $\mathbf{0}$ with an independent variance of $(\lambda\tau)^{-1}$ of each element of this vector. As the prior covariance is proportional to the real noise variance τ^{-1}, a smaller variance will cause stronger shrinkage due to a more informative prior.

What if the noise distribution is not Gaussian? Would that invalidate the approach taken by RLS and the Kalman filter? Fortunately, the Gauss-Markov Theorem (for example, [97]) states that the least squares estimate is optimal independent of the shape of the noise distribution, as long as its variance is constant over all observations. Nonetheless, adding the assumption of Gaussian noise and acquiring a Gaussian model for the weight vector allows us to specify the predictive density. Without these assumptions, we would be unable make any statements about this density, and are subsequently also unable to provide a measure for the prediction confidence.

In summary, demonstrating the formal equivalence between the RLS algorithm and the Kalman filter for a stationary system state has significantly increased the understanding of the assumptions underlying the RLS method and provides intuitive interpretations for matching and recency-weighting by relating them to an increased uncertainty about the observations.

5.3.7 Incremental Noise Precision Estimation

So far, the discussion of the incremental methods has focused on estimating the weight vector that solves (5.5). Let us now consider how we can estimate the noise precision by incrementally solving (5.6).

For batch learning it was already demonstrated that (5.11) and (5.13) provide a biased and unbiased noise precision estimate that solves (5.6). The same solutions are valid when using an incremental approach, and thus, after N observations,

$$\hat{\tau}_N^{-1} = c_N^{-1} \|\mathbf{X}_N \hat{\mathbf{w}}_N - \mathbf{y}_N\|_{\mathbf{M}_N}^2 \tag{5.62}$$

provides a biased estimate of the noise precision, and

$$\hat{\tau}_N^{-1} = (c_N - D_{\mathcal{X}})^{-1} \|\mathbf{X}_N \hat{\mathbf{w}}_N - \mathbf{y}_N\|_{\mathbf{M}_N}^2 \tag{5.63}$$

is the unbiased estimate. Ideally, $\hat{\mathbf{w}}_N$ is the weight vector that satisfies the Principle of Orthogonality, but if gradient-based methods are utilised, we are forced to rely on the current (possibly quite wrong) estimate.

Let us firstly derive a gradient-based method for estimating the noise precision, which is the one applied in XCS. Following that, a much more accurate approach is introduced that can be used alongside the RLS algorithm to track the exact noise precision estimate after (5.63) for each additional observation.

Estimation by Gradient Descent

The problem of computing (5.62) can be reformulated as finding the minimum of

$$\sum_{n=1}^{N} m(\mathbf{x}_n)\left(\tau^{-1} - (\hat{\mathbf{w}}_N^T \mathbf{x}_n - y_n)^2\right)^2. \tag{5.64}$$

That the minimum of the above with respect to τ is indeed (5.62) can be easily shown by the solution of setting its gradient with respect to τ to zero.

This minimisation problem can now be solved with any gradient-based method. Applying the LMS algorithm, the resulting update equation is given by

$$\hat{\tau}_{N+1}^{-1} = \hat{\tau}_N^{-1} + \gamma m(\mathbf{x}_{N+1})\left((\hat{\mathbf{w}}_{N+1}^T \mathbf{x}_{N+1} - y_{N+1})^2 - \hat{\tau}_N^{-1}\right). \tag{5.65}$$

While this method provides a computationally cheap approach to estimating the noise precision, it is flawed in several ways: firstly, it suffers under some circumstances from slow convergence speed, just as any other gradient-based method. Secondly, at each step, the method relies on the updated weight vector estimate, but does not take into account that changing the weight vector also modifies past estimates and with it the squared estimation error. Finally, by minimising (5.64) we are computing the biased estimate (5.62) rather than the unbiased estimate (5.63). The following method address all of these problems.

Estimation by Direct Tracking

Assume that the sequence of weight vector estimates $\{\hat{\mathbf{w}}_1, \hat{\mathbf{w}}_2, \dots\}$ satisfies the Principle of Orthogonality, which we can achieve by utilising the RLS algorithm. In the following, a method for incrementally updating $\|\mathbf{X}_N \hat{\mathbf{w}}_N - \mathbf{y}_N\|_{\mathbf{M}_N}^2$ is derived, which then allows for accurate tracking of the unbiased noise precision estimate (5.63).

At first, let us derive a simplified expression for $\|\mathbf{X}_N \hat{\mathbf{w}}_N - \mathbf{y}_N\|_{\mathbf{M}_N}^2$: based on the Corollary to the Principle of Orthogonality (5.17) and $-\mathbf{y}_N = -\mathbf{X}_N \hat{\mathbf{w}}_N + (\mathbf{X}_N \hat{\mathbf{w}}_N - \mathbf{y}_N)$ we get

$$\begin{aligned} \mathbf{y}_N^T \mathbf{M}_N \mathbf{y}_N &= \hat{\mathbf{w}}_N^T \mathbf{X}_N^T \mathbf{M}_N \mathbf{X}_N \hat{\mathbf{w}}_N - 2\hat{\mathbf{w}}_N^T \mathbf{X}_N^T \mathbf{M}_N (\mathbf{X}_N \hat{\mathbf{w}}_N - \mathbf{y}_N) \\ &\quad + (\mathbf{X}_N \hat{\mathbf{w}}_N - \mathbf{y}_N)^T \mathbf{M}_N (\mathbf{X}_N \hat{\mathbf{w}}_N - \mathbf{y}_N) \\ &= \hat{\mathbf{w}}_N^T \mathbf{X}_N^T \mathbf{M}_N \mathbf{X}_N \hat{\mathbf{w}}_N + \|\mathbf{X}_N \hat{\mathbf{w}}_N - \mathbf{y}_N\|_{\mathbf{M}_N}^2, \end{aligned} \tag{5.66}$$

which, for the sum of squared errors, results in

$$\|\mathbf{X}_N \hat{\mathbf{w}}_N - \mathbf{y}_N\|_{\mathbf{M}_N}^2 = \mathbf{y}_N^T \mathbf{M}_N \mathbf{y}_N - \hat{\mathbf{w}}_N^T \mathbf{X}_N^T \mathbf{M}_N \mathbf{X}_N \hat{\mathbf{w}}_N. \tag{5.67}$$

Expressing $\|\mathbf{X}_{N+1} \hat{\mathbf{w}}_{N+1} - \mathbf{y}_{N+1}\|_{\mathbf{M}_{N+1}}^2$ in terms of $\|\mathbf{X}_N \hat{\mathbf{w}}_N - \mathbf{y}_N\|_{\mathbf{M}_N}^2$ requires combining (5.31), (5.32) and (5.67), and the use of $\mathbf{\Lambda}_N \hat{\mathbf{w}}_N = \mathbf{X}_N^T \mathbf{M}_N \mathbf{y}_N$ after (5.30), which, after some algebra, results in the following:

Theorem 5.7 (Incremental Sum of Squared Error Update). *Let the sequence of weight vector estimates $\{\hat{\mathbf{w}}_1, \hat{\mathbf{w}}_2, \dots\}$ satisfy the Principle of Orthogonality (5.16). Then*

$$\begin{aligned}&\|\mathbf{X}_{N+1}\hat{\mathbf{w}}_{N+1} - \mathbf{y}_{N+1}\|^2_{M_{N+1}} \\ &\quad = \|\mathbf{X}_N\hat{\mathbf{w}}_N - \mathbf{y}_N\|^2_{M_N} + m(\mathbf{x}_{N+1})(\hat{\mathbf{w}}_N^T\mathbf{x}_{N+1} - y_{N+1})(\hat{\mathbf{w}}_{N+1}^T\mathbf{x}_{N+1} - y_{N+1})\end{aligned} \tag{5.68}$$

holds.

An almost equal derivation reveals that the sum of squared errors for the recency-weighted RLS variant is given by

$$\begin{aligned}&\|\mathbf{X}_{N+1}\hat{\mathbf{w}}_{N+1} - \mathbf{y}_{N+1}\|^2_{M_{N+1}} \\ &\quad = \lambda^{m(\mathbf{x}_{N+1})}\|\mathbf{X}_N\hat{\mathbf{w}}_N - \mathbf{y}_N\|^2_{M_N} \\ &\qquad + m(\mathbf{x}_{N+1})(\hat{\mathbf{w}}_N^T\mathbf{x}_{N+1} - y_{N+1})(\hat{\mathbf{w}}_{N+1}^T\mathbf{x}_{N+1} - y_{N+1}),\end{aligned} \tag{5.69}$$

where, when compared to (5.68), the current sum of squared errors is additionally discounted.

In summary, the unbiased noise precision estimate can be tracked by directly solving (5.63), where the match count is updated by

$$c_{N+1} = c_N + m(\mathbf{x}_{N+1}), \tag{5.70}$$

and the sum of squared errors is updated by (5.68). As Theorem 5.7 states, (5.68) is only valid if the Principle of Orthogonality holds. However, using the computationally cheaper RLS implementation that involves (5.35) introduces an initial bias and hence violates the Principle of Orthogonality. Nonetheless, if δ in $\mathbf{\Lambda}_0^{-1} = \delta\mathbf{I}$ is set to a very large positive scalar, this bias is negligible, and hence (5.68) is still applicable with only minor inaccuracy.

Example 5.8 (Noise Precision Estimation for Averaging Classifiers). Consider averaging classifiers, such that $x_n = 1$ for all $n > 0$. Given the use of gradient-based methods to estimate the weight vector violates the Principle of Orthogonality, and hence (5.65) has to be used estimate the noise precision, resulting in

$$\hat{\tau}_{N+1}^{-1} = \hat{\tau}_N^{-1} + m(\mathbf{x}_{N+1})\left((\hat{w}_{N+1} - y_{N+1})^2 - \hat{\tau}_N^{-1}\right). \tag{5.71}$$

Alternatively, we can use the RLS algorithm (5.46) for averaging classifiers, and use (5.68) to accurately track the noise precision by

$$\hat{\tau}_{N+1}^{-1} = \hat{\tau}_N^{-1} + m(\mathbf{x}_{N+1})(\hat{w}_N - y_{N+1})(\hat{w}_{N+1} - y_{N+1}). \tag{5.72}$$

Note that while the computational cost of both approaches is equal (in its application to averaging classifiers), the second approach is vastly superior in its weight vector and noise precision estimation accuracy and should therefore be always preferred.

Squared Error or Absolute Error?

XCSF (of which XCS is a special case) initially applied the NLMS method (5.29) [237], and later the RLS algorithm by (5.34) and (5.35) [142, 143] to estimate the weight vector. The classifier estimation error is tracked by the LMS update

$$\hat{\tau}_{N+1}^{-1} = \hat{\tau}_N^{-1} + m(\mathbf{x}_{N+1}) \left(|\hat{\mathbf{w}}_{N+1}^T \mathbf{x}_{N+1} - y_{N+1}| - \hat{\tau}_N^{-1} \right), \tag{5.73}$$

to – after N observations – perform stochastic incremental gradient descent on the error function

$$\sum_{n=1}^{N} m(\mathbf{x}_n) \left(\tau^{-1} - |\hat{\mathbf{w}}_N^T \mathbf{x}_n - y_n| \right)^2 . \tag{5.74}$$

Therefore, the error that is estimated is the mean absolute error

$$c_N^{-1} \sum_{n=1}^{N} m(\mathbf{x}_n) \left| \hat{\mathbf{w}}_N^T \mathbf{x}_n - y_n \right| , \tag{5.75}$$

rather than the MSE (5.62). Thus, XCSF does not estimate the error that its weight vector estimate aims at minimising, and does not justify this inconsistency – probably because the errors that are minimised have never before been explicitly expressed. While there is no systematic study that compares using (5.62) rather than (5.75) as the classifier error estimate in XCSF, we have recommended in [155] to use the MSE for the reason of consistency and easier tracking by (5.68), and – as shown here – to provide its probabilistic interpretation as the noise precision estimate $\hat{\tau}$ of the linear model.

5.3.8 Summarising Incremental Learning Approaches

Various approaches to estimating the weight vector and noise precision estimate of the linear model (5.3) have been introduced. While the gradient-based models, such as LMS or NLMS, are computationally cheap, they require problem-dependent tuning of the step size and might feature slow convergence to the optimal estimates. RLS and Kalman filter approaches, on the other hand, scale at best with $\mathcal{O}(D_\mathcal{X}^2)$, but are able to accurately track both the optimal weight vector estimate and its associated noise precision estimate simultaneously.

Table 5.1 gives a summary of all the methods introduced in this chapter (omitting the recency-weighted variants), together with their computational complexity. As can be seen, this complexity is exclusively dependent on the size of the input vectors for use by the classifier model (in contrast to their use for matching). Given that we have averaging classifiers, we have $D_\mathcal{X} = 1$, and thus, all methods have equal complexity. In this case, the RLS algorithm with direct noise precision tracking should always be applied. For higher-dimensional input spaces, the choice of the algorithm depends on the available computational resources, but the RLS approach should always be given a strong preference.

Table 5.1. A summary of batch and incremental methods presented in this chapter for training the linear regression model of a single classifier. The notation and initialisation values are explained throughout the chapter.

Batch Learning	
$\hat{\mathbf{w}} = (\mathbf{X}^T\mathbf{M}\mathbf{X})^{-1}\mathbf{X}^T\mathbf{M}\mathbf{y}$ or $\hat{\mathbf{w}} = (\sqrt{\mathbf{M}}\mathbf{X})^+\sqrt{\mathbf{M}}\mathbf{y}$ $\hat{\tau}^{-1} = (c - D_{\mathcal{X}})^{-1}\Vert\mathbf{X}\hat{\mathbf{w}} - \mathbf{y}\Vert_M^2$ with $c = \mathrm{Tr}(\mathbf{M})$	
Incremental Weight Vector Estimate	Complexity
LMS $\hat{\mathbf{w}}_{N+1} = \hat{\mathbf{w}}_N + \gamma_{N+1} m(\mathbf{x}_{N+1})\mathbf{x}_{N+1}(y_{N+1} - \hat{\mathbf{w}}_N^T\mathbf{x}_{N+1})$	$\mathcal{O}(D_{\mathcal{X}})$
NLMS $\hat{\mathbf{w}}_{N+1} = \hat{\mathbf{w}}_N + \gamma_{N+1} m(\mathbf{x}_{N+1})\frac{\mathbf{x}_{N+1}}{\Vert\mathbf{x}_{N+1}\Vert^2}(y_{N+1} - \hat{\mathbf{w}}_N^T\mathbf{x}_{N+1})$	$\mathcal{O}(D_{\mathcal{X}})$
RLS (Inverse Covariance Form) $\hat{\mathbf{w}}_{N+1} = \hat{\mathbf{w}}_N + m(\mathbf{x}_{N+1})\mathbf{\Lambda}_{N+1}^{-1}\mathbf{x}_{N+1}(y_{N+1} - \hat{\mathbf{w}}_N^T\mathbf{x}_{N+1})$, $\mathbf{\Lambda}_{N+1} = \mathbf{\Lambda}_N + m(\mathbf{x}_{N+1})\mathbf{x}_{N+1}\mathbf{x}_{N+1}^T$	$\mathcal{O}(D_{\mathcal{X}}^3)$
RLS (Covariance Form) $\hat{\mathbf{w}}_{N+1} = \hat{\mathbf{w}}_N + m(\mathbf{x}_{N+1})\mathbf{\Lambda}_{N+1}^{-1}\mathbf{x}_{N+1}(y_{N+1} - \hat{\mathbf{w}}_N^T\mathbf{x}_{N+1})$, $\mathbf{\Lambda}_{N+1}^{-1} = \mathbf{\Lambda}_N^{-1} - m(\mathbf{x}_{N+1})\frac{\mathbf{\Lambda}_N^{-1}\mathbf{x}_{N+1}\mathbf{x}_{N+1}^T\mathbf{\Lambda}_N^{-1}}{1+m(\mathbf{x}_{N+1})\mathbf{x}_{N+1}^T\mathbf{\Lambda}_N^{-1}\mathbf{x}_{N+1}}$	$\mathcal{O}(D_{\mathcal{X}}^2)$
Kalman Filter (Covariance Form) $\zeta_{N+1} = m(\mathbf{x}_{N+1})\mathbf{\Lambda}_N^{-1}\mathbf{x}_{N+1}\left(m(\mathbf{x}_{N+1})\mathbf{x}_{N+1}^T\mathbf{\Lambda}_N^{-1}\mathbf{x}_{N+1} + \tau_{N+1}^{-1}\right)^{-1}$, $\hat{\mathbf{w}}_{N+1} = \hat{\mathbf{w}}_N + \zeta_{N+1}\left(y_{N+1} - \hat{\mathbf{w}}_N^T\mathbf{x}_{N+1}\right)$, $\mathbf{\Lambda}_{N+1}^{-1} = \mathbf{\Lambda}_N^{-1} - \zeta_{N+1}\mathbf{x}_{N+1}^T\mathbf{\Lambda}_N^{-1}$	$\mathcal{O}(D_{\mathcal{X}}^2)$
Kalman Filter (Inverse Covariance Form) $\mathbf{\Lambda}_{N+1}\hat{\mathbf{w}}_{N+1} = \mathbf{\Lambda}_N\hat{\mathbf{w}}_N + m(\mathbf{x}_{N+1})\tau_{N+1}\mathbf{x}_{N+1}y_{N+1}$, $\mathbf{\Lambda}_{N+1} = \mathbf{\Lambda}_N + m(\mathbf{x}_{N+1})\tau_{N+1}\mathbf{x}_{N+1}\mathbf{x}_{N+1}^T$, $\hat{\mathbf{w}}_{N+1} = \mathbf{\Lambda}_{N+1}(\mathbf{\Lambda}_{N+1}\hat{\mathbf{w}}_{N+1})^{-1}$	$\mathcal{O}(D_{\mathcal{X}}^3)$
Incremental Noise Precision Estimate	Complexity
LMS (for biased estimate (5.62)) $\hat{\tau}_{N+1}^{-1} = \hat{\tau}_N^{-1} + m(\mathbf{x}_{N+1})\left((\hat{\mathbf{w}}_{N+1}^T\mathbf{x}_{N+1} - y_{N+1})^2 - \hat{\tau}_N^{-1}\right)$	$\mathcal{O}(D_{\mathcal{X}})$
Direct tracking (for unbiased estimate (5.63)) Only valid in combination with RLS/Kalman filter in Inverse Covariance Form or in Covariance Form with insignificant prior $\Vert\mathbf{X}_{N+1}\hat{\mathbf{w}}_{N+1} - \mathbf{y}_{N+1}\Vert_{M_{N+1}}^2 = \Vert\mathbf{X}_N\hat{\mathbf{w}}_N - \mathbf{y}_N\Vert_{M_N}^2$ $+m(\mathbf{x}_{N+1})(\hat{\mathbf{w}}_N^T\mathbf{x}_{N+1} - y_{N+1})(\hat{\mathbf{w}}_{N+1}^T\mathbf{x}_{N+1} - y_{N+1})$, $c_{N+1} = c_N + m(\mathbf{x}_{N+1})$, $\hat{\tau}_{N+1}^{-1} = (c_{N+1} - D_{\mathcal{X}})^{-1}\Vert\mathbf{X}_{N+1}\hat{\mathbf{w}}_{N+1} - \mathbf{y}_{N+1}\Vert_{M_{N+1}}^2$	$\mathcal{O}(D_{\mathcal{X}})$

5.4 Empirical Demonstration

Having described the advantage of utilising the RLS algorithm to estimating the weight vector and tracking the noise variance simultaneously, this section gives a brief empirical demonstration of its superiority over gradient-based methods. The two experiments show on one hand that the speed of convergence of the

LMS and NLMS algorithm is lower than for the RLS algorithm and depends on the values of the input, and on the other hand that direct tracking of the noise variance is more accurate than estimating it by the LMS method.

5.4.1 Experimental Setup

The following classifier setups are used:

NLMS Classifier. This classifier uses the NLMS algorithm (5.29) to estimate the weight vector, starting with $\hat{\mathbf{w}}_0 = \mathbf{0}$, and a constant step size of $\gamma = 0.2$. For one-dimensional input spaces, $D_{\mathcal{X}} = 1$, with $x_n = 1$ for all $n > 0$, the NLMS algorithm is equivalent to the LMS algorithm (5.25), in which variable step sizes according to the MAM update [220] are used,

$$\gamma_N = \begin{cases} 1/c_N & \text{if } c_N \leq 1/\gamma, \\ \gamma & \text{otherwise,} \end{cases} \tag{5.76}$$

which is equivalent to bootstrapping the estimate by RLS (see Example 5.6). The noise variance is estimated by the LMS algorithm (5.63), with an initial $\tau_0^{-1} = 0$, and a step size that follows the MAM update (5.76). Thus, the NLMS classifier uses the same techniques for weight vector and noise variance estimation as XCS(F), with the only difference that the correct variance rather than the mean absolute error (5.75) is estimated (see also Sect. 5.3.7). Hence, the performance of NLMS classifiers reflects the performance of classifiers in XCS(F).

RLSLMS Classifier. The weight vector is estimated by the RLS algorithm, using (5.34) and (5.35), with initialisation $\hat{\mathbf{w}}_0 = \mathbf{0}$ and $\mathbf{\Lambda}_0^{-1} = 1000\mathbf{I}$. The noise variance is estimated by the LMS algorithm, just as for the NLMS Classifier. This setup conforms to XCSF classifiers with RLS as first introduced by Lanzi et al. [142, 143].

RLS Classifier. As before, the weight vector is estimated by the RLS algorithm (5.34) and (5.35), with initialisation $\hat{\mathbf{w}}_0 = \mathbf{0}$ and $\mathbf{\Lambda}_0^{-1} = 1000\mathbf{I}$. The noise variance is estimated by tracking the sum of squared errors according to (5.68) and evaluating (5.63) for the unbiased variance estimate.

In both experiments, all three classifiers are used for the same regression task, with the assumption that they match all inputs, that is, $m(\mathbf{x}_n) = 1$ for all $n > 0$. Their performance of estimating the weight vector is measured by the MSE of their model evaluated with respect to the target function f over 1000 inputs that are evenly distributed over the function's domain, using (5.11). The quality of the estimate noise variance is evaluated by its squared error when compared to the unbiased noise variance estimate (5.13) of a linear model trained by (5.8) over 1000 observations that are evenly distributed over the function's domain.

For the first experiment, averaging classifiers with $x_n = 1$ for all $n > 0$ are used to estimate weight and noise variance of the noisy target function $f_1(x) = 5 + \mathcal{N}(0, 1)$. Hence, the correct weight estimate is $\hat{w} = 5$, with noise variance $\hat{\tau}^{-1} = 1$. As the function output is independent of its input, its domain does not

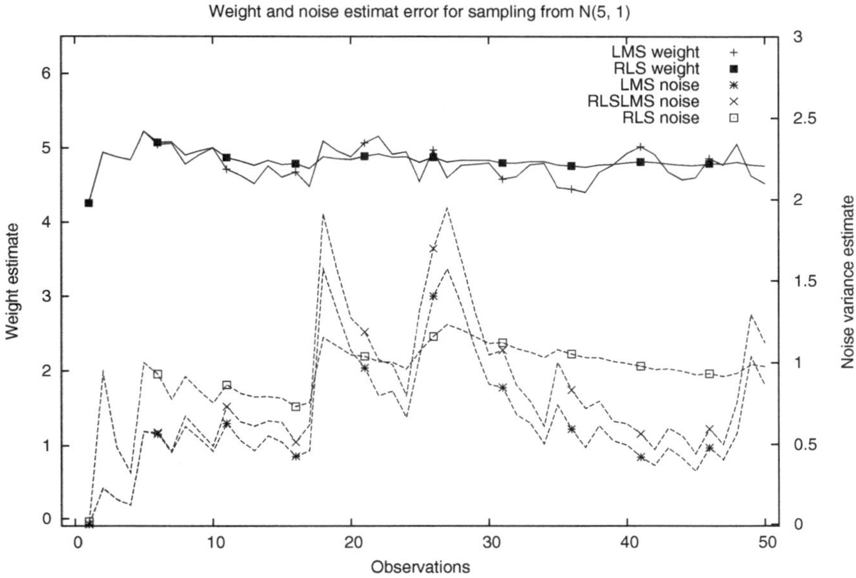

Fig. 5.1. The graph shows the weight estimate (on the left scale) and noise variance estimate (on the right scale) of different averaging classifiers when being presented with observations sampled from $\mathcal{N}(5, 1)$. The weight estimate of the RLSLMS classifier is not shown, as it is equivalent to the estimate of the RLS classifier.

need to be defined. The target function of the second experiment is the sinusoid $f_2(\mathbf{x}_n) = \sin(i_n)$ with inputs $\mathbf{x}_n = (1, i_n)$, hence, using classifiers that model straight lines. The experiment is split into two parts, where in the first part, the function is modelled over the domain $i_n \in [0, \pi/2)$, and in the second part over $i_n \in [pi/2, \pi)$. The classifiers are trained incrementally, by presenting them with observations that are uniformly sampled from the target function's domain.

Statistical significance of difference in the classifiers' performances of estimating the weight vector and noise variance is evaluated by comparing the sequence of model MSEs and squared noise variance estimation errors respectively, after each additional observations, and over 20 experimental runs. These sequences violate the standard analysis of variances (ANOVA) assumption of homogeneity of covariances, and thus the randomised ANOVA procedure [184], specifically designed for such cases, was used. It is based on estimating the sampling distribution of the null hypothesis ("all methods feature the same performance") by sampling the standard two-way ANOVA F-values from randomly reshuffled performance curves between the methods, where we use a samples size of 5000. The two factors are the type of classifier that is used, and the number of observations that the classifier has been trained on, where performance is measured by the model or noise variance error. Significant differences are only reported between classifier types, and Tukey's HSD post hoc test is employed to determine the direction of the effect.

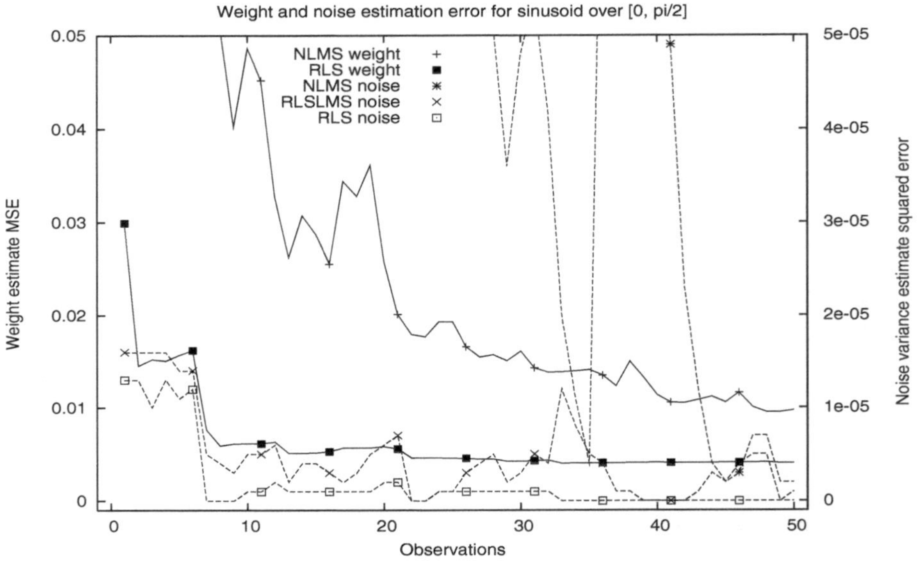

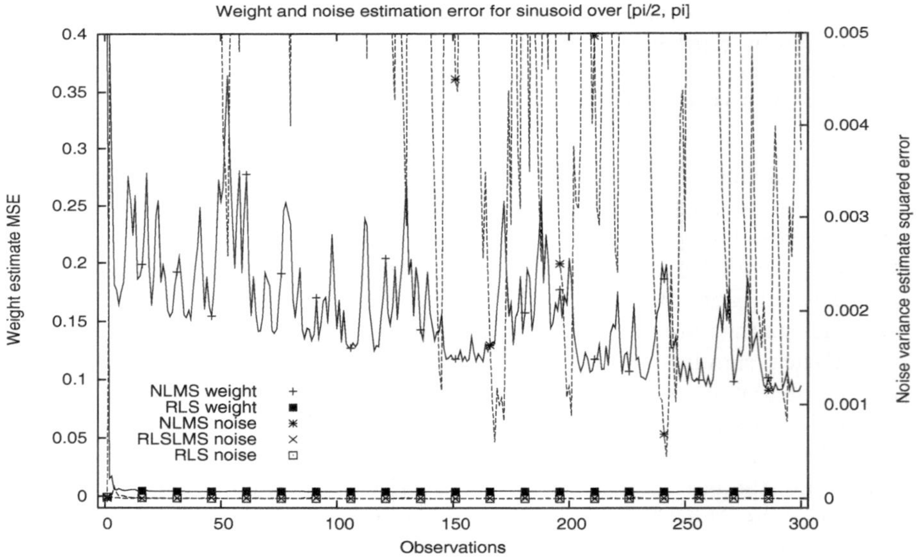

Fig. 5.2. The graphs show the MSE of the weight vector estimate (on the left scale) and squared noise variance estimate error (on the right scale) of different classifiers when approximating a sinusoid. The classifiers are presented with input $\mathbf{x}_n = (1, i_n)^T$ and output $y_n = \sin(i_n)$. In the upper graph, the sinusoid was sampled from the range $i_n \in [0, \pi/2]$, and in the lower graph the samples are taken from the range $i_n \in [\pi/2, \pi]$. The MSE of the weight vector estimate for the RLSLMS classifier is not show, as it is equivalent to the MSE of the RLS classifier.

Figures 5.1 and 5.2 show one run of training the classifiers on f_1 and f_2 respectively. Figure 5.1 illustrates how the weight and noise variance estimate differs for different classifiers when trained on the same 50 observations. Figure 5.2, on the other hand, does not display the estimates itself, but rather shows the error of the weight vector and noise variance estimates. Let us firstly focus on the ability of the different classifiers to estimate the weight vector.

5.4.2 Weight Vector Estimate

In the following, the RLSLMS classifier will be ignored due to its equivalence to the RLS classifier when estimating the weight vector. Figure 5.1 shows that while both the NLMS and the RLS algorithm estimate the weight to be about $\hat{w} = 5$, the RLS algorithm is more stable in its estimate. In fact, comparing the model MSEs by the randomised ANOVA procedure reveals that this error is significantly lower for the RLS method (randomised ANOVA: $F_{\text{alg}}(2, 2850) = 38.0$, $F^*_{\text{alg},.01} = 25.26$, $p < .01$). Figure 5.1 also clearly illustrates that utilising the MAM causes the weight estimates to be initially equivalent to the RLS estimates, until $1/\gamma = 5$ observations are reached. As the input to the averaging classifier is always $x_n = 1$, the speed of convergence of the LMS classifier is independent of these inputs.

The second experiment, on the other hand, demonstrates how ill-conditioned inputs cause the convergence speed of the NLMS algorithm to deteriorate. The upper graph of Figure 5.2 shows that while the weight estimate is close to optimal after 10 observations for the RLS classifier, the NLMS classifier requires more than 50 observations to reach a similar performance, when modelling f_2 over $i_n \in [0, \pi/2)$. Even worse, changing the sampling range to $i_n \in [\pi/2, \pi)$ causes the NLMS performance to drop such that it still features an MSE of around 0.1 after 300 observations, while the performance of the RLS classifier remains unchanged, as shown by the lower graph of Figure 5.2. This drop can be explained by the increasing eigenvalues of $c_N^{-1}\mathbf{X}_N^T\mathbf{M}_N\mathbf{X}_N$ that reduce the speed of convergence (see Sect. 5.25). The minimal MSE of a linear model is in both cases approximately 0.00394, and the difference in performance between the NLMS and the RLS classifier is in both cases significant (randomised ANOVA for $i_n \in [0, \pi/2]$: $F_{\text{alg}}(2, 2850) = 973.0$, $F^*_{\text{alg},.001} = 93.18$, $p < .001$; randomised ANOVA for $i_n \in [\pi/2, \pi]$: $F_{\text{alg}}(2, 17100) = 88371.5$, $F^*_{\text{alg},.001} = 2190.0, p < .001$).

5.4.3 Noise Variance Estimate

As the noise variance estimate depends by (5.63) on a good estimate of the weight vector, classifiers that perform poorly on estimating the weight vector can be expected to not perform any better when estimating the noise variance. This suggestion is confirmed when considering the noise variance estimate of the NLMS classifier in Fig. 5.1 that fluctuates heavily around the correct value of 1. While the RLSLMS classifier has the equivalent weight estimate to the RLS classifier, its noise variance estimate fluctuates almost as heavily as that of the NLMS classifier, as it also uses LMS to perform this estimate. Thus, while a good weight vector estimate is a basic requirement for estimating the noise variance, the applied LMS method seems to perform even worse when estimating the noise variance than when estimating the weight. As can be seen in Fig. 5.1, direct tracking of the noise variance in combination with the RLS algorithm for a stable weight estimate gives the least noise and accurate estimate. Indeed, while there is no significant difference in the squared estimation error between the NLMS and RLSLMS classifier (randomised ANOVA: $F_{\text{alg}}(2, 2850) = 53.68$, $F^*_{\text{alg},.001} = 29.26$, $p < .001$; Tukey's HSD: $p > .05$), the RLS classifier features a significantly better estimate than both of the other classifier types (Tukey's HSD: for both NLMS and RLSLMS $p < .01$).

Conceptually, the same pattern is observed in the second experiment, as shown in Fig. 5.2. However, in this case, the influence of a badly estimated weight vector becomes clearer, and is particularly visible for the NLMS classifier. Recall that this figure shows the estimation errors rather than the estimates itself, and hence, the upper graph shows that the NLMS classifier only provides estimates that are comparable to the RLSLMS and RLS classifier after 30 observations. The performance of NLMS in the case of ill-conditioned inputs is even worse; its estimation performance never matches that of the classifiers that utilise the RLS algorithm for their weight vector estimate. In contrast to the first experiment there is no significant difference between the noise variance estimation error of the RLSLMS and RLS classifiers, but in both cases they are significantly better than the NLMS classifier (for $i_n \in [0, \pi/2]$: randomised ANOVA: $F_{\text{alg}}(2, 2850) = 171.41$, $F^*_{\text{alg},.001} = 32.81$, $p < .001$; Tukey's HSD: NMLS vs. RLSLMS and RLS $p < .01$, RLSLMS vs. RLS $p > .05$; for $i_n \in [\pi/2, \pi]$: randomised ANOVA: $F_{\text{alg}}(2, 17100) = 4268.7$, $F^*_{\text{alg},.001} = 577.89$, $p < .001$; Tukey's HSD: NLMS vs. RLS and RLSLMS $p < .01$, RLSLMS vs. RLS $p > .05$).

In summary, both experiments in combination demonstrate that to provide a good noise variance estimate, the method needs to estimate the weight vector well, and that direct tracking of this estimate is better than its estimation by the LMS algorithm.

5.5 Classification Models

After having extensively covered the training of linear regression classifier models, let us turn our focus on classification models. In this case, we assume that input space and output space to be $\mathcal{X} = \mathbb{R}^{D_\mathcal{X}}$ and $\mathcal{Y} = \{0, 1\}^{D_\mathcal{Y}}$, where $D_\mathcal{Y}$ is the number of classes of the problem. An output vector $\mathbf{y}$ representing class j is 0 in all its elements except for $y_j = 1$.

Taking the generative point-of-view, a classifier is assumed to have generated an observation of some class with a certain probability, independent of the associated input, resulting in the classifier model

$$p(\mathbf{y}|\mathbf{x}, \mathbf{w}) = \prod_{j=1}^{D_\mathcal{Y}} w_j^{y_j}, \qquad \text{with } \sum_{j=1}^{D_\mathcal{Y}} w_j. \tag{5.77}$$

Therefore, the probability of the classifier having generated class j is given by w_j, which is the jth element of its parameter vector $\mathbf{w} \in \mathbb{R}^{D_\mathcal{Y}}$.

5.5.1 A Quality Measure for Classification

Good classifiers are certain about which classes they are associated with. This implies that one aims at finding classifiers that have a high probability associated with a single class, and low probability for all other classes.

For a two-class problem, the relation $w_2 = 1 - w_1$ is required to hold to satisfy $\sum_j w_j = 1$. In such a case, the model's variance $\text{var}(\mathbf{y}|\mathbf{w}) = w_1(1 - w_1)$ is a good measure of the model's quality as it is $\text{var}(\mathbf{y}|\mathbf{w}) = 0$ for $w_1 = 0$ or $w_2 = 0$, and has its maximum $\text{var}(\mathbf{y}|\mathbf{w}) = 0.25$ at $w_1 = 0.5$, which is the point of maximum uncertainty.

The same principle can be extended to multi-class problems, by taking the product of the elements of $\mathbf{w}$, denoted τ^{-1}, and given by

$$\tau^{-1} = \prod_{i=1}^{D_\mathcal{Y}} w_j. \tag{5.78}$$

In the three-class case, for example, the worst performance occurs at $w_1 = w_2 = w_3 = 1/3$, at which point τ^{-1} is maximised. Note that τ^{-1} is, unlike for linear regression, formally *not* the precision estimate.

As τ^{-1} is easily computed from $\mathbf{w}$, its estimate does not need to be maintained separately. Thus, the description of batch and incremental learning approaches deals exclusively with the estimation of $\mathbf{w}$.

5.5.2 Batch Approach for Classification

Recall that the aim of a classifier is to solve (4.24), which, together with (5.77) results in the constrained optimisation problem

$$\max_w \sum_{n=1}^{N} m(\mathbf{x}_n) \sum_{j=1}^{D_\mathcal{Y}} y_{nj} \ln w_j, \tag{5.79}$$

$$\text{subject to} \sum_{j=1}^{D_\mathcal{Y}} w_j = 1.$$

Using the Lagrange multiplier λ to express the constraint $1 - \sum_j w_j = 0$, the aim becomes to maximise

$$\sum_{n=1}^{N} m(\mathbf{x}_n) \sum_{j=1}^{D_\mathcal{Y}} y_{nj} \ln w_j + \lambda \left(1 - \sum_{j=1}^{D_\mathcal{Y}} w_j \right). \tag{5.80}$$

Differentiating the above with respect to w_j for some j, setting it to 0, and solving for w_j results in the estimate

$$\hat{w}_j = \lambda^{-1} \sum_{n=1}^{N} m(\mathbf{x}_n) y_{nj}. \tag{5.81}$$

Solving for λ and using $\sum_j \hat{w}_j = 1$ and $\sum_j y_{nj} = 1$ for all N, we get $\lambda = \sum_n m(\mathbf{x}_n) = c$, which is the match count after N observations. As a result, $\mathbf{w}$ is after N observations by the principle of maximum likelihood given by

$$\hat{\mathbf{w}} = c^{-1} \sum_{n=1}^{N} m(\mathbf{x}_n)\mathbf{y}_n, \tag{5.82}$$

Thus, the jth element of $\hat{w}$, representing the probability of the classifier having generated an observation of class j, is the number of matched observations of this class divided by the total number of observations – a straightforward frequentist measure.

5.5.3 Incremental Learning for Classification

Let $\hat{\mathbf{w}}_N$ be the estimate of $\mathbf{w}$ after N observations. Given the new observation $(\mathbf{x}_{N+1}, \mathbf{y}_{N+1})$, the aim of the incremental approach is to find a computationally efficient approach to update $\hat{\mathbf{w}}_N$ to reflect this new knowledge. By (5.82), $c_{N+1}\hat{\mathbf{w}}_{N+1}$ is given by

$$\begin{aligned} c_{N+1}\hat{\mathbf{w}}_{N+1} &= \sum_{n=1}^{N+1} m(\mathbf{x}_n)\mathbf{y}_n \\ &= \sum_{n=1}^{N} m(\mathbf{x}_n)\mathbf{y}_n + m(\mathbf{x}_{N+1})\mathbf{y}_{N+1} \\ &= (c_{N+1} - m(\mathbf{x}_{N+1}))\hat{\mathbf{w}}_N + m(\mathbf{x}_{N+1})\mathbf{y}_{N+1}. \end{aligned} \tag{5.83}$$

Dividing the above by c_{N+1} results in the final incremental update

$$\hat{\mathbf{w}}_{N+1} = \hat{\mathbf{w}}_N - c_{N+1}^{-1} m(\mathbf{x}_{N+1}) \left(\hat{\mathbf{w}}_N - \mathbf{y}_{N+1}\right). \tag{5.84}$$

This update tracks (5.82) accurately, is of complexity $\mathcal{O}(D_\mathcal{Y})$, and only requires the parameter vector $\hat{\mathbf{w}}$ and the match count c to be stored. Thus, it is accurate and efficient.

Example 5.9 (Classifier Model for Classification). Figure 5.3 shows the data of a classification task with two distinct classes. Observations of classes 1 and 2 are shown by circles and squares, respectively. The larger rectangles indicate the matched areas of the input space of the three classifiers c_1, c_2, and c_3. Based on these data, the number of matched observations of each class as well as $\hat{\mathbf{w}}$ and $\hat{\tau}$ are shown for each classifier in Table 5.2.

Recall that the elements of $\hat{\mathbf{w}}$ represent the estimated probabilities of having generated an observation of a specific class. The estimates in Table 5.2 show that Classifier c_3 is most certain about modelling class 2, while Classifier c_2 is most uncertain about which class it models. These values are also reflected in $\hat{\tau}^{-1}$, which is highest for c_2 and lowest for c_3. Thus, c_3 is the "best" classifier, while c_2 is the "worst" – an evaluation that reflects what can be observed in Fig. 5.3.

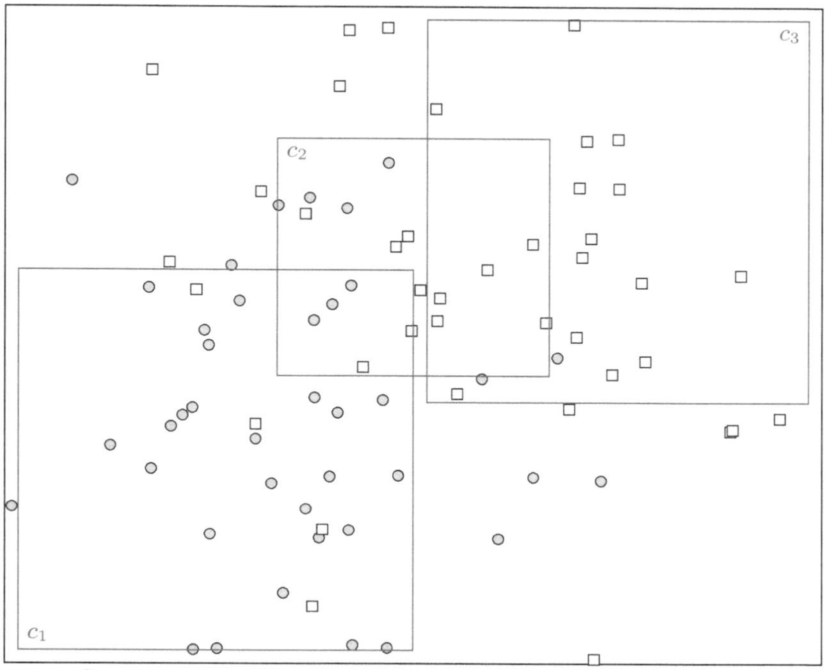

Fig. 5.3. Classification data of two different classes and three classifiers. The circles represent class 1, and the squares are samples of class 2. The larger rectangles within the input space are the matched ares of the three classifiers c_1, c_2, and c_3.

Table 5.2. Estimates resulting from the classification task illustrated in Fig. 5.3. The table gives the number of observations of each class matched per classifier. Additionally, it shows the parameter estimate $\hat{\mathbf{w}}$ and the measure $\hat{\tau}^{-1}$ of each classifier's prediction quality, evaluated by (5.82) and (5.78) respectively.

Classifier	Class 1	Class 2	$\hat{\mathbf{w}}_k^T$	$\hat{\tau}_k^{-1}$
c_1	27	5	$(0.84, 0.16)$	0.134
c_2	7	10	$(0.41, 0.59)$	0.242
c_3	2	19	$(0.09, 0.91)$	0.082

5.6 Discussion and Summary

The aim of a local model representing a classifier is to maximise its likelihood, as follows from the probabilistic LCS model of the previous chapter. In this chapter, several batch and incremental learning approaches for training linear regression models and classification models have been described and compared.

With respect to linear regression, the maximum likelihood estimate of the weight vector was shown to be a weighted least squares problem (5.5), that by itself is a well known problem with a multitude of approaches that goes far beyond the ones described in this chapter. Nonetheless, it is usually not stated as such in the LCS literature, and neither approached from first principles. Additional novelties in the LCS context are a probabilistic interpretation of the linear model and its noise structure, the resulting explicit formulation of the predictive density, and rigorous batch and incremental estimates of the noise variance.

The weight update of the original XCS conforms to (5.25) with $x_n = 1$ for $n > 0$ and hence aims at minimising the squared error (5.5). Later, XCS was modified to act as regression model [240], and extended to XCSF to model straight lines [241] by using the NLMS update (5.29), again without explicitly stating a single classifier's aim. In a similar manner, the classifier model was extended to a full linear model [141][4].

Simultaneously, and similar to the discussion in Sect. 5.3.4, the convergence of gradient-based methods was identified as a problem [142, 143], with a discussion based on steepest gradient descent rather than the NLMS method. As an alternative, the RLS algorithm was proposed to estimate the weight vector, but the aim of a classifier was specified without considering matching, and matching was implemented by only updating the classifier's parameter if that classifier matches the current input. While this is a valid procedure from the algorithmic perspective, it does not make matching explicit in the classifier's aim, and cannot deal with matching to a degree. The aim formulation (5.5), in contrast, provides both features and thereby leads to a better understanding and greater flexibility of the classifier model.

While XCSF weight estimation research did not stop at linear models [156, 175], the presented work was not extend beyond their realm to avoid the introduction of multiple local optima that make estimating the globally optimal weight vector significantly more complicated. In addition, there is always the trade-off between the complexity of the local models and the global model to consider: if more powerful local models are used, less of them are necessary to provide the same level of complexity of the global model, but the increased complexity and power makes their model usually harder to understand. For these reasons, linear classifier models provide a good trade-off between ease of training and power of the model, that are still relatively simple to interpret.

In contrast to the large amount of research activity seeking to improve the weight vector estimation method in XCS, its method of estimating the classifier model quality based on the absolute rather than the squared error was left untouched since the initial introduction of XCS until we questioned its validity in on the basis of the identified model aim [78], as also discussed in Sect. 5.3.7. The modified error measure not only introduces consistency, but also allows accurate tracking of the noise precision estimate with the method developed in Sect. 5.3.7, as previously shown [78]. Used as a drop-in replacement for the mean absolute error measure in XCSF, Loiacono et al. have shown that it, indeed, improves

---

[4] Despite the title "Extending XCSF Beyond Linear Approximation" of [141], the underlying model is still linear.

the generalisation capabilities as it provides a more accurate and stable estimate of the model quality of a classifier and subsequently a fitness estimate with the same qualities [155].

Nonetheless, the linear regression training methods introduced in this chapter are by no means to be interpreted as the ultimate methods to use to train the classifier models. Alternatively, one can use the procedure deployed in this chapter to adapt other parameter estimation techniques to their use in LCS. Still, currently the RLS algorithm is the best known incremental method to track the optimal weight estimate under the given assumptions, while simultaneously accurately estimating the noise variance. Hence, given that one aims at minimising the squared error (5.5), it should be the method of choice.

As an alternative to the squared error that corresponds to the assumption of Gaussian noise, one can consistently aim at estimating the weight vector that minimises the mean absolute error (5.75) [157]. However, this requires a modification of the assumptions about the distributions of the different linear model variables. Additionally, there is currently no known method to incrementally track the optimal weight estimate, as RLS does for the squared error measure. This also means that (5.68) cannot be used to track the model error, and slower gradient-based alternatives have to applied.

With respect to classification, the training of an appropriate LCS model has been discussed for both batch and incremental training. The method differs from current XCS-based LCS, such as UCS [161], in that it does not require augmentation of the input space by a separate class label (see Sect. 3.1.3), and evaluating classifiers based on how accurate its associated class is represented within its matched area of the input space. Instead, no assumptions are made about which class is modelled by a classifier, and the probability of having generated the observations of either class is estimated. This estimate can additionally be used to measure the quality of a classifier, based on the idea that good classifiers predict a single class with high probability. This concept has been firstly applied in an XCS-like context by Dam, Abbass, and Lokan in a Bayesian formulation for two-class classification problems with the result of improved performance and faster learning [67]. Further evaluation and extensions to multi-class problems are still pending.

A later chapter reconsiders the probabilistic structure of both the linear regression and classification models, and shows how the development of a probabilistic approach allows the model to be embedded in a fully Bayesian framework that also lends itself to application to multi-dimensional output spaces in the regression case. Before that, let us in the following chapter consider another LCS component that, contrary to the weight vector estimate of XCS, has received hardly any attention in LCS research: how the local models provided by the classifiers are combined to form a global model.

6 Mixing Independently Trained Classifiers

An essential part of the introduced model and of LCS in general that hardly any research has been devoted to is how to combine the local models provided by the classifiers to produce a global model. More precisely, given an input and the output prediction of all matched classifiers, the task is to combine these predictions to form a global prediction. This task will be called the *mixing problem*, and some model that provides an approach to this task a *mixing model*.

Whilst some early LCS (for example, SCS [95]) aimed at choosing a single "best" classifier to provide the global prediction, in modern Michigan-style LCS, predictions of matching classifiers have been mixed to give the "system prediction", that is, what will be called the global prediction. In XCS, for example, Wilson [237] defined the mixing model as follows:

> "There are several reasonable ways to determine [the global prediction] $P(a_i)$. We have experimented primarily with a fitness-weighted average of the prediction of classifiers advocating a_i. Presumably, one wants a method that yields the system's "best guess" as to the payoff [...] to be received if a_i is chosen",

and maintains this model for all XCS derivatives without any further discussion. As will be discussed in Sect. 6.2.5, the fitness he is referring to is a complex heuristic measure of the quality of a classifier. While the aim is *not* to redefine the fitness of a classifier in XCS, it is questioned if it is really the best measure to use when mixing the local classifier predictions. The mixing model has been changed in YCS [33], a simplified version of XCS and accuracy-based LCS in general, such that the classifier update equations can be formulated by difference equations, and by Wada et al. [223] to linearise the underlying model for the purpose of correcting XCS for use with reinforcement learning (see Sects. 4.5 and 9.3.6). In either case the motivation for changing the mixing model differs from the motivation in this chapter, which is to improve the performance of the model itself, rather than to simplify it or to modify its formulation for the use in reinforcement learning.

A formal treatment of the mixing problem requires a formal statement of the aim that is to be reached. In a previous, related study [83] this aim was defined

J. Drugowitsch: Des. & Anal. of Learn. Class. Sys.: A Prob. Approach, SCI 139, pp. 101–121, 2008.
springerlink.com

by the minimisation of the mean squared error of the global prediction with respect to the target function, given a fixed set of fully trained classifiers. As will be discussed in Sect. 6.4, this aim does not completely conform to the LCS model that was introduced in Chap. 4.

Rather than using the mean squared error as a measure of the quality of a mixing model, this chapter follows pragmatically the approach that was introduced with the probabilistic LCS model: each classifier k provides a localised probabilistic input/output mapping $p(y|\mathbf{x}, \boldsymbol{\theta}_k)$, and the value of a binary latent random variance z_{nk} determines if classifier k generated the nth observation. Each observation is generated by one and only one matching classifier, and so the vector $\mathbf{z}_n = (z_{n1}, \dots, z_{nK})^T$ has a single element with value 1, with all other elements being 0. As the values of the latent variables are unknown, they are modelled by the probabilistic model $g_k(\mathbf{x}) \equiv p(z_{nk} = 1|\mathbf{x}_n, \mathbf{v}_k)$, which is the mixing model. The aim is to find a mixing model that is sufficiently easy to train and maximises the data likelihood (4.9), given by

$$l(\boldsymbol{\theta}; \mathcal{D}) = \sum_{n=1}^{N} \ln \sum_{k=1}^{K} g_k(\mathbf{x}_n) p(y_n|\mathbf{x}_n, \boldsymbol{\theta}_k). \tag{6.1}$$

One possibility for such a mixing model was already introduced in Chap. 4 as a generalisation of the gating network used in the Mixtures-of-Experts model, and is given by the matching-augmented softmax function (4.22). Further alternatives will be introduced in this chapter.

The approach is called "pragmatic", as by maximising the data likelihood, the problem of overfitting is ignored, together with the identification of a good model structure that is essential to LCS. Nonetheless, the methods introduced here will reappear in only sightly modified form once these issues are dealt with, and discussing them here provides a better understanding in later chapters. Additionally, XCS implicitly uses an approach similar to maximum likelihood to train its classifiers and mixing models, and deals with overfitting only at the level of localising the classifiers in the input space (see App. B). Therefore, the methods and approaches discussed here can be used as a drop-in replacement for the XCS mixing model and for related LCS.

To summarise, we assume to have a set of K fully trained classifier, each of which provides a localised probabilistic model $p(y|\mathbf{x}, \boldsymbol{\theta}_k)$. The aim is to find a mixing model that provides the generative probability $p(z_{nk} = 1|\mathbf{x}_n, \mathbf{v}_k)$, that is, the probability that classifier k generated observation n, given input $\mathbf{x}_n$ and mixing model parameters $\mathbf{v}_k$, that maximises the data likelihood (6.1). Additional requirements are a sufficiently easy training and a good scaling of the method with the number of classifiers.

We will firstly concentrate on the model that was introduced in Chap. 4, and provide two approaches to training this model. Due to the thereafter discussed weaknesses of these training procedures, a set of formally inspired and computationally cheap heuristics are introduced. Some empirical studies show that these heuristics perform competitively when compared to the optimum. The chapter concludes by comparing the approach of maximising the likelihood to a closely related previous study [83], to linear LCS models, and to models that treat classifiers and mixing model as separate components by design.

6.1 Using the Generalised Softmax Function

By relating the probabilistic structure of LCS to the Mixtures-of-Experts model in Chap. 4, the probability of classifier k generating the nth observation is given by the generalised softmax function (4.22), that is,

$$g_k(\mathbf{x}_n) = \frac{m_k(\mathbf{x}_n)\exp(\mathbf{v}_k^T\phi(\mathbf{x}_n))}{\sum_{j=1}^K m_j(\mathbf{x}_n)\exp(\mathbf{v}_j^T\phi(\mathbf{x}_n))}, \tag{6.2}$$

where $\mathbf{V} = \{\mathbf{v}_k\}$ is the set of mixing model parameters $\mathbf{v}_k \in \mathbb{R}^{D_V}$, and $\phi(\mathbf{x})$ is a transfer function that maps the input space $\mathcal{X}$ into some D_V-dimensional real space $\mathbb{R}^{D_V}$. In LCS, this function is usually $\phi(\mathbf{x}) = 1$ for all $\mathbf{x} \in \mathcal{X}$, with $D_V = 1$, but to stay general, we do not make any assumptions about the form of ϕ.

Assuming knowledge of the predictive densities of all classifiers $p(y|\mathbf{x}, \boldsymbol{\theta}_k)$, the data likelihood (6.1) is maximised by the expectation-maximisation algorithm by finding the values for $\mathbf{V}$ that maximise (4.13), given by

$$\sum_{n=1}^N \sum_{k=1}^K r_{nk} \ln g_k(\mathbf{x}_n). \tag{6.3}$$

In the above equation, r_{nk} stands for the responsibility of classifier k for observation n, given by (4.12), that is

$$r_{nk} = \frac{g_k(\mathbf{x}_n)p(y_n|\mathbf{x}_n, \boldsymbol{\theta}_k)}{\sum_{j=1}^K g_j(\mathbf{x_n})p(y_n|\mathbf{x}_n, \boldsymbol{\theta}_j)}. \tag{6.4}$$

Thus, we want to fit the mixing model to the data by minimising the cross-entropy $-\sum_n \sum_k r_{nk} \ln g_k(\mathbf{x}_n)$ between the responsibilities and the generative mixing model.

6.1.1 Batch Learning by Iterative Reweighted Least Squares

The softmax function is a generalised linear model, and specialised tools have been developed to fit such models [165]. Even though a generalisation of this function is used, the same tools are applicable, as shown in this section. In particular, the Iterative Reweighted Least Squares (IRLS) will be employed to find the mixing model parameters.

The IRLS can be derived by applying the Newton-Raphson iterative optimisation scheme [19] that, for minimising an error function $E(\mathbf{V})$, takes the form

$$\hat{\mathbf{V}}^{(\text{new})} = \hat{\mathbf{V}}^{(\text{old})} - \mathbf{H}^{-1}\nabla E(\mathbf{V}), \tag{6.5}$$

where $\mathbf{H}$ is the Hessian matrix whose elements comprise the second derivatives of $E(\mathbf{V})$, and $\nabla E(\mathbf{V})$ is the gradient vector of $E(\mathbf{V})$ with respect to $\mathbf{V}$. Even though not immediately obvious, its name derives from a reformulation of the update procedure that reveals that, at each update step, the algorithm solves a weighted least squares problem where the weights change at each step [19].

As we want to maximise (6.3), our function to minimise is the cross-entropy

$$E(\mathbf{V}) = -\sum_{n=1}^{N}\sum_{k=1}^{K} r_{nk} \ln g_k(\mathbf{x}_n). \tag{6.6}$$

The gradient of g_k with respect to $\mathbf{v}_j$ is

$$\nabla_{v_j} g_k(\mathbf{x}) = g_k(x)(\mathbf{I}_{kj} - g_j(\mathbf{x}))\phi(\mathbf{x}), \tag{6.7}$$

and, thus, the gradient of $E(\mathbf{V})$ evaluates to

$$\nabla_V E(\mathbf{V}) = \begin{pmatrix} \nabla_{v_1} E(\mathbf{V}) \\ \vdots \\ \nabla_{v_K} E(\mathbf{V}) \end{pmatrix}, \qquad \nabla_{v_j} E(\mathbf{V}) = \sum_{n=1}^{N} (g_j(\mathbf{x}_n) - r_{nj})\phi(\mathbf{x}_n), \tag{6.8}$$

where we have used $\sum_k g_k(\mathbf{x}) = 1$. The Hessian matrix

$$\mathbf{H} = \begin{pmatrix} \mathbf{H}_{11} & \cdots & \mathbf{H}_{1K} \\ \vdots & \ddots & \vdots \\ \mathbf{H}_{K1} & \cdots & \mathbf{H}_{KK} \end{pmatrix}, \tag{6.9}$$

is constructed by evaluating its $D_V \times D_V$ blocks

$$\mathbf{H}_{kj} = \mathbf{H}_{jk} = \sum_{n=1}^{N} g_k(\mathbf{x}_n)(\mathbf{I}_{kj} - g_j(\mathbf{x}_n))\phi(\mathbf{x}_n)\phi(\mathbf{x}_n)^T, \tag{6.10}$$

that result from $\mathbf{H}_{kj} = \nabla_{v_k}\nabla_{v_j} E(\mathbf{V})$.

To summarise the IRLS algorithm, given N observations $\mathcal{D} = \{\mathbf{X}, \mathbf{Y}\}$, and knowledge of the classifier parameters $\{\boldsymbol{\theta}_1, \dots, \boldsymbol{\theta}_K\}$ to evaluate $p(y|\mathbf{x}, \boldsymbol{\theta}_k)$, we can incrementally improve the estimate $\hat{\mathbf{V}}$ by repeatedly performing (6.5), starting with arbitrary initial values for $\hat{\mathbf{V}}$. As the Hessian matrix $\mathbf{H}$ given by (6.9) is positive definite [19], the error function $E(\mathbf{V})$ is convex, and the IRLS algorithm will approach is unique minimum, although, not monotonically [119]. Thus, $E(\mathbf{V})$ after (6.6) will decrease, and can be used to monitor convergence of the algorithm.

Note, however, that by (6.5), a single step of the algorithm requires computation of the gradient $\nabla_V E(\mathbf{V})$ of size KD_V, the $KD_V \times KD_V$ Hessian matrix $\mathbf{H}$, and the inversion of the latter. Due to this inversion, a single iteration of the

IRLS algorithm is of complexity $\mathcal{O}(N(KD_V)^3)$, which prohibits its application in LCS, where we require algorithms that preferably scale linearly with the number of classifiers. Nonetheless, it is of significant theoretical value, as it provides the values for $\mathbf{V}$ that maximise (6.3) and can therefore act as a benchmark for other mixing models and their associated methods.

6.1.2 Incremental Learning by Least Squares

Following a similar but slightly modified derivation to the one give by Jordan and Jacobs [121], we can incrementally approximate the maximum of (6.3) by a recursive least squares procedure that is of lower complexity than the IRLS algorithm. Due to the convexity of $E(\mathbf{V})$, its unique minimum is found when its gradient is $\nabla_V E(\mathbf{V}) = \mathbf{0}$, that is, when $\hat{\mathbf{V}}$ satisfies

$$\sum_{n=1}^{N} (g_k(\mathbf{x}_n) - r_{nk})\phi(\mathbf{x}_n) = \mathbf{0}, \qquad k = 1, \dots, K. \tag{6.11}$$

Substituting (6.2) for g_k, we want to solve

$$\sum_{n=1}^{N} m_k(\mathbf{x}_n) \left(\frac{\exp(\hat{\mathbf{v}}_k^T \phi(\mathbf{x}_n))}{\sum_{j=1}^{K} m_j(\mathbf{x}_n) \exp(\hat{\mathbf{v}}_j^T \phi(\mathbf{x}_n))} - \frac{r_{nk}}{m_k(\mathbf{x}_n)} \right) \phi(\mathbf{x}_n) = \mathbf{0} \tag{6.12}$$

Thus, the difference between the left-hand term and the right-hand term inside the brackets is to be minimised, weighted by $m_k(\mathbf{x}_n)$, such that

$$m_k(\mathbf{x}_n) \frac{\exp(\hat{\mathbf{v}}_k^T \phi(\mathbf{x}_n))}{\sum_{j=1}^{K} m_j(\mathbf{x}_n) \exp(\hat{\mathbf{v}}_j^T \phi(\mathbf{x}_n))} \approx m_k(\mathbf{x}_n) \frac{r_{nk}}{m_k(\mathbf{x}_n)}, \tag{6.13}$$

holds for all n. Solving the above for $\hat{\mathbf{v}}_k^T \phi(\mathbf{x}_n)$, its desired target values is

$$\ln \frac{r_{nk}}{m_k(\mathbf{x}_n)} - \ln C_n, \tag{6.14}$$

where $C_n = \sum_j m_j(\mathbf{x}_n) \exp(\hat{\mathbf{v}}_j^T \phi(\mathbf{x}_n))$ is the normalising term that is common to all $\hat{\mathbf{v}}_k^T \phi(\mathbf{x}_n)$ and can therefore be omitted, as it disappears when $\hat{\mathbf{v}}_k^T \phi(\mathbf{x}_n)$ is converted to $g_k(\mathbf{x}_n)$. Therefore, the target for $\hat{\mathbf{v}}_k^T \phi(\mathbf{x}_k)$ is $\ln \frac{r_{nk}}{m_k(\mathbf{x}_n)}$, weighted by $m_k(\mathbf{x}_n)$. This allows us to reformulate the problem of finding values for $\hat{\mathbf{V}}$ that maximise (6.3) as the K linear least squares problems of minimising

$$\sum_{n=1}^{N} m_k(\mathbf{x}_n) \left(\hat{\mathbf{v}}_k^T \phi(\mathbf{x}_n) - \ln \frac{r_{nk}}{m_k(\mathbf{x}_n)} \right)^2, \qquad k = 1, \dots, K. \tag{6.15}$$

Even though $r_{nk} = 0$ if $m_k(\mathbf{x}_n) = 0$, and therefore $\frac{r_{nk}}{m_k(\mathbf{x}_n)}$ is undefined in such a case, this does not cause any problems, as in such a case the weight is equally zero which makes computing the target superfluous. Also note that each of these problems operate on an input space of dimensionality D_V, and hence, using the least squares methods introduced in the previous chapter, have either complexity $\mathcal{O}(NKD_V^3)$ for the batch solution or $\mathcal{O}(KD_V^2)$ for each step of the incremental solution. Given that we usually have $D_V = 1$ in LCS, this is certainly an appealing property.

When minimising (6.15) it is essential to consider that the values for r_{nk} by (6.4) depend on the current $\hat{\mathbf{v}}_k$ of all classifiers. Consequently, when performing batch learning, it is not sufficient to solve all K least squares problems only once, as the corresponding targets change with the updated values of $\hat{\mathbf{V}}$. Thus, again one needs to repeatedly update the estimate $\hat{\mathbf{V}}$ until the cross-entropy (6.6) converges.

On the other hand, using recursive least squares to provide an incremental approximation of $\hat{\mathbf{V}}$ we need to honour the non-stationarity of the target values by using the recency-weighted RLS variant. Hence, according to Sect. 5.3.5 the update equations take the form

$$\begin{aligned}\hat{\mathbf{v}}_{kN+1} &= \lambda^{m_k(\mathbf{x}_n)}\hat{\mathbf{v}}_{kN} \\ &\quad + m_k(\mathbf{x}_{N+1})\mathbf{\Lambda}_{kN+1}^{-1}\phi(\mathbf{x}_{N+1})\left(\ln\frac{r_{nk}}{m_k(\mathbf{x}_n)} - \hat{\mathbf{v}}_{kN}^T\phi(\mathbf{x}_{N+1})^T\right),\end{aligned} \tag{6.16}$$

$$\begin{aligned}\mathbf{\Lambda}_{kN+1}^{-1} &= \lambda^{-m(\mathbf{x}_{N+1})}\mathbf{\Lambda}_{kN}^{-1} \\ &\quad - m(\mathbf{x}_{N+1})\lambda^{-m(\mathbf{x}_{N+1})}\frac{\mathbf{\Lambda}_{kN}^{-1}\phi(\mathbf{x}_{N+1})\phi(\mathbf{x}_{N+1})^T\mathbf{\Lambda}_{kN}^{-1}}{\lambda^{m_k(\mathbf{x}_n)} + m_k(\mathbf{x}_{N+1})\phi(\mathbf{x}_{N+1})^T\mathbf{\Lambda}_{kN}^{-1}\phi(\mathbf{x}_{N+1})},\end{aligned} \tag{6.17}$$

where the $\hat{\mathbf{v}}_k$'s and $\mathbf{\Lambda}_k^{-1}$'s are initialised to $\hat{\mathbf{v}}_{k0} = \mathbf{0}$ and $\mathbf{\Lambda}_{k0}^{-1} = \delta\mathbf{I}$ for all k, with δ being a large scalar. In [121], Jordan and Jacobs initially set $\lambda = 0.99$ and increased a fixed fraction (0.6) of the remaining distance to 1.0 every 1000 updates. This seems a sensible approach to start with, but further empirical experience is required to make definite recommendations.

As pointed out by Jordan and Jacobs [121], approximating the values of $\hat{\mathbf{V}}$ by least squares does not result in the same parameter estimates as when using the IRLS algorithm, due to the use of least squares rather than maximum likelihood. In fact, the least squares approach can be seen as an approximation to the maximum likelihood solution under the assumption that the residual in (6.15) in small, which is equivalent to assuming that the LCS model can fit the underlying regression surface and that the noise is small. Nonetheless, they demonstrate empirically that the least squares approach provides good results even when the residual is large in the early stages of training [121]. In any case, in terms of complexity it is a very appealing alternative to the IRLS algorithm.

6.2 Heuristic-Based Mixing Models

While the IRLS algorithm minimises (6.6), it does not scale well with the number of classifiers. The least squares approximation, on the other hand, scales well, but minimises (6.15) instead of (6.6), which does not always give good results, as will be demonstrated in Sect. 6.3. As an alternative, this section introduces some heuristic mixing models that scale linearly with the number of classifiers, just like the least squares approximation, and feature better performance.

Before discussing different heuristics, let us define the requirements on g_k: to preserve their probabilistic interpretation, we require $g_k(\mathbf{x}) \geq 0$ for all k and $\mathbf{x}$, and $\sum_k g_k(\mathbf{x}) = 1$ for all $\mathbf{x}$. In addition, we need to honour matching, which means that if $m_k(\mathbf{x}) = 0$, we need to have $g_k(\mathbf{x}) = 0$. These requirements are met if we define

$$g_k(\mathbf{x}) = \frac{m_k(\mathbf{x})\gamma_k(\mathbf{x})}{\sum_{j=1}^K m_j(\mathbf{x})\gamma_j(\mathbf{x})}, \tag{6.18}$$

where $\{\gamma_k : \mathcal{X} \rightarrow \mathbb{R}^+\}$ is a set of K functions returning positive scalars, that implicitly rely on the mixing model parameters $\mathbf{V}$. Thus, the mixing model defines a weighted average, where the weights are specified on one hand by the matching functions, and on the other hand by the functions γ_k. The heuristics differ among each other only in how they define the γ_k's.

Note that the generalised softmax function (6.2) also performs mixing by weighted average, as it conforms to (6.18) with $\gamma_k(\mathbf{x}) = \exp(\mathbf{v}_k^T\mathbf{x})$ and mixing model parameters $\mathbf{V} = \{\mathbf{v}_k\}$. The weights it assigns to each classifier are determined by the log-linear model $\exp(\mathbf{v}_k^T\mathbf{x})$, which needs to be trained separately, depending on the responsibilities that express the goodness-of-fit of the classifier models for the different inputs. In contrast, all heuristic models that are introduced here rely on measures that are part of the classifiers' linear regression models and do not need to be fitted separately. As they do not have any adjustable parameters, they all have $\mathbf{V} = \emptyset$. The heuristics assume classifiers to use regression rather than classification models. For the classification case, similar heuristics are easily found by using the observations of the following section, that are valid for any form of classifier model, to guide the design of these heuristics.

6.2.1 Properties of Weighted Averaging Mixing

Let $\hat{f}_k : \mathcal{X} \rightarrow \mathbb{R}$ be given by $\hat{f}_k(\mathbf{x}) = \mathbb{E}(y|\mathbf{x}, \boldsymbol{\theta}_k)$, that is, the estimator of classifier k defined by the mean of the conditional distribution of the output given the input and the classifier parameters. Equally, let $\hat{f} : \mathcal{X} \rightarrow \mathbb{R}$ be the global model estimator, given by $\hat{f}(\mathbf{x}) = \mathbb{E}(y|\mathbf{x}, \theta)$. As by (4.8) we have $p(y|\mathbf{x}, \theta) = \sum_k g_k(\mathbf{x})p(y|\mathbf{x}, \boldsymbol{\theta}_k)$, the global estimator is related to the local estimators by

$$\hat{f}(\mathbf{x}) = \int_{\mathcal{Y}} y \sum_k g_k(\mathbf{x})p(y|\mathbf{x}, \boldsymbol{\theta}_k)\mathrm{d}y = \sum_k g_k(\mathbf{x})\hat{f}_k(\mathbf{x}), \tag{6.19}$$

and, thus, is also a weighted average of the local estimators. From this follows that $\hat{f}$ is bounded from below and above by the lowest and highest estimate of the local models, respectively, that is

$$\min_k \hat{f}_k(\mathbf{x}) \leq \hat{f}(\mathbf{x}) \leq \max_k \hat{f}_k(\mathbf{x}), \quad \forall \mathbf{x} \in \mathcal{X}. \tag{6.20}$$

In general, we aim at minimising the deviation of the global estimator $\hat{f}$ from the target function f that describes the data-generating process. If we measure this deviation by the difference measure $h(f(\mathbf{x}) - \hat{f}(\mathbf{x}))$, where h is some convex function $h : \mathbb{R} \rightarrow \mathbb{R}^+$, mixing by a weighted average allows for the derivation of an upper bound on this difference measure:

Theorem 6.1. *Given the global estimator* $\hat{f} : \mathcal{X} \rightarrow \mathbb{R}$, *that is formed by a weighted averaging of* K *local estimators* $\hat{f}_k : \mathcal{X} \rightarrow \mathbb{R}$ *by* $\hat{f}(\mathbf{x}) = \sum_k g_k(\mathbf{x})\hat{f}_k(\mathbf{x})$, *such that* $g_k(\mathbf{x}) \geq 0$ *for all* $\mathbf{x}$ *and* k, *and* $\sum_k g_k(\mathbf{x}) = 1$ *for all* $\mathbf{x}$, *the difference between the target function* $f : \mathcal{X} \rightarrow \mathbb{R}$ *and the global estimator is bounded from above by*

$$h\left(\hat{f}(\mathbf{x}) - f(\mathbf{x})\right) \leq \sum g_k(\mathbf{x}) h\left(\hat{f}_k(\mathbf{x}) - f(\mathbf{x})\right), \quad \forall \mathbf{x} \in \mathcal{X}, \tag{6.21}$$

where $h : \mathbb{R} \rightarrow \mathbb{R}^+$ *is a convex function. More specifically, we have*

$$\left(\hat{f}(\mathbf{x}) - f(\mathbf{x})\right)^2 \leq \sum g_k(\mathbf{x}) \left(\hat{f}_k(\mathbf{x}) - f(\mathbf{x})\right)^2, \quad \forall \mathbf{x} \in \mathcal{X}, \tag{6.22}$$

and

$$\left|\hat{f}(\mathbf{x}) - f(\mathbf{x})\right| \leq \sum g_k(\mathbf{x}) \left|\hat{f}_k(\mathbf{x}) - f(\mathbf{x})\right|, \quad \forall \mathbf{x} \in \mathcal{X}. \tag{6.23}$$

Proof. For any $\mathbf{x} \in \mathcal{X}$, we have

$$\begin{aligned}
h\left(\hat{f}(\mathbf{x}) - f(\mathbf{x})\right) &= h\left(\sum_k g_k(\mathbf{x})\hat{f}_k(\mathbf{x}) - f(\mathbf{x})\right) \\
&= h\left(\sum_k g_k(\mathbf{x}) \left(\hat{f}_k(\mathbf{x}) - f(\mathbf{x})\right)\right) \\
&\leq \sum_k g_k(\mathbf{x}) h\left(\hat{f}_k(\mathbf{x}) - f(\mathbf{x})\right),
\end{aligned}$$

where we have used $\sum_k g_k(\mathbf{x}) = 1$, and the inequality is Jensen's Inequality (for example, [231]), based on the convexity of h and the weighted average property of g_k. Having proven (6.21), (6.22) and (6.23) follow from the convexity of $h(a) = a^2$ and $h(a) = |a|$, respectively.

Therefore, the error of the global estimator can be minimised by assigning high weights, that is, high values of $g_k(\mathbf{x})$, to classifiers whose error of the local estimator is small. Observing in (6.18) that the value of $g_k(\mathbf{x})$ is directly proportional to the value of $\gamma_k(\mathbf{x})$, a good heuristic will assign high values to $\gamma_k(\mathbf{x})$ if the error of the local estimator can be expected to be small. The design of all heuristics is based on this intuition.

The probabilistic formulation of the LCS model results in a further bound, this time on the variance of the output prediction:

Theorem 6.2. *Given the density* $p(y|\mathbf{x}, \boldsymbol{\theta})$ *for output* y *given input* $\mathbf{x}$ *and parameters* $\boldsymbol{\theta}$, *formed by the* K *classifier model densities* $p(y|\mathbf{x}, \boldsymbol{\theta}_k)$ *by* $p(y|\mathbf{x}, \boldsymbol{\theta}_k) = \sum_k g_k(\mathbf{x})p(y|\mathbf{x}, \boldsymbol{\theta}_k)$, *such that* $g_k(\mathbf{x}) \geq 0$ *for all* $\mathbf{x}$ *and* k, *and* $\sum_k g_k(\mathbf{x}) = 1$ *for* *all* $\mathbf{x}$*, the variance of* y *is bounded from above by the weighted average of the variance of the local models for* y*, that is*

$$var(y|\mathbf{x}, \boldsymbol{\theta}) = \sum_k g_k(\mathbf{x})^2 var(y|\mathbf{x}, \boldsymbol{\theta}_k) \leq \sum_k g_k(\mathbf{x}) var(y|\mathbf{x}, \boldsymbol{\theta}_k), \quad \forall \mathbf{x} \in \mathcal{X}. \quad (6.24)$$

Proof. To show the above, we again take the view that each observation was generated by one and only one classifier, and introduce the indicator variable I as a conceptual tool that takes the value k if classifier k generated the observation, giving $g_k(\mathbf{x}) \equiv p(I = k|\mathbf{x})$, where we are omitting the parameters of the mixing models implicit in g_k. We also use $p(y|\mathbf{x}, \boldsymbol{\theta}_k) \equiv p(y|\mathbf{x}, I = k)$ to denote the model provided by classifier k. Thus, we have $p(y|\mathbf{x}, \boldsymbol{\theta}) = \sum_k p(I = k|\mathbf{x})p(y|\mathbf{x}, I = k)$, and, analogously, $\mathbb{E}(y|\mathbf{x}, \boldsymbol{\theta}) = \sum_k p(I = k|\mathbf{x})\mathbb{E}(y|\mathbf{x}, I = k)$. However, similarly to the basic relation $\text{var}(aX + bY) = a^2\text{var}(X) + b^2\text{var}(Y) + 2ab\text{cov}(X, Y)$, we have for the variance

$$\text{var}(y|\mathbf{x}, \boldsymbol{\theta}) = \sum_k p(I = k)^2 \text{var}(y|\mathbf{x}, I = k) + 0, \quad (6.25)$$

where the covariance terms are zero as the classifier models are conditionally independent given I. This confirms the equality in (6.24). The inequality is justified by observing that the variance is non-negative, and $0 \leq g_k(\mathbf{x}) \leq 1$ and so $g_k(\mathbf{x})^2 \leq g_k(\mathbf{x})$.

Here, not only a bound but also an exact expression for the variance of the combined prediction is provided. This results in a different view on the design criteria for possible heuristics: we want to assign weights that are in some way inversely proportional to the classifier prediction variance. As the prediction variance indicates the expected prediction error, this design criterion conforms to the one that is based on Theorem 6.1.

Neither Theorem 6.1 nor Theorem 6.2 assume that the local models are linear. In fact, they apply to any case where a global model results from a weighted average of a set of local models. Thus, they can also be used in LCS when the classifier models are classification model, or non-linear model (for example, [156, 175]).

Example 6.3 (Mean and Variance of a Mixture of Gaussians). Consider 3 classifiers that, for some input $\mathbf{x}$ provide the predictions $p(y|\mathbf{x}, \boldsymbol{\theta}_1) = \mathcal{N}(y|0.2, 0.1^2)$, $p(y|\mathbf{x}, \boldsymbol{\theta}_2) = \mathcal{N}(y|0.5, 0.05^2)$, and $p(y|\mathbf{x}, \boldsymbol{\theta}_3) = \mathcal{N}(y|0.7, 0.2^2)$. Using the mixing weights inversely proportional to their variance, that is $g_1(\mathbf{x}) = 0.20$, $g_2(\mathbf{x}) = 0.76$, and $g_3(\mathbf{x}) = 0.04$, our global estimator $\hat{f}(\mathbf{x})$, determined by (6.19), results in $\hat{f}(\mathbf{x}) = 0.448$. Let us assume that the target function value is given by $f(\mathbf{x}) = 0.5$, resulting in the squared prediction error $(f(\mathbf{x}) - \hat{f}(\mathbf{x}))^2 \approx 0.002704$. This error is correctly upper-bounded by (6.22), that results in $(f(\mathbf{x}) - \hat{f}(\mathbf{x}))^2 \leq 0.0196$. The correctness of (6.24) is demonstrated by taking 10^6 samples from the predictive distributions of the different classifiers, resulting in the sample vectors $\mathbf{s}_1$, $\mathbf{s}_2$, and $\mathbf{s}_3$, each of size 10^6. Thus, we can produce a sample vector of the global prediction by $\mathbf{s} = \sum_k g_k(\mathbf{x})\mathbf{s}_k$, which has the sample variance 0.00190. This conforms to – and thus empirically validates – the variance after (6.24), which results in $\mathrm{var}(y|\mathbf{x}, \boldsymbol{\theta}) = 0.00191 \leq 0.0055$.

6.2.2 Inverse Variance

The unbiased noise variance estimate of a linear regression classifier k is, after (5.13), given by

$$\hat{\tau}_k^{-1} = (c_k - D_{\mathcal{X}})^{-1} \sum_{n=1}^{N} m_k(\mathbf{x}_n) \left(\hat{\mathbf{w}}_k^T \mathbf{x}_n - y_n \right)^2 , \tag{6.26}$$

and is therefore approximately the mean sum of squared prediction errors. If this estimate is small, the squared prediction error is, on average, known to be small and we can expect the predictions to have a low error. Hence, inverse variance mixing is defined by using mixing weights that are inversely proportional to the noise variance estimates of the according classifiers. More formally, $\gamma_k(\mathbf{x}) = \hat{\tau}_k$ in (6.18) for all $\mathbf{x}$. The previous chapter has shown how to estimate the noise variance of a classifier by batch or incremental learning.

6.2.3 Prediction Confidence

If the classifier model is probabilistic, its prediction can be given by a probabilistic density. Knowing this density allows for the specification of an interval on the output into which 95% of the observations are likely to fall, known as the 95% confidence interval. The width of this interval therefore gives a measure of how certain we are about the prediction made by this classifier. This is the underlying idea of mixing by prediction confidence.

More formally, the predictive density of the linear classifier model is given for classifier k by marginalising $p(y, \boldsymbol{\theta}_k|\mathbf{x}) = p(y|\mathbf{x}, \boldsymbol{\theta}_k)p(\boldsymbol{\theta}_k)$ over the parameters $\boldsymbol{\theta}_k$, and results in

$$p(y|\mathbf{x}) = \mathcal{N}\left(y | \hat{\mathbf{w}}_k^T \mathbf{x}, \hat{\tau}_k^{-1}(\mathbf{x}^T \boldsymbol{\Lambda}_k^{-1} \mathbf{x} + 1) \right) , \tag{6.27}$$

as already introduced in Sect. 5.3.6. The 95% confidence interval – indeed that of any percentage – is directly proportional to the standard deviation of this density, which is the square root of its variance. Thus, to assign higher weights to classifiers with a higher confidence prediction, that is, a prediction with a smaller confidence interval, $\gamma_k(\mathbf{x})$ is set to

$$\gamma_k(\mathbf{x}) = \left(\hat{\tau}_k^{-1}(\mathbf{x}^T \boldsymbol{\Lambda}_k^{-1} \mathbf{x} + 1) \right)^{-1/2} . \tag{6.28}$$

Compared to mixing by inverse variance, this measure additionally takes the uncertainty of the weight vector estimate into account and is consequently dependent on the input. Additionally, it relies on the assumption of Gaussian noise and a Gaussian weight vector model, which might not hold – in particular when the number of observations that the classifier is trained on is small. Therefore, despite using more information than mixing by inverse variance, it cannot be guaranteed to perform better.

6.2.4 Maximum Prediction Confidence

The global model density is by (4.8) given by a mixture of the densities of the local models. As for the local models, the spread of the global prediction determines a confidence interval on the global model. Minimising the spread of the global prediction maximises its confidence. Due to mixing by weighted average, the spread of the global density if bounded from below and above by the smallest and the largest spread of the contributing classifiers. Thus, in order to minimise the spread of the global prediction, we only consider the predictive density of the classifier with the smallest predictive spread.

Using this concept, mixing to maximise the prediction confidence is formalised by setting $\gamma_k(\mathbf{x})$ to 1 only for the classifier with the lowest prediction spread, that is,

$$\gamma_k(\mathbf{x}) = \begin{cases} 1 & \text{if } k = \text{argmax}_k\, m_k(\mathbf{x}) \left(\hat{\tau}_k^{-1}(\mathbf{x}^T \mathbf{\Lambda}_k^{-1}\mathbf{x} + 1)\right)^{-1/2}, \\ 0 & \text{otherwise.} \end{cases} \tag{6.29}$$

Note the addition of $m_k(\mathbf{x})$ to ensure that the *matching* highest confidence classifier is picked.

As for mixing by confidence, using only the classifier with the highest prediction confidence relies on several assumptions that might by violated. Thus, maximum confidence mixing can be expected to perform worse than mixing by inverse variance in cases where these assumptions are violated. In such cases it might even fare worse than mixing by confidence, as it relies on these assumptions more heavily.

6.2.5 XCS

While none of the approaches discussed before are currently used in any LCS, the mixing model used XCS(F) is here – for the sake of comparison – described in the same formal framework. Mixing in XCS(F) has not changed since it was firstly specified in [237], despite its multiple other changes and improvements. Additionally, the mixing model in XCS(F) is closely linked to the fitness of a classifier as used by the genetic algorithm, and is thus overly complex. Due to the algorithmic description of an incremental method, the aims of XCS(F) are usually not explicitly specified. Nonetheless, all mixing parameters in XCS(F) are updated by the LMS method, for which the formally equivalent, but more intuitive, batch approaches have already been discussed in the previous chapter.

Recall, that the LMS algorithm for single-dimensional constant inputs is specified by (5.25) to update some scalar estimate $\hat{w}$ of an output y after observing the $(N+1)$th output by

$$\hat{w}_{N+1} = \hat{w}_N + \gamma_{N+1}(y_{N+1} - \hat{w}_N), \tag{6.30}$$

where γ_{N+1} is some scalar step size. As shown in Example 5.2, this update equation aims at minimising a sum of squared errors (5.5), whose minimum is achieved by

$$\hat{w} = c_k^{-1} \sum_{n=1}^{N} m(\mathbf{x}_n) y_n, \tag{6.31}$$

given all N observations. Hence, (6.31) is the batch formulation for the solution that the incremental (6.30) approximates.

Applying this relation to the XCS update equations for the mixing parameters, the mixing model employed by XCS(F) can be described as follows: The *error* ϵ_k of classifier k in XCS(F) is the mean absolute prediction error of its local models, and is given by

$$\epsilon_k = c_k^{-1} \sum_{n=1}^{N} m(\mathbf{x}_n) \left| y_n - \hat{\mathbf{w}}_k^T \mathbf{x}_n \right| . \tag{6.32}$$

The classifier's *accuracy* is some inverse function $\kappa(\epsilon_k)$ of the classifier error. This function was initially given by an exponential [237], but was later [239, 57] redefined to

$$\kappa(\epsilon) = \begin{cases} 1 & \text{if } \epsilon < \epsilon_0, \\ \alpha \left(\frac{\epsilon}{\epsilon_0} \right)^{-\nu} & \text{otherwise,} \end{cases} \tag{6.33}$$

where the constant scalar ϵ_0 is known as the *minimum error*, the constant α is a scaling factor, and the constant ν is a mixing power factor [57]. The accuracy is constantly 1 up to the error ϵ_0 and then drops off steeply, with the shape of the drop determined by α and ν. The *relative accuracy* is a classifier's accuracy for a single input normalised by the sum of the accuracies of all classifiers matching that input. The *fitness* is the relative accuracy of a classifier averaged over all inputs that it matches, that is

$$F_k = c_k^{-1} \sum_{n=1}^{N} \frac{m_k(\mathbf{x}_n)\kappa(\epsilon_k)}{\sum_{j=1}^{K} m_j(\mathbf{x}_n)\kappa(\epsilon_j)} \tag{6.34}$$

This fitness is the measure of a classifier's prediction quality, and hence γ_k is input-independently given by $\gamma_k(\mathbf{x}) = F_k$.

Note that the magnitude of a relative accuracy depends on both the error of a classifier, and on the error of the classifiers that match the same input. This makes the fitness of classifier k dependent on inputs that are matched by classifiers that share inputs with classifier k, but are not necessarily matched by this classifier. This might be a good measure for the fitness of a classifier (where prediction quality is not all that counts), but it does not perform too well as a measure of the prediction quality of a classifier.

6.3 Empirical Comparison

In order to compare how well the different heuristics perform with respect to the aim of maximising (6.1), their performance is evaluated on a set of four regression tasks. The results show that i) mixing by inverse variance outperforms the other heuristic methods, ii) also performs better than the least squares approximation, and iii) mixing as done in XCS(F) performs worse than all other methods.

In all experiments a set of K linear regression classifiers is created such that the number of classifiers matching each input is about the same for all inputs. These classifiers are trained on all available observations by batch learning, before the mixing models are applied and their performance measured by the likelihood (6.1). This setup was chosen for several reasons: firstly, mixing is only required if several classifiers match the same input, which is provided by the generated set of classifiers. Secondly, the classifiers are trained before the mixing models are applied, as we want to only compare the mixing models based on the same set of classifiers, and not how training of classifiers and mixing them interacts. Finally, the likelihood measure is used to compare the performance of the mixing models, rather than some form of squared error or similar, as the aim in this chapter is to discuss methods that maximise this likelihood, rather than any other measure.

6.3.1 Experimental Design

Regression Tasks. The mixing models are evaluated on four regression tasks $f : \mathbb{R} \to \mathbb{R}$, given in Table 6.1. The input range is $[0, 1]$, and the output is shifted and scaled such that $-0.5 \leq f(x) \leq 0.5$. 1000 observations $(i_n, f(i_n))$ are taken from the target function f at regular intervals, from 0 to 1, to give the output vector $\mathbf{y} = (f(i_1), \dots, f(i_{1000}))^T$. The input matrix for averaging classifiers is given by $\mathbf{X} = (1, \dots, 1)^T$, and for classifiers that model straight lines by a 1000×2 matrix $\mathbf{X}$ with the nth row given by $(1, i_n)$.

Table 6.1. The set of functions used for evaluating the performance of the different mixing models. The functions are taken from Donoho and Johnstone [73], and have been previously used in Booker [23] in an LCS-related study. The functions are samples over the range $[0, 1]$ and their outputs are normalised to $-0.5 \leq f(x) \leq 0.5$.

Function	*Definition*
Blocks	$f(x) = \sum h_j K(x - x_j), \quad K(x) = (1 + \mathrm{sgn}(x))/2,$ $(x_j) = (0.1, 0.13, 0.15, 0.23, 0.25, 0.40, 0.44, 0.65, 0.76, 0.78, 0.81),$ $(h_j) = (4, -5, 3, -4, 5, -4.2, 2.1, 4.3, -3.1, 5.1, -4.2).$
Bumps	$f(x) = \sum h_j K((x - x_j)/w_j), \quad K(x) = (1 + \lvert x \rvert^4)^{-1},$ $(x_j) = x_{\mathrm{Blocks}},$ $(h_j) = (4, 5, 3, 4, 5, 4.2, 2.1, 4.3, 3.1, 5.1, 4.2),$ $(w_j) = (0.005, 0.005, 0.006, 0.01, 0.01, 0.03, 0.01, 0.01, 0.005, 0.008, 0.005).$
Doppler	$f(x) = (x(1 - x))^{1/2} \sin(2\pi(1 + 0.05)/(x + 0.05))$
Heavisine	$f(x) = 4 \sin 4\pi x - \mathrm{sgn}(x - 0.3) - \mathrm{sgn}(0.72 - x)$

Classifier Generation and Training. For each experimental run K classifiers are created, where K depends on the experiment. Each classifier matches an interval $[l_k, u_k]$ of the input space, that is $m_k(i_n) = 1$ if $l_k \leq i_n \leq u_k$, and $m_k(i_n) = 0$ otherwise. Even coverage such that about an equal number of classifiers matches each input is achieved by splitting the input space into 1000 bins, and localising the classifiers one by one in a "Tetris"-style way: the average width in bins of the matched interval of a classifier needs to be $1000c/K$ such that on average c classifiers match each bin. The interval width of a new classifier is sampled from $\mathcal{B}(1000, (1000c/K)/1000)$, where $\mathcal{B}(n, p)$ is a binomial distribution for n trials and a success probability of p. The minimal width is limited from below by 3, such that each classifier is at least trained on 3 observations. The new classifier is then localised such that the number of classifiers that match the same bins is minimal. If several such locations are possible, one is chosen uniformly at random. Having positioned all K classifier, they are trained by batch learning using (5.9) and (5.13). The number of classifiers that match each input is in all experiments set to $c = 3$.

Mixing Models. The performance of the following mixing models is compared: the IRLS algorithm (*IRLS*) and its least-squares approximation (*LS*) on the generalised softmax function with $\phi(\mathbf{x}) = 1$ for all $\mathbf{x}$, the inverse variance (*InvVar*) heuristics, the mixing by confidence (*Conf*) and mixing by maximum confidence (*MaxConf*) heuristics, and mixing by XCS(F) (*XCS*). When classifiers model straight lines, the IRLS algorithm (*IRLSf*) and its least-squares approximation (*LSf*) with a transfer function $\phi(\mathbf{x}) = (1, i_n)^T$ are used additionally, to allow for an additional soft-linear partitioning beyond the realm of matching (see the discussion in Sect. 4.3.5 for more information). Training by the IRLS algorithm is performed incrementally according to Sect. 6.1.1, until the change in cross-entropy (6.6) between two iterations is smaller than 0.1%. The least-squares approximation is performed repeatedly in batches rather than as described in Sect. 6.1.2, by using (5.9) to find the $\mathbf{v}_k$'s that minimise (6.15). Convergence is assumed when the change in (6.6) between two batch updates is smaller than 0.05% (this value is smaller than for the IRLS algorithm, as the least squares approximation takes smaller steps). The heuristic mixing models do not require any separate training and are applied such as described in Sect. 6.2. For XCS, the standard setting $\epsilon_0 = 0.01$, $\alpha = 0.1$, and $\nu = 5$, as recommended by Butz and Wilson [57], are used.

Evaluating the Performance. Having generated and trained a set of classifiers, each mixing model is trained with the same set to make their performance directly comparable. It is measured by evaluating (6.1), where $p(y_n|\mathbf{x}_n, \boldsymbol{\theta}_k)$ is computed by (5.3), using the same observations that the classifiers where trained on, and the g_k's are provided by the different mixing models. As the IRLS algorithm maximises the data likelihood (6.1) when using the generalised softmax function as the mixing model, its performance is used as a benchmark that the other models are compared to. Their performance is reported as a fraction of the likelihood of the IRLS algorithm with $\phi(\mathbf{x}) = 1$.

Statistical Analysis. A two-way analysis of variance (ANOVA) is used to determine if the performance of the different mixing models differ significantly, with the first factor being the type of mixing model (IRLS, IRLSf, LS, LSf, InvVar, Conf, MaxConf, XCS) and the second factor being the combination of regression task and type of classifier (Blocks, Bumps, Doppler, Heavisine, either with averaging classifiers, or classifiers that model straight lines). The direction of the difference is determined by Tukey's HSD post-hoc test. As the optimal likelihood as measured by IRLS varies strongly with different sets of classifiers, the performance is measured as a fraction of the optimal likelihood for a particular classifier set rather than the likelihood itself.

6.3.2 Results

The first experiment compares the performance of all mixing models when using $K = 50$ classifiers. For all functions and both averaging classifiers and classifiers that model straight lines, 50 experimental runs were performed per function[1]. To show the different test functions, and to give the reader an intuitive idea how mixing is performed, Figures 6.1 to 6.4 show the predictions of the different methods of a single run when using classifiers that model straight lines. The mean likelihoods over these 50 runs as a fraction of the mean likelihood of the IRLS

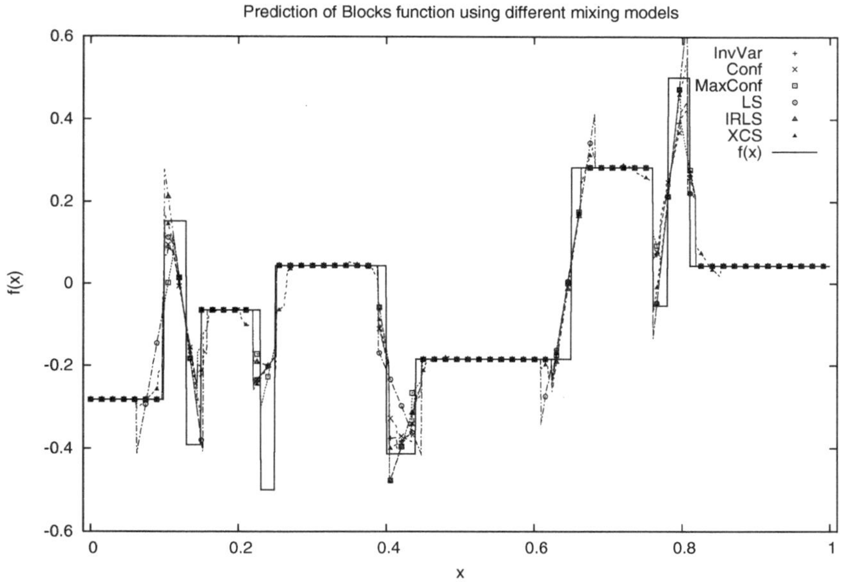

Fig. 6.1. Resulting predictions of a single run, using different mixing models for the Blocks function. See the text for an explanation of the experimental setup.

[1] In our experience, performing the experiments with fewer runs provided insufficient data to permit significance tests to reliably detect the differences.

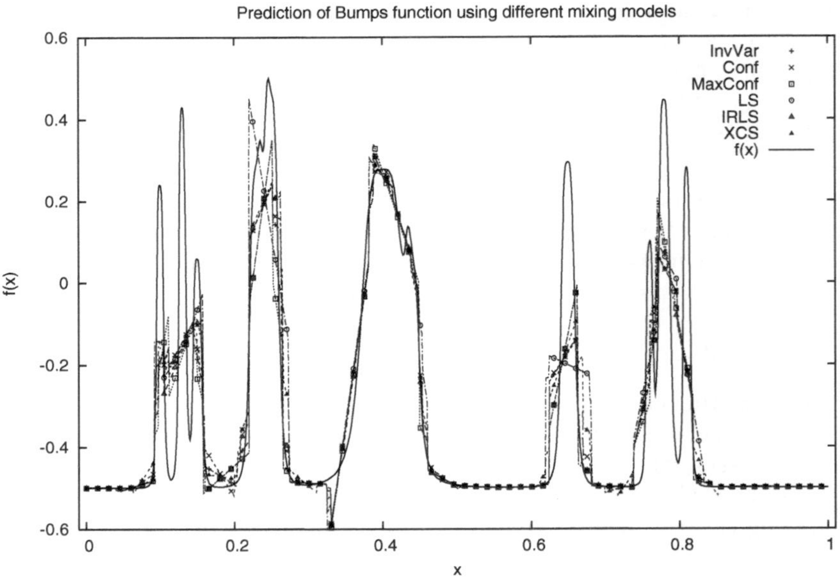

Fig. 6.2. Resulting predictions of a single run, using different mixing models for the Bumps function. See the text for an explanation of the experimental setup.

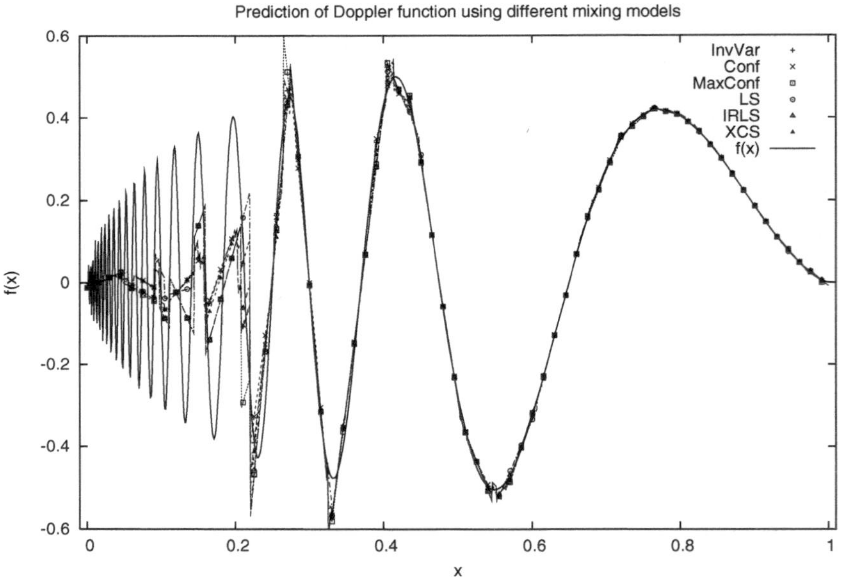

Fig. 6.3. Resulting predictions of a single run, using different mixing models for the Doppler function. See the text for an explanation of the experimental setup.

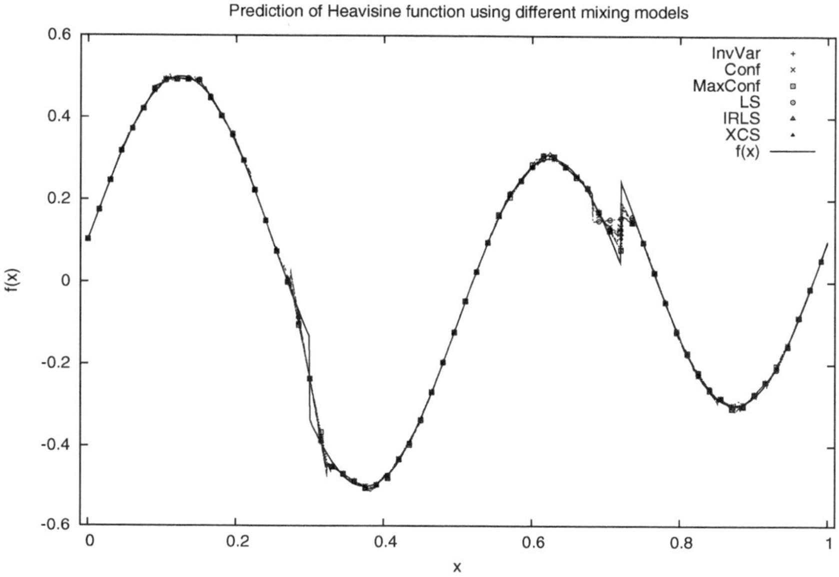

Fig. 6.4. Resulting predictions of a single run, using different mixing models for the Heavisine function. See the text for an explanation of the experimental setup.

Table 6.2. The mean likelihoods of the different mixing models, as a fraction of the mean likelihood of IRLS, averaged over 50 experimental runs per function. A *lin* added to the function name indicates the use of classifiers that model straight lines rather than averaging classifiers. For averaging classifiers, IRLS and IRLSf, and LS and LSf are equivalent, and so their results are combined. The results written in bold indicate that there is no significant difference to the best-performing mixing model for this function. Those results that are significantly worse than the best mixing model but not significantly worse than the best model in their group are written in italics. Statistical significance was determined by Tukey's HSD post-hoc test at the 0.01 level.

Function	*Likelihood of Mixing Model as Fraction of IRLS*							
	IRLS	IRLSf	LS	LSf	InvVar	Conf	MaxConf	XCS
Blocks	**1.00000**		**0.99473**		**0.99991**	**0.99988**	**0.99973**	**0.99877**
Bumps	**1.00000**		*0.94930*		**0.98442**	**0.97740**	*0.96367*	*0.94678*
Doppler	**1.00000**		*0.94930*		**0.98442**	**0.97740**	*0.96367*	*0.94678*
Heavisine	**1.00000**		*0.96289*		**0.96697**	*0.95123*	*0.95864*	*0.95807*
Blocks lin	**1.00000**	**1.00014**	**0.99141**	**0.99559**	**0.99955**	**0.99929**	**0.99956**	**0.99722**
Bumps lin	**1.00000**	**0.99720**	*0.94596*	*0.94870*	**0.98425**	**0.97494**	**0.97797**	*0.94107*
Doppler lin	**1.00000**	**0.99856**	*0.94827*	**0.98628**	**0.98723**	**0.97818**	**0.98172**	*0.94395*
Heavisine lin	**1.00000**	**0.99523**	**0.98480**	**0.96854**	**0.98448**	**0.97347**	**0.99005**	**0.95739**

method are shown in Table 6.2. An ANOVA reveals that there is a significant performance difference between the different methods ($F(7, 2744) = 43.0688$, $p = 0.0$). Comparing the means shows that the method that performs best is IRLS, followed by IRLSf, InvVar, MaxConf, Conf, LSf, LS, and last, XCS. The

Table 6.3. p-values for Tukey's HSD post-hoc comparison of the different mixing methods. The performance values were gathered in 50 experimental runs per function, using both averaging classifiers and classifiers that model straight lines. The p-values reported are for a post-doc comparison only considering the factor that determines the mixing method. The methods are ordered by performance, with the leftmost and bottom method being the best-performing one. The p-values in italics indicate that no significant difference between the methods at the 0.01 level was detected.

	IRLS	IRLSf	InvVar	MaxConf	Conf	LSf	LS	XCS
XCS	0.0000	0.0000	0.0000	0.0000	0.0000	*0.0283*	*0.5131*	-
LS	0.0000	0.0000	0.0000	0.0000	0.0000	*0.8574*	-	
LSf	0.0000	0.0000	0.0000	0.0095	*0.0150*	-		
Conf	0.0000	0.0000	*0.1044*	*0.9999*	-			
MaxConf	0.0000	0.0000	*0.1445*	-				
InvVar	0.0001	0.0002	-					
IRLSf	*0.8657*	-						
IRLS	-							

p-values of Tukey's HSD post-hoc test are given in Table 6.3. They show that the performance difference between all methods is significant at the 0.01 level, except for the ones that are written in italics.

The same experiment where preformed with $K \in \{20, 100, 400\}$, classifiers, yielding qualitatively similar results. This shows that the presented performance differences are not sensitive to the number of classifiers used.

6.3.3 Discussion

As can be seen from the results, IRLS is in almost all cases significantly better, and in no case significantly worse than any other methods that were applied. IRLSf uses more information than IRLS to mix the classifier predictions, and thus can be expected to perform better. As can be seen from Table 6.2, however, it frequently features worse performance, though not significantly. This worse performance can be attributed to the used stopping criterion that is based on the relative change of the likelihood between two successive iterations. This likelihood increases more slowly when using IRLSf, which leads the stopping criterion to abort learning earlier for IRLSf than IRLS, causing it to perform worse.

InvVar is the best method of the introduced heuristics and constantly outperforms LS and LSf. Even though it does not perform significantly better than Conf and MaxConf, its mean is higher and the method relies on less assumptions. Thus, it should be the preferred method amongst the heuristics that were introduced.

As expected, XCS features a worse performance than all other methods, which can be attribute to the fact that the performance measure of the local model is influenced by the performance of the local models that match the same inputs. This might introduce some smoothing, but it remains questionable if such smoothing is ever advantageous. This doubt is justified by observing that XCS performs worst even on the smoothest function in the test set, which is the Heavisine function.

Overall, these experiments confirm empirically that IRLS performs best. However, due to its high complexity and bad scaling properties, it is not recommendable for applications that require the use of a large number of classifiers. While the least squares approximation could be used as an alternative in such cases, the results suggest that InvVar provides better results. Additionally, it is easier to implement than LS and LSf, and requires no incremental update. Thus, it should be the preferred method to use.

6.4 Relation to Previous Work and Alternatives

A closely related previous study has investigated mixing models for LCS with the aim of minimising the mean squared error of the global prediction rather than maximising its likelihood [83]. Formally, the aim was to find a mixing model that minimises

$$\sum_{n=1}^{N} \left(\hat{f}(\mathbf{x}_n) - f(\mathbf{x}_n) \right)^2, \tag{6.35}$$

where f is the target function, and $\hat{f}(\mathbf{x}_n)$ is the global output prediction for input $\mathbf{x}_n$. This problem statement can be derived from a model that assumes the relation between f and $\hat{f}$ to be $\hat{f}(\mathbf{x}) = f(\mathbf{x}) + \epsilon$, where $\epsilon \sim \mathcal{N}(0, \sigma^2)$ is a zero-mean constant variance Gaussian that represents the random noise. The maximum likelihood estimate for the parameters of $\hat{f}$ is found by maximising $\sum_n \ln \mathcal{N}(f(\mathbf{x}_n)|\hat{f}(\mathbf{x}_n), \sigma^2)$, which is equivalent to minimising (6.35).

In the LCS model with linear regression classifiers, introduced in Chap. 4, on the other hand, zero-mean constant variance Gaussian noise is assumed on each *local* model $p(y|\mathbf{x}, \boldsymbol{\theta}_k)$ rather than the global model $p(y|\mathbf{x}, \boldsymbol{\theta})$. These models are related by $p(y|\mathbf{x}, \boldsymbol{\theta}) = \sum_k g_k(\mathbf{x}) p(y|\mathbf{x}, \boldsymbol{\theta}_k)$, and as $g_k(\mathbf{x})$ might change with $\mathbf{x}$, the noise variance of the global model is very likely not constant. As a result, the maximum likelihood estimate for the LCS model as introduced in Chap. 4 does not conform to minimising (6.35). Nonetheless, the results based on minimising (6.35) are qualitatively the same as they show that amongst the heuristics InvVar features competitive performance, is usually better than Conf and MaxConf, and always outperforms XCS.

Modelling the noise on the local model level rather than the global model level is required to train the classifiers independently. It also makes explicit the need for a mixing model. In contrast, one could – as in Sect. 4.5 – assume a linear LCS model that features noise at the global level, such that an output y given some input $\mathbf{x}$ is modelled by

$$p(y|\mathbf{x}, \boldsymbol{\theta}) = \mathcal{N}\left(y \Big| \sum_{k=1}^{K} g_k(\mathbf{x}) \mathbf{w}_k^T \mathbf{x}, \tau^{-1} \right), \tag{6.36}$$

where $g_k(\mathbf{x})$ is some function of the matching functions $m_k(\mathbf{x})$, independent of $\boldsymbol{\theta}$. In such a case, one could interpret the values of $g_k(\mathbf{x})$ to form the mixing model but it is less clear how to separate the global model into local classifier models. Maximising the likelihood for such a model results in the least-squares problem (6.35) with $\hat{f}(\mathbf{x}; \boldsymbol{\theta}) = \sum_k g_k(\mathbf{x})\mathbf{w}_k^T \mathbf{x}$, the solution to which has been discussed in the previous chapter.

To the other extreme, one could from the start assume that the classifiers are trained independently, such that each of them provides the model c_k with predictive density $p(\mathbf{y}|\mathbf{x}, c_k)$. The global model is formed by marginalising over the local models,

$$p(\mathbf{y}|\mathbf{x}) = \sum_{k=1}^{K} p(\mathbf{y}|\mathbf{x}, c_k) p(c_k|\mathbf{x}), \tag{6.37}$$

where $p(c_k|\mathbf{x})$ is the probability of the model of classifier k being the "true" model, given a certain input $\mathbf{x}$. This term can be used to introduce matching, by setting $p(c_k|\mathbf{x}) = 0$ if $m_k(\mathbf{x}) = 0$. Averaging over models by their probability is known as Bayesian Model Averaging [107], which might initially look like resulting in the same formulation as the model derived from the generalised MoE model. The essential difference, however, is that $p(\mathbf{y}|\mathbf{x}, c_k)$ is independent of the model parameters $\boldsymbol{\theta}_k$ as it marginalises over them,

$$p(\mathbf{y}|\mathbf{x}, c_k) = \int p(\mathbf{y}|\mathbf{x}, \boldsymbol{\theta}_k, c_k) p(\boldsymbol{\theta}_k|c_k) \mathrm{d}\boldsymbol{\theta}_k. \tag{6.38}$$

Therefore, it cannot be directly compared to the mixing models introduced in this chapter, and should be treated as a different LCS model, closely related to ensemble learning. Further research is required to see if such an approach leads to viable LCS formulations.

6.5 Summary and Outlook

This chapter dealt with an essential LCS component that directly emerges from the introduced LCS model and is largely ignored by LCS research: how to combine a set of localised models, provided by the classifiers, to provide a global prediction. The aim of this "mixing problem" was defined by maximising the data likelihood (6.1) of the previously introduced LCS model.

As was shown, the IRLS algorithm is a possible approach to finding the globally optimal mixing parameters $\mathbf{V}$ to the generalised softmax mixing model, but it suffers from high complexity, and can therefore act as nothing more than a benchmark to compare other approaches to. The least squares approximation, on the other hand, scales well but lacks the desired performance, as shown in experiments.

As an alternative, heuristics that are inspired by formal properties of mixing by weighted average have been introduced. Not only do they scale well with the number of classifiers as they do not have any adjustable parameters other than the classifier parameters, but they also perform better than mixing by the least squares approximation. In particular, mixing by inverse variance makes the least assumptions of the introduced heuristics, and is also the best-performing one (though not significantly) and therefore our recommended choice. The heuristics were designed for linear regression classifier models, but the same concepts apply to designing heuristics for classification models.

The mixing model in XCS was never designed to maximise the data likelihood, and therefore the comparison to other heuristics might not seem completely fair. However, it was shown previously [83] that it also performs worst with respect to the mean squared error measure, and thus is not a good choice for a mixing model. Rather, mixing by inverse variance should be used as a drop-in replacement in XCS, but this recommendation is more strongly based on previous experiments [83] (see Sect. 6.4) rather than the empirical results presented here.

This chapter completes the discussion of how to find the LCS model parameters $\boldsymbol{\theta}$ by the principle of maximum likelihood for a fixed model structure $\mathcal{M}$. The next step is to provide a framework that lets us in addition find a good model structure, that is, a good set of classifiers. The taken approach is unable to identify good model structures at the model structure level $\mathcal{M}$ alone, but requires the reformulation of the probabilistic model itself to avoid overfitting even when finding the model parameters for a fixed model structure. This requires a deviation from the principle of maximum likelihood, which, however, does not completely invalidate the work that was presented in the last two chapters. Rather, the new update equations for parameter learning are up to small modifications similar to the ones that provide maximum likelihood estimates. Investigating these differences provides valuable insight into how exactly model selection infiltrates the parameter learning process.

7 The Optimal Set of Classifiers

This chapter deals with the question of what it means for a set of classifiers to be optimal in the light of the available data, and how to provide a formal solution to this problem. As such, it tackles the core task of LCS, whose ultimate aim is it to find such a set.

Up until now there is no general definition of what LCS ought to learn. Rather, there is an intuitive understanding of what a desirable set of classifiers should look like, and LCS algorithms are designed around such an understanding. However, having LCS that perform according to intuition in simple problems where the desired solution is known does not mean that they will do so in more complex tasks. Furthermore, how do we know that our intuition does not betray us?

While there are a small number of studies on what LCS want to learn and how that can be measured [130, 133, 135], they concentrate exclusively on the case where the input is encoded as a binary string, and even then they list several possible approaches rather than providing a single conclusive answer. However, considering the complexity of the problem at hand, it is understandable that approaching it is anything but trivial. The solution structure is strongly dependent on the chosen representation, but what is the best representation? Do we want the classifiers to partition the input space such that each of them independently provides a part of the solution, or do we expect them to cooperate? Should we prefer default hierarchies, where predictions of more general classifiers, that is, classifiers that match larger areas of the input space, are overridden by more specific ones, in a tree-like structure? Are the predictions of the classifiers supposed to be completely accurate, or do we allow for some error? And these are just a few questions to consider.

Rather than listing all possible questions and going through them one by one, the problem is here approached from another side, based on how LCS were characterised in Chapter 3: a fixed set of classifiers, that is, a fixed model structure $\mathcal{M}$, provides a certain hypothesis about the data-generating process that generated the observed data $\mathcal{D}$. With this in mind, "What do LCS want to learn?" becomes "Which model structure $\mathcal{M}$ explains the available data $\mathcal{D}$ best?". But, what exactly does "best" mean? Fortunately, evaluating the suitability of a

J. Drugowitsch: Des. & Anal. of Learn. Class. Sys.: A Prob. Approach, SCI 139, pp. 123–164, 2008.
springerlink.com

model with respect to the available data is a common task in machine learning, known as *model selection*. Hence, the complex problem of defining the optimal set of classifiers can be reduced to identifying a suitable model, and to applying it. This is what will be done for the rest of this chapter.

Firstly, let us consider the question of optimality, and, in general, which model properties are desirable. Using Bayesian model selection to identify good sets of classifiers, the LCS model is reformulated as a fully Bayesian model for regression. Classification is handled in a later section. Subsequently, a longer, more technical section demonstrates how variational Bayesian inference is applied to find closed-form approximations to posterior distributions. This also results in a closed-form expression for the quality of a particular model structure that allows us to compare the suitability of different LCS model structures to explain the available data. As such, this chapter provides the first general (that is, representation-independent) definition of optimality for a set of classifiers, and with it an answer to the question what LCS want to learn.

7.1 What Is Optimal?

Let us consider two extremes: N classifiers, such that each observation is matched by exactly one classifier, or a single classifier that matches all inputs. In the first case, each classifier replicates its associated observation completely accurately, and so the whole set of classifiers is a completely accurate representation of the data; it has an optimal goodness-of-fit. Methods that minimise the empirical risk, such as maximum likelihood or squared error minimisation, would evaluate such a set as being optimal. Nonetheless, it does not provide any generalisation in noisy data, as it does not differentiate between noise and the pattern in the data. In other words, having one classifier per observation does not provide us with any additional information than the data itself, and thus is not a desired solution.

Using a single classifier that matches all inputs, on the other hand, is the simplest LCS model structure, but has a very low expressive power. That is, it can only express very simple pattern in the data, and will very likely have a bad goodness-of-fit. Thus, finding a good set of classifiers involves balancing the goodness-of-fit of this set and its complexity, which determines its expressive power. This trade-off must be somehow expressed in each method that avoids overfitting.

7.1.1 Current LCS Approaches

XCS has the ability to find a set of classifiers that generalises over the available data [237, 238], and so has YCS [33] and CCS [153, 154]. This means that they do not simply minimise the overall model error but have some built-in model selection capability, however crude it might be.

Let us first consider XCS: its ability to generalise is brought about by a combination of the accuracy-definition of a classifier and the operation of its genetic algorithm. A classifier is considered as being accurate if its mean absolute prediction error over all matched observations is below the *minimum error*[1] threshold ϵ_0. The genetic algorithm provides accurate classifiers that match larger areas of the input space with more reproductive opportunity. However, overly general classifiers, that is, classifiers that match overly large areas of the input space, will feature a mean absolute error that is larger than ϵ_0, and are not accurate anymore. Thus, the genetic algorithm "pushes" towards more general classifiers, but only until they reach ϵ_0 [53]. In combination with the competition between classifiers that match the same input, XCS can be said to aim at finding the smallest non-overlapping set of accurate classifiers. From this perspective we could define an optimal set of classifiers that is dependent on ϵ_0. However, such a definition is not very appealing, as i) it is based on an algorithm, rather than having an algorithm that is based on the definition; ii) it is based solely on intuition; iii) the best set of classifiers is fully determined by the setting of ϵ_0 that might depend on the task at hand; and iv) ϵ_0 is the same for the whole input space, and so XCS cannot cope with tasks where the noise varies for different areas of the input space.

YCS [33] was developed by Bull as a simplified version of XCS such that its classifier dynamics can be modelled by difference equations. While it still measures the mean absolute prediction error of each classifier, it defines the fitness as being inversely proportional to this error, rather than using any accuracy concept based on some error threshold. Additionally, its genetic algorithm differs from the one used in XCS in that it selects classifiers from the whole set rather than only from the set that matches the current input. Having a fitness that is inverse to the error will make the genetic algorithm assign a higher reproductive opportunity to low-error classifiers that match many inputs. How low this error has to be depends on the error of other competing classifiers in the set, and on the maximum number of classifiers allowed, as that number determines the number of classifiers that the genetic algorithm aims at assigning to each input. Due to these dependencies it is difficult to define which set of classifiers YCS aims at finding, particularly as it depends on the dynamics of the genetic algorithm and the interplay of several system parameters. Its pressure towards more general classifiers comes from those classifiers matching more inputs and thus updating their error estimates more quickly, which gives them an initial higher fitness than more specific classifiers. However, this pressure is implicit and weaker than in XCS, which is easily seen in Fig. 1(a) of [33], where general and specific, but equally accurate, classifiers peacefully and stably co-exist in the population. It can only be stated that YCS supports classifiers that match larger areas of the input space, but only up until their errors get too large when compared to other classifiers in the set.

CCS [153, 154], in contrast, has a very clear definition of what types of classifiers win the competition in a classification task: it aims at maximally general

[1] The term *minimum error* for ϵ_0 is a misnomer, as it specifies the maximum error that classifier can have to still be accurate. Thus, ϵ_0 should be called the *maximum admissible error* or similar.

and maximally accurate classifiers by combining a generality measures, given by the proportion of overall examples correctly classified, and an error measures that is inversely proportional to the number of correct positive classifications over all classification attempts of a rule[2]. The trade-off between generality and error is handled by a constant γ that needs to be tuned. Thus, as in XCS, it is dependent on a system parameter that is to be set by the user. Additionally, in its current form, CCS aims at evolving rules that are completely accurate, and is thus unable to cope with noisy data [153, 154]. The set of classifiers it aims for can be described as the smallest set of classifiers that has the best trade-off between error and generality, as controlled by the parameter γ.

7.1.2 Model Selection

Due to the shortcomings of the previously discussed LCS, these will not be consider when defining the optimal set of classifiers. Rather, existing concepts from current model selection methods will be used. Even though most of these methods have different philosophical background, they all result in the principle of minimising a combination of the model error and a measure of the model complexity. To provide good model selection it is essential to use a good model complexity measure, and it has been shown that, generally, methods that consider the distribution of the data when judging the model complexity outperform methods that do not [125]. Furthermore, it is also of advantage to use the full training data rather than an independent test set [13].

Bayesian model selection meets these requirements and has additionally already been applied to the Mixtures-of-Expert model [227, 20, 216]. This makes it an obvious choice as a model selection criterion for LCS. A short discussion of alternative model selection criteria that might be applicable to LCS is provided in Sect. 7.6, later in this chapter.

7.1.3 Bayesian Model Selection

Given a model structure $\mathcal{M}$ and the data $\mathcal{D}$, Bayesian model selection is based on finding the probability density of the model structure given the data by Bayes' rule

$$p(\mathcal{M}|\mathcal{D}) \propto p(\mathcal{D}|\mathcal{M})p(\mathcal{M}), \tag{7.1}$$

where $p(\mathcal{M})$ is the prior over the set of possible model structures. The "best" model structure given the data is the one with the highest probability density $p(\mathcal{M}|\mathcal{D})$.

The data-dependent term $p(\mathcal{D}|\mathcal{M})$ is a likelihood known as the *evidence* for model structure $\mathcal{M}$, and is for a parametric model with parameters $\boldsymbol{\theta}$ evaluated by

---

[2] In [153, 154], the generality measure is called the *accuracy*, and the ratio of positive correct classifications over the total number of classification attempts is the *error*, despite it being some inverse measure of the error.

$$p(\mathcal{D}|\mathcal{M}) = \int_{\boldsymbol{\theta}} p(\mathcal{D}|\boldsymbol{\theta}, \mathcal{M}) p(\boldsymbol{\theta}|\mathcal{M}) \mathrm{d}\boldsymbol{\theta}, \tag{7.2}$$

where $p(\mathcal{D}|\boldsymbol{\theta}, \mathcal{M})$ is the data likelihood for a given model structure $\mathcal{M}$, and $p(\boldsymbol{\theta}|\mathcal{M})$ are the parameter priors given the same model structure. Thus, in order to perform Bayesian model selection, one needs to have a prior over the model structure space $\{\mathcal{M}\}$, a prior over the parameters given a model structure, and an efficient way of computing the model evidence (7.2).

As expected from a good model selection method, an implicit property of Bayesian model selection is that it penalises overly complex models [159]. This can be intuitively explained as follows: probability distributions that are more widely spread generally have lower peaks as the area underneath their density function is always 1. While simple model structures only have a limited capability of expressing data sets, more complex model structures are able to express a wider range of different data sets. Thus, their prior distribution will be more widely spread. As a consequence, conditioning a simple model structure on some data that it can express will cause its distribution to have a larger peak than a more complex model structure than is also able to express this data. This shows that, in cases where a simple model structure is able to explain the same data as a more complex model structure, Bayesian model selection will prefer the simpler model structure.

Example 7.1 (Bayesian Model Selection Applied to Polynomials). As in Example 3.1, consider a set of 100 observation from the 2nd degree polynomial $f(x) = 1/3 - x/2 + x^2$ with additive Gaussian noise $\mathcal{N}(0, 0.1^2)$ over the range $x \in [0, 1]$. Assuming ignorance of the data-generating process, the acquired model is a polynomial of unknown degree d. As was shown in Example 3.1, minimising the empirical risk leads to overfitting, as increasing the degree of the polynomial and with it the model complexity reduces this risk. Minimising the expected risk, on the other hands leads to correctly identifying the "true" model, but this risk is

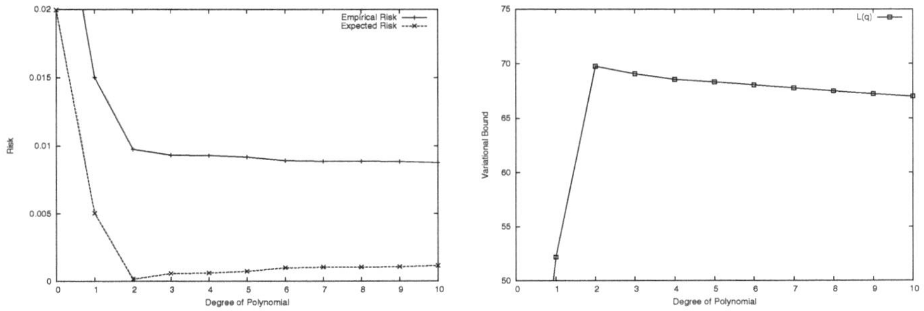

(a) (b)

Fig. 7.1. Expected and empirical risk, and the variational bound of the fit of polynomials of various degree to 100 noisy observations of a 2nd-order polynomial. (a) shows how the expected and empirical risk change with the degree of the polynomial. (b) shows the same for the variational bound. More information is given in Example 7.1.

usually not directly accessible. The graph from Fig. 3.1(b) that shows how both risk measures change with d is reproduced in Fig. 7.1(a) for convenience.

Using a Bayesian model of the data-generating process, one can assess the probability of the data supporting the polynomial having a particular degree by Bayesian model selection. The model acquired for this task is the same that is later introduced for linear regression classifiers and thus will not be discussed in detail. Variational Bayesian inference, as described Sect. 7.3.1, is used to evaluate a lower "variational" bound $\mathcal{L}(q)$ on the model log-probability, that is $\mathcal{L}(q) \leq \ln p(\mathcal{D}|\mathcal{M}) + \text{const.} = \ln p(\mathcal{M}|\mathcal{D}) + \text{const.}$ under the assumption of a uniform model prior $p(\mathcal{M})$. As shown in Fig. 7.1(b), $\mathcal{L}(q)$ is highest for $d = 2$, which demonstrates that Bayesian model selection correctly identifies the data-generating model.

7.1.4 Applying Bayesian Model Selection to Finding the Best Set of Classifiers

Applied to LCS, the model structure is, as previously described, defined by the number of classifiers K and their matching functions $\mathbf{M} = \{m_k : \mathcal{X} \rightarrow [0,1]\}$, giving $\mathcal{M} = \{K, \mathbf{M}\}$. In order to find the best set of classifiers, we need to maximise its probability density with respect to the data (7.1), which is equivalent to maximising its logarithm

$$\ln p(\mathcal{M}|\mathcal{D}) = \ln p(\mathcal{D}|\mathcal{M}) + \ln p(\mathcal{M}) + \text{const.}, \tag{7.3}$$

where the constant term captures the normalising constant and can be ignored when comparing the different model structures, as it is shared between them.

Evaluating the log-evidence $\ln p(\mathcal{D}|\mathcal{M})$ in (7.3) requires us to firstly specify a parameter prior $p(\boldsymbol{\theta}|\mathcal{M})$, and then to evaluate (7.2) to get the evidence of $\mathcal{M}$. Unfortunately, the LCS model described in Chap. 4 is not fully Bayesian and needs to be reformulated before the evidence can be evaluated. Additionally, the resulting probabilistic model structure does not provide a closed-form solution to (7.2). Thus, the rest of this chapter is devoted to i) introducing a fully Bayesian LCS model, and ii) applying an approximation method called *Variational Bayesian inference* that gives us a closed-form expression for the evidence. Before we do so, let us discuss the prior $p(\mathcal{M})$ on the model structure itself, and why the requirement of specifying parameter and model structure priors is not an inherit weakness of the method.

7.1.5 The Model Structure Prior $p(\mathcal{M})$

Specifying the prior for $p(\mathcal{M})$ lets us express our belief about which model structures are best at representing the data, prior to knowledge of the data. Recall that $\mathcal{M} = \{\mathbf{M}, K\}$ and thus $p(\mathcal{M})$ can be decomposed into $p(\mathcal{M}) = p(\mathbf{M}|K)p(K)$. Our belief about the number of classifiers K is that this number is certainly always finite, which requires $p(K) \rightarrow 0$ with $K \rightarrow \infty$. The beliefs about the set of matching functions of $\mathbf{M}$ given some K is less clear. Let us only observe that $\mathbf{M}$ contains K matching functions such that the set of possible $\mathbf{M}$ grows exponentially with K.

The question of how to best specify $p(\mathcal{M})$, and if there even is a "best" prior on $\mathcal{M}$, is not completely clear and requires further investigation. For now, $p(\mathcal{M}) \propto 1/K$, or

$$\ln p(\mathcal{M}) = -\ln K! + \text{const.} \tag{7.4}$$

is used for illustrative purposes. This prior can be interpreted as the prior $p(K) = (\mathrm{e}-1)^{-1} 1/K!$ on the number of classifiers, where $\mathrm{e} \equiv \exp(1)$, and a uniform $p(\mathbf{M}|K)$ that is absorbed by the constant term. Such a prior satisfies $p(K) \rightarrow 0$ for $K \rightarrow \infty$ and expresses that we expect the number of classifiers in the model to be small[3].

7.1.6 The Myth of No Prior Assumptions

A prior in the Bayesian sense is specified by a prior probability distribution and expresses what is known about a random variable in the absence of some evidence. For parametric models, the prior usually expresses what the model parameters are expected to be, in the absence of any observations. As such, it is part of the assumptions that are made about the data-generating process. Combining the information of the prior and the data gives the posterior.

Having the need to specify prior distributions could be considered as a weakness of Bayesian model selection, or even Bayesian statistics. Similarly, it could also be seen as a weakness of the presented approach to define the best set of classifiers. This view is justified by the idea that there exist other methods that do not make any prior assumptions. But is this really the case?

Let us investigate the class of linear models as described in Chap. 5. Due to linking the recursive least squares algorithm to ridge regression in Sect. 5.3.5 and the Kalman filter in Sect. 5.3.6, it was shown that the ridge regression problem

$$\min_{w} \left(\|\mathbf{X}\mathbf{w} - \mathbf{y}\|^2 + \lambda \|\mathbf{w}\|^2 \right) \tag{7.5}$$

is equivalent to conditioning a multivariate Gaussian prior $\boldsymbol{\omega}_0 \sim \mathcal{N}(\mathbf{0}, (\lambda\tau)^{-1}\mathbf{I})$ on the available data $\{\mathbf{X}, \mathbf{y}\}$, where τ is the noise precision of the linear model with respect to the data. Such a prior means that we assume each element of the weight vector to be independent — due to the zero off-diagonal elements of the diagonal covariance matrix — and zero-mean Gaussian with variance $(\lambda\tau)^{-1}$. That is, we assume the elements most likely to be zero, but they can also have other values with a likelihood that decreases with their deviation from zero.

Setting $\lambda = 0$ reduces (7.5) to a standard linear least squares problem without any prior assumptions — as it seems — besides the linear relation between the

[3] As pointed out by Dr. Dan Richardson, University of Bath, the prior $p(K) \propto 1/K!$ has $\mathbb{E}(K) < 2$ and thus expresses the belief that the number of classifiers is expected to be on average less than 2. He proposed the alternative prior $p(K) = \exp(-V)V^K/K!$, where V is a constant related to volume, and $\mathbb{E}(K)$ increases with V.

input and the output and the constant noise variance. Let us have a closer look at how $\lambda = 0$ influences $\boldsymbol{\omega}_0$: As $\lambda \to 0$ causes $(\lambda\tau)^{-1} \to \infty$, one can interpret the prior $\boldsymbol{\omega}_0$ to be the multivariate Gaussian $\mathcal{N}(\mathbf{0}, \infty\mathbf{I})$ (ignoring the problems that come with the use of ∞). As a Gaussian with increasing variance approaches the uniform distribution, the elements of the weight vectors are now equally likely to take any possible value of the real line. Even though such a prior seems unbiased at first, let us not forget that the uniform density puts most of its weight on large values due to its uniform tails [70]. Thus, as linear least squares is equivalent to ridge regression with $\lambda = 0$, its prior assumptions on the values of the weight vector elements is that they are uncorrelated but most likely take very large values. Large weight vector values, however, are usually a sign of non-smooth functions. Thus, linear least squares implicitly assumes that the function it models is not smooth.

As discussed in Sect. 3.1.1, a smooth function is a prerequisite for generalisation. Thus, we do actually assume smoothness of the function, and therefore ridge regression with $\lambda > 0$ is more appropriate than plain linear least squares. The prior that is associated with ridge regression is known as a *shrinkage prior* [102], as it causes the weight vector elements to be smaller than without using this prior. Ridge regression itself is part of a family of *regularisation* methods that add the assumption of function smoothness to guide parameter learning in otherwise ill-defined circumstances [213].

In summary, even methods that seemingly make no assumptions about the parameter values are biased by implicit priors, as was shown by comparing ridge regression to linear least squares. In any case, it is important to be aware of these priors, as they are part of the assumptions that a model makes about the data-generating process. Thus, when introducing the Bayesian LCS model, special emphasis is put on how the introduced parameter priors express our assumptions.

7.2 A Fully Bayesian LCS for Regression

The Bayesian LCS model for regression is equivalent to the one introduced as a generalisation of the Mixtures-of-Experts model in Chap. 4, with the differences that here, classifiers are allowed to perform multivariate rather than univariate regression, and that priors and associated hyperpriors are assigned to all model parameters. As such, it is a generalisation of the previous model as it completely subsumes it. A similar model for classification will be briefly discussed in Sect. 7.5. For now the classifiers are not assumed to be trained independently. This independence will be re-introduced at a later stage, analogous to Sect. 4.4.

Table 7.1 gives a summary of the Bayesian LCS model, and Fig. 7.2 shows its variable dependency structure as a directed graph. The model is besides the additional matching similar to the Bayesian MoE model by Waterhouse et al. [227, 226], to the Bayesian mixture model of Ueda and Ghahramani [216], and to the Bayesian MoE model of Bishop and Svensén [20]. Each of its components will now be described in more detail.

Table 7.1. Bayesian LCS model, with all its components. For more details on the model see Sect. 7.2.

Data, Model Structure, and Likelihood	
	N observations $\{(\mathbf{x}_n, \mathbf{y}_n)\}$, $\mathbf{x}_n \in \mathcal{X} = \mathbb{R}^{D_\mathcal{X}}$, $\mathbf{y}_n \in \mathcal{Y} = \mathbb{R}^{D_\mathcal{Y}}$ Model structure $\mathcal{M} = \{K, \mathbf{M}\}, k = 1, \dots, K$ K classifiers Matching functions $\mathbf{M} = \{m_k : \mathcal{X} \to [0,1]\}$
Likelihood	$p(\mathbf{Y}\|\mathbf{X}, \mathbf{W}, \boldsymbol{\tau}, \mathbf{Z}) = \prod_{n=1}^N \prod_{k=1}^K p(\mathbf{y}_n\|\mathbf{x}_n, \mathbf{W}_k, \tau_k)^{z_{nk}}$
Classifiers	
Variables	Weight matrices $\mathbf{W} = \{\mathbf{W}_k\}$, $\mathbf{W}_k \in \mathbb{R}^{D_\mathcal{Y}} \times \mathbb{R}^{D_\mathcal{X}}$ Noise precisions $\boldsymbol{\tau} = \{\tau_k\}$ Weight shrinkage priors $\boldsymbol{\alpha} = \{\alpha_k\}$ Noise precision prior parameters a_τ, b_τ α-hyperprior parameters a_α, b_α
Model	$p(\mathbf{y}\|\mathbf{x}, \mathbf{W}_k, \tau_k) = \mathcal{N}(\mathbf{y}\|\mathbf{W}_k\mathbf{x}, \tau_k^{-1}\mathbf{I}) = \prod_{j=1}^{D_\mathcal{Y}} \mathcal{N}(y_j\|\mathbf{w}_{kj}^T\mathbf{x}, \tau_k^{-1})$
Priors	$p(\mathbf{W}_k, \tau_k\|\alpha_k) = \prod_{j=1}^{D_\mathcal{Y}} \left(\mathcal{N}(\mathbf{w}_{kj}\|\mathbf{0}, (\alpha_k\tau_k)^{-1}\mathbf{I})\mathrm{Gam}(\tau_k\|a_\tau, b_\tau)\right)$ $p(\alpha_k) = \mathrm{Gam}(\alpha_k\|a_\alpha, b_\alpha)$
Mixing	
Variables	Latent variables $\mathbf{Z} = \{\mathbf{z}_n\}$, $\mathbf{z}_n = (z_{n1}, \dots, z_{nK})^T \in \{0,1\}^K$, 1-of-$K$ Mixing weight vectors $\mathbf{V} = \{\mathbf{v}_k\}$, $\mathbf{v}_k \in \mathbb{R}^{D_V}$ Mixing weight shrinkage priors $\boldsymbol{\beta} = \{\beta_k\}$ β-hyperprior parameters a_β, b_β
Model	$p(\mathbf{Z}\|\mathbf{X}, \mathbf{V}, \mathbf{M}) = \prod_{n=1}^N \prod_{k=1}^K g_k(\mathbf{x}_n)^{z_{nk}}$ $g_k(\mathbf{x}) \equiv p(z_k = 1\|\mathbf{x}, \mathbf{v}_k, m_k) = \frac{m_k(\mathbf{x})\exp(\mathbf{v}_k^T\phi(\mathbf{x}))}{\sum_{j=1}^K m_j(\mathbf{x})\exp(\mathbf{v}_j^T\phi(\mathbf{x}))}$
Priors	$p(\mathbf{v}_k\|\beta_k) = \mathcal{N}(\mathbf{v}_k\|\mathbf{0}, \beta_k^{-1}\mathbf{I})$ $p(\beta_k) = \mathrm{Gam}(\beta_k\|a_\beta, b_\beta)$

7.2.1 Data, Model Structure, and Likelihood

To evaluate the evidence of a certain model structure $\mathcal{M}$, the data $\mathcal{D}$ and the model structure $\mathcal{M}$ need to be known. The data $\mathcal{D}$ consists of N observations, each given by an input/output pair $(\mathbf{x}_n, \mathbf{y}_n)$. The input vector $\mathbf{x}_n$ is an element of the $D_\mathcal{X}$-dimensional real input space $\mathcal{X} = \mathbb{R}^{D_\mathcal{X}}$, and the output vector $\mathbf{y}_n$ is an element of the $D_\mathcal{Y}$-dimensional real output space $\mathcal{Y} = \mathbb{R}^{D_\mathcal{Y}}$. Hence, $\mathbf{x}_n$ has $D_\mathcal{X}$, and $\mathbf{y}_n$ has $D_\mathcal{Y}$ elements. The input matrix $\mathbf{X}$ and output matrix $\mathbf{Y}$ are defined according to (3.4).

The data is assumed to be standardised by a linear transformation such that all $\mathbf{x}$ and $\mathbf{y}$ have mean $\mathbf{0}$ and a range of 1. The purpose of this standardisation is the same as the one given by Chipman, George and McCulloch [62], which is to make it easier to intuitively gauge parameter values. For example, with the data being standardised, a weight value of 2 can be considered large as a half range increase in $\mathbf{x}$ would result in a full range increase in $\mathbf{y}$.

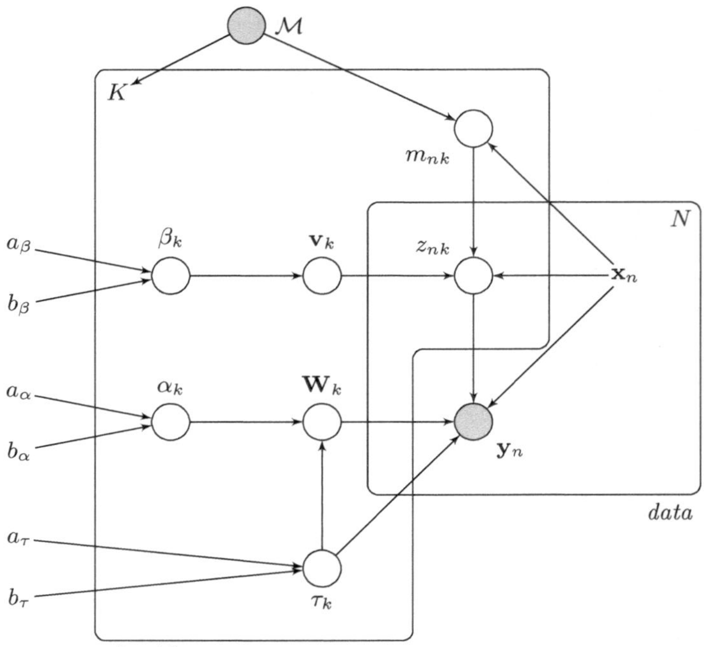

Fig. 7.2. Directed graphical model of the Bayesian LCS model. See the caption of Fig. 4.1 for instructions on how to read this graph. Note that to train the model, both the data $\mathcal{D}$ and the model structure $\mathcal{M}$ are assumed to be given. Hence, the $\mathbf{y}_n$'s and $\mathcal{M}$ are observed random variables, and the $\mathbf{x}_n$'s are constants.

The model structure $\mathcal{M} = \{K, \mathbf{M}\}$ specifies on one hand that K classifiers are used, and on the other hand, where these classifiers are localised. Each classifier k has an associated matching function $m_k : \mathcal{X} \rightarrow [0, 1]$, that returns for each input the probability of classifier k matching this input, as described in Sect. 4.3.1. Each input is assumed to be matched by at least one classifier, such that for each input $\mathbf{x}_n$ we have $\sum_k m_k(\mathbf{x}_n) > 0$. This needs to be the case in order to be able to model all of the inputs. As the model structure is known, all probability distributions are implicitly conditional on $\mathcal{M}$.

The data likelihood is specified from the generative point-of-view by assuming that each observation was generated by one and only one classifier. Let $\mathbf{Z} = \{\mathbf{z}_n\}$ be the N latent binary vectors $\mathbf{z}_n = (z_{n1}, \dots, z_{nK})^T$ of size K. We have $z_{nk} = 1$ if classifier k generated observation n, and $z_{nk} = 0$ otherwise. As each observation is generated by a single classifier, only a single element of each $\mathbf{z}_n$ is 1, and all other elements are 0. Under the standard assumption of independent and identically distributed data, that gives the likelihood

$$p(\mathbf{Y}|\mathbf{X}, \mathbf{W}, \boldsymbol{\tau}, \mathbf{Z}) = \prod_{n=1}^{N} \prod_{k=1}^{K} p(\mathbf{y}_n|\mathbf{x}_n, \mathbf{W}_k, \tau_k)^{z_{nk}}, \quad (7.6)$$

where $p(\mathbf{y}_n|\mathbf{x}_n, \mathbf{W}_k, \boldsymbol{\tau})$ is the model for the input/output relation of classifier k, parametrised by $\mathbf{W} = \{\mathbf{W}_k\}$ and $\boldsymbol{\tau} = \{\tau_k\}$. Let us continue with the classifier model, and then the model for the latent variables $\mathbf{Z}$.

7.2.2 Multivariate Regression Classifiers

The classifier model for classifier k is given by

$$\begin{aligned} p(\mathbf{y}|\mathbf{x}, \mathbf{W}_k, \tau_k) &= \mathcal{N}(\mathbf{y}|\mathbf{W}_k\mathbf{x}, \tau_k^{-1}\mathbf{I}) \\ &= \prod_{j=1}^{D_\mathcal{Y}} \mathcal{N}(y_j|\mathbf{w}_{jk}^T\mathbf{x}, \tau_k^{-1}) \\ &= \prod_{j=1}^{D_\mathcal{Y}} \left(\frac{\tau_k}{2\pi}\right)^{1/2} \exp\left(-\frac{\tau_k}{2}(y_j - \mathbf{w}_{kj}^T\mathbf{x})^2\right), \end{aligned} \tag{7.7}$$

where y_j is the jth element of $\mathbf{y}$, $\mathbf{W}_k$ is the $D_\mathcal{Y} \times D_\mathcal{X}$ weight matrix, and τ_k is the scalar noise precision. $\mathbf{w}_{kj}^T$ is the jth row vector of the weight matrix $\mathbf{W}_k$.

This model assumes that each element of the output $\mathbf{y}$ is linearly related to $\mathbf{x}$ with coefficients $\mathbf{w}_{kj}$, that is, $y_j \approx \mathbf{w}_{kj}^T\mathbf{x}$. Additionally, it assumes the elements of the output vector to be independent and feature zero-mean Gaussian noise with constant variance τ_k^{-1}. Note that the noise variance is assumed to be the same for each element of this output. It would be possible to assign each output element its own noise variance estimate, but this model variation was omitted for the sake of simplicity. If we have $D_\mathcal{Y} = 1$, we return to the univariate regression model (5.3) that is described at length in Chap. 5.

7.2.3 Priors on the Classifier Model Parameters

Each element of the output is assumed to be related to the input by a smooth function. As a consequence, the elements of the weight matrix $\mathbf{W}_k$ are assumed to be small which is expressed by assigning shrinkage priors to each row vector $\mathbf{w}_{kj}$ of the weight matrix $\mathbf{W}_k$. Additionally, the noise precision is assumed to be larger, but not much larger than 0, and in no case infinite, which is given by the prior $\text{Gam}(\tau_k|a_\tau, b_\tau)$ on the noise precision. Thus, the prior on $\mathbf{W}_k$ and τ_k is given by

$$\begin{aligned} p(\mathbf{W}_k, \tau_k|\alpha_k) &= \prod_{j=1}^{D_\mathcal{Y}} p(\mathbf{w}_{kj}, \tau_k|\alpha_k) \\ &= \prod_{j=1}^{D_\mathcal{Y}} \left(\mathcal{N}(\mathbf{w}_{kj}|\mathbf{0}, (\alpha_k\tau_k)^{-1}\mathbf{I})\text{Gam}(\tau_k|a_\tau, b_\tau)\right) \\ &= \prod_{j=1}^{D_\mathcal{Y}} \left(\left(\frac{\alpha_k\tau_k}{2\pi}\right)^{D_\mathcal{X}/2} \frac{b_\tau^{a_\tau}\tau_k^{(a_\tau-1)}}{\Gamma(a_\tau)} \exp\left(-\frac{\alpha_k\tau_k}{2}\mathbf{w}_{kj}^T\mathbf{w}_{kj} - a_\tau\tau_k\right)\right), \end{aligned} \tag{7.8}$$

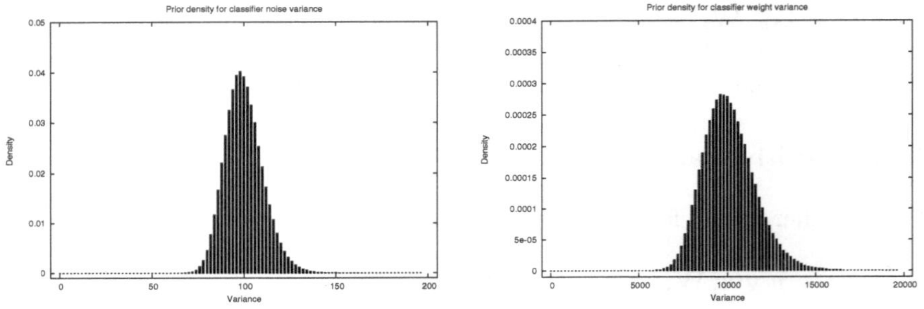

(a) (b)

Fig. 7.3. Histogram plot of the density of the (a) noise variance, and (b) variance of the weight vector prior. The plot in (a) was generated by sampling from τ_k^{-1} and shows that the prior on the variance is very flat, with the highest peak at a density of around 0.04 and a variance of about 100. The plot in (b) was generated by sampling from $(\alpha_k \tau_k)^{-1}$ and shows an even broader density for the variance of the zero mean weight vector prior, with its peak at around 0.00028 at a variance of about 10000.

where $\Gamma(\cdot)$ is the gamma function, α_k parametrises the variance of the Gaussian, and a_τ and b_τ are the parameters of the Gamma distribution. This prior distribution is known as *normal inverse-gamma*, as the inverse variance parameter of the Gaussian is distributed according to a Gamma distribution. Its use is advantageous, as conditioning it on a Gaussian results again in a normal inverse-gamma distribution, that is, it is a *conjugate prior* of the Gaussian distribution.

The prior assumes that elements of the weight vectors $\mathbf{w}_{jk}$ are independent and most likely zero, which is justified by the standardised data and the lack of further information. Its likelihood of deviating from zero is parametrised by α_k. τ_k is added to the variance term of the normal distribution for mathematical convenience, as it simplifies the computation of the posterior and predictive density.

The noise precision is distributed according to a Gamma distribution, which we will parametrise similar to Bishop and Svensén [20] by $a_\tau = 10^{-2}$ and $b_\tau = 10^{-4}$ to keep the prior sufficiently broad and uninformative, as shown in Fig. 7.3(a). An alternative approach would be to set the prior on τ_k to express the belief that the variance of the localised models will be most likely smaller than the variance of a single global model of the same form. We will not follow this approach, but more information on how to set the distribution parameters in such a case can be found in work on Bayesian Treed Models by Chipman, George and McCulloch [62].

We could specify a value for α_k by again considering the relation between the local models and global model, as done by Chipman, George and McCulloch [62]. However, we rather follow the approach of Bishop and Svensén [20], and treat α_k as a random variable that is modelled in addition to $\mathbf{W}_k$ and τ_k. It is assigned a conjugate Gamma distribution

$$p(\alpha_k) = \text{Gam}(\alpha_k | a_\alpha, b_\alpha) = \frac{b_\alpha^{a_\alpha} \alpha_k^{(a_\alpha - 1)}}{\Gamma(a_\alpha)} \exp(-a_\alpha \alpha_k), \tag{7.9}$$

which is kept sufficiently broad and uninformative by setting $a_\alpha = 10^{-2}$ and $b_\alpha = 10^{-4}$. The combined effect of τ_k and α_k on the weight vector prior variance is shown in Fig. 7.3(b).

7.2.4 Mixing by the Generalised Softmax Function

As in Chap. 4, the latent variables are modelled by the generalised softmax function (4.22), given by

$$g_k(\mathbf{x}) \equiv p(z_k = 1|\mathbf{x}, \mathbf{v}_k) = \frac{m_k(\mathbf{x}) \exp(\mathbf{v}_k^T \phi(\mathbf{x}))}{\sum_{j=1}^K m_j(\mathbf{x}) \exp(\mathbf{v}_j^T \phi(\mathbf{x}))}. \tag{7.10}$$

It assumes that, given that classifier k matched input $\mathbf{x}$, the probability of classifier k generating observation n is related to $\phi(\mathbf{x})$ by a log-linear function $\exp(\mathbf{v}_k^T \phi(\mathbf{x}))$, parametrised by $\mathbf{v}_k$. The transfer function $\phi : \mathcal{X} \rightarrow \mathbb{R}^{D_V}$ maps the input into a D_V-dimensional real space, and therefore the vector $\mathbf{v}_k$ is of size D_V and also an element of that space. In LCS, we usually have $D_V = 1$ and $\phi(\mathbf{x}) = 1$ for all $\mathbf{x} \in \mathcal{X}$, but to stay general, no assumptions about ϕ and D_V will be made.

Making use of the 1-of-K structure of $\mathbf{z}$, its joint probability is given by

$$p(\mathbf{z}|\mathbf{x}, \mathbf{V}) = \prod_{k=1}^K g_k(\mathbf{x})^{z_k}. \tag{7.11}$$

Thus, the joint probability of all $\mathbf{z}_n$ becomes

$$p(\mathbf{Z}|\mathbf{X}, \mathbf{V}) = \prod_{n=1}^N \prod_{k=1}^K g_k(\mathbf{x}_n)^{z_{nk}}, \tag{7.12}$$

which fully specifies the model for $\mathbf{Z}$.

7.2.5 Priors on the Mixing Model

Due to the normalisation, the mixing function g_k is over-parametrised, as it would be sufficient to specify $K-1$ vectors $\mathbf{v}_k$ and leave $\mathbf{v}_K$ constant [165]. This would make the values for all $\mathbf{v}_k$'s to be specified in relation to the constant $\mathbf{v}_K$, and causes problems if classifier K is removed from the current set. Thus, g_k is rather left over-parametrised, and it is assumed that all $\mathbf{v}_k$'s are small, which is expressed by the shrinkage prior

$$\begin{aligned} p(\mathbf{v}_k|\beta_k) &= \mathcal{N}(\mathbf{v}_k|\mathbf{0}, \beta_k^{-1}\mathbf{I}) \\ &= \left(\frac{\beta_k}{2\pi}\right)^{D_V/2} \exp\left(-\frac{\beta_k}{2}\mathbf{v}_k^T \mathbf{v}_k\right). \end{aligned} \tag{7.13}$$

Thus, the elements of $\mathbf{v}_k$ are assumed to be independent and zero-mean Gaussian with precision β_k.

Rather than specifying a value for β_k, it is again modelled by the Gamma hyperprior

$$p(\beta_k) = \text{Gam}(\beta_k | a_\beta, b_\beta) = \frac{b_\beta^{a_\beta} \beta_k^{(a_\beta - 1)}}{\Gamma(a_\beta)} \exp(-a_\beta \beta_k), \tag{7.14}$$

with hyper-parameters set to $a_\beta = 10^{-2}$ and $b_\beta = 10^{-4}$ to get a broad and uninformative prior for the variance of the mixing weight vectors. The shape of the prior is the same as for τ_k^{-1}, which is shown in Fig. 7.3(a).

7.2.6 Joint Distribution over Random Variables

Assuming knowledge of $\mathbf{X}$ and $\mathcal{M}$, the joint distribution over all random variables decomposes into

$$\begin{aligned} p(\mathbf{Y}, \mathbf{U} | \mathbf{X}) = & \, p(\mathbf{Y} | \mathbf{X}, \mathbf{W}, \boldsymbol{\tau}, \mathbf{Z}) p(\mathbf{W}, \boldsymbol{\tau} | \boldsymbol{\alpha}) p(\boldsymbol{\alpha}) \\ & \times p(\mathbf{Z} | \mathbf{X}, \mathbf{V}) p(\mathbf{V} | \boldsymbol{\beta}) p(\boldsymbol{\beta}), \end{aligned} \tag{7.15}$$

where $\mathbf{U}$ collectively denotes the hidden variables $\mathbf{U} = \{\mathbf{W}, \boldsymbol{\tau}, \boldsymbol{\alpha}, \mathbf{Z}, \mathbf{V}, \boldsymbol{\beta}\}$. This decomposition is also clearly visible in Fig. 7.2, where the dependency structure between the different variables and parameters is graphically illustrated. All priors are independent for different k's, and so we have

$$p(\mathbf{W}, \boldsymbol{\tau} | \boldsymbol{\alpha}) = \prod_{k=1}^{K} p(\mathbf{W}_k, \tau_k | \alpha_k), \tag{7.16}$$

$$p(\boldsymbol{\alpha}) = \prod_{k=1}^{K} p(\alpha_k), \tag{7.17}$$

$$p(\mathbf{V} | \boldsymbol{\beta}) = \prod_{k=1}^{K} p(\mathbf{v}_k | \beta_k), \tag{7.18}$$

$$p(\boldsymbol{\beta}) = \prod_{k=1}^{K} p(\beta_k). \tag{7.19}$$

By inspecting (7.6) and (7.12) it can be seen that, similar to the priors, both $p(\mathbf{Y} | \mathbf{X}, \mathbf{W}, \boldsymbol{\tau}, \mathbf{Z})$ and $p(\mathbf{Z} | \mathbf{X}, \mathbf{V})$ factorise over k, and therefore the joint distribution (7.15) factorises over k as well. This property will be used when deriving the required expressions to compute the evidence $p(\mathcal{D} | \mathcal{M})$.

7.3 Evaluating the Model Evidence

This rather technical section is devoted to deriving an expression for the model evidence $p(\mathcal{D} | \mathcal{M})$ for use in (7.3). Evaluating (7.2) does not yield a closed-form expression. Hence, we will make use of an approximation technique known as *variational Bayesian inference* [119, 19] that provides us with such a closed-form expression.

Alternatively, sampling techniques, such as Markov Chain Monte Carlo (MCMC) methods, could be utilised to get an accurate posterior and model evidence. However, the model structure search is expensive and requires a quick evaluation of the model evidence for a given model structure, and therefore the computational burden of sampling techniques makes approximating the model evidence by variational methods a better choice.

For the remainder of this chapter, all distributions are treated as being implicitly conditional on $\mathbf{X}$ and $\mathcal{M}$, to keep the notation simple. Additionally, the range for sums and products will not always be specified explicitly, as they are usually obvious from their context.

7.3.1 Variational Bayesian Inference

The aim of Bayesian inference and model selection is, on one hand, to find a variational distribution $q(\mathbf{U})$ that approximates the true posterior $p(\mathbf{U}|\mathbf{Y})$ and, on the other hand, to get the model evidence $p(\mathbf{Y})$. Variational Bayesian inference is based on the decomposition [19, 118]

$$\ln p(\mathbf{Y}) = \mathcal{L}(q) + \mathrm{KL}(q\|p), \tag{7.20}$$

$$\mathcal{L}(q) = \int q(\mathbf{U}) \ln \frac{p(\mathbf{U},\mathbf{Y})}{q(\mathbf{U})} \mathrm{d}\mathbf{U}, \tag{7.21}$$

$$\mathrm{KL}(q\|p) = -\int q(\mathbf{U}) \ln \frac{p(\mathbf{U}|\mathbf{Y})}{q(\mathbf{U})} \mathrm{d}\mathbf{U}, \tag{7.22}$$

which holds for any choice of q. As the Kullback-Leibler divergence $\mathrm{KL}(q\|p)$ is always non-negative, and zero if and only if $p(\mathbf{U}|\mathbf{Y}) = q(\mathbf{U})$ [232], the variational bound $\mathcal{L}(q)$ is a lower bound on $\ln p(\mathbf{Y})$ and only equivalent to the latter if $q(\mathbf{U})$ is the true posterior $p(\mathbf{U}|\mathbf{Y})$. Hence, the posterior can be approximated by maximising the lower bound $\mathcal{L}(q)$, which brings the variational distribution closer to the true posterior and at the same time yields an approximation of the model evidence by $\mathcal{L}(q) \leq \ln p(\mathbf{Y})$.

Factorial Distributions

To make this approach tractable, we need to choose a family of distributions $q(\mathbf{U})$ that gives an analytical solution. A frequently used approach (for example, [20, 227]) that is sufficiently flexible to give a good approximation to the true posterior is to use the set of distributions that factorises with respect to disjoint groups $\mathbf{U}_i$ of variables

$$q(\mathbf{U}) = \prod_i q_i(\mathbf{U}_i), \tag{7.23}$$

which allows maximising $\mathcal{L}(q)$ with respect to each group of hidden variables separately while keeping the other ones fixed. This results in

$$\ln q_i^*(\mathbf{U}_i) = \mathbb{E}_{i \neq j}\left(\ln p(\mathbf{U},\mathbf{Y})\right) + \text{const.}, \tag{7.24}$$

when maximising with respect to $\mathbf{U}_i$, where the expectation is taken with respect to all hidden variables except for $\mathbf{U}_i$, and the constant term is the logarithm of the normalisation constant of q_i^* [19, 118]. In our case, we group the variables according to their priors by $\{\mathbf{W}, \boldsymbol{\tau}\}$, $\{\boldsymbol{\alpha}\}$, $\{\mathbf{V}\}$, $\{\boldsymbol{\beta}\}$, $\{\mathbf{Z}\}$.

Handling the Softmax Function

If the model has a conjugate-exponential structure, (7.24) gives an analytical solution with a distribution form equal to the prior of the corresponding hidden variable. However, in our case the generalised softmax function (7.10) does not conform to this conjugate-exponential structure, and needs to be dealt with separately. A possible approach is to replace the softmax function by an exponential lower bound on it, which consequently introduces additional variational variables with respect to which $\mathcal{L}(q)$ also needs to be maximised. This approach was followed By Bishop and Svensén [20] and Jaakkola and Jordan [119] for the logistic sigmoid function, but currently there is no known exponential lower bound function on the softmax besides a conjectured one by Gibbs [93][4]. Alternatively, we can follow the approach taken by Waterhouse et al. [227, 226], where $q_V^*(\mathbf{V})$ is approximated by a Laplace approximation. Due to the lack of better alternatives, this approach is chosen, despite such an approximation invalidating the lower bound nature of $\mathcal{L}(q)$.

Update Equations and Model Posterior

To get the update equations for the parameters of the variational distribution, we need to evaluate (7.24) for each group of hidden variables in $\mathbf{U}$ separately, similar to the derivations by Waterhouse et al. [226] and Ueda and Ghahramani [216]. This provides us with an approximation for the posterior $p(\mathbf{U}|\mathbf{Y})$ and will be shown in the following sections.

Approximating the model evidence $p(\mathbf{Y})$ requires a closed-form expression for $\mathcal{L}(q)$ by evaluating (7.21), where many terms of the variational update equations can be reused, as will be shown after having derived the update equations.

7.3.2 Classifier Model $q^*_{W,\tau}(W, \tau)$

The maximum of $\mathcal{L}(q)$ with respect to $\mathbf{W}$ and $\boldsymbol{\tau}$ is given by evaluating (7.24) for $q_{W,\tau}$, which, by using (7.15), (7.16) and (7.6), results in

$$\begin{aligned}\ln q^*_{W,\tau}(\mathbf{W}, \boldsymbol{\tau}) &= \mathbb{E}_Z(\ln p(\mathbf{Y}|\mathbf{W}, \boldsymbol{\tau}, \mathbf{Z})) + \mathbb{E}_\alpha(\ln 0p(\mathbf{W}, \boldsymbol{\tau}|\boldsymbol{\alpha})) + \text{const.} \\ &= \sum_k \sum_n \mathbb{E}_Z(z_{nk} \ln p(\mathbf{y}_n|\mathbf{W}_k, \tau_k)) \\ &\quad + \sum_k \mathbb{E}_\alpha(\ln p(\mathbf{W}_k, \tau_k|\alpha_k)) + \text{const.,} \end{aligned} \tag{7.25}$$

[4] A more general bound was recently developed by Wainwright, Jaakkola and Willsky [225], but its applicability still needs to be evaluated.

where the constant represents all terms in (7.15) that are independent of $\mathbf{W}$ and $\boldsymbol{\tau}$, and $\mathbb{E}_Z$ and $\mathbb{E}_\alpha$ are the expectations evaluated with respect to $\mathbf{Z}$ and $\boldsymbol{\alpha}$ respectively. This expression shows that $q^*_{W,\tau}$ factorises with respect to k, which allows us to handle the $q_{W,\tau}(\mathbf{W}_k, \tau_k)$'s separately, by solving

$$\ln q^*_{W,\tau}(\mathbf{W}_k, \tau_k) = \sum_n \mathbb{E}_Z(z_{nk} \ln p(\mathbf{y}_n|\mathbf{W}_k, \tau_k)) + \mathbb{E}_\alpha(\ln p(\mathbf{W}_k, \tau_k|\alpha_k)) + \text{const.} \tag{7.26}$$

Using the classifier model (7.7), we get

$$\begin{aligned}
&\sum_n \mathbb{E}_Z(z_{nk} \ln p(\mathbf{y}_n|\mathbf{W}_k, \tau_k)) \\
&\quad= \sum_n \mathbb{E}_Z(z_{nk}) \ln \prod_j \mathcal{N}(y_{nj}|\mathbf{w}_{kj}^T \mathbf{x}_n, \tau_k^{-1}) \\
&\quad= \sum_n r_{nk} \sum_j \left(\frac{1}{2} \ln \tau_k - \frac{\tau_k}{2}(y_{nj} - \mathbf{w}_{kj}^T \mathbf{x}_n)^2 \right) + \text{const.} \\
&\quad= \frac{D_\mathcal{Y}}{2} \left(\sum_n r_{nk} \right) \ln \tau_k + \text{const.} \\
&\qquad - \frac{\tau_k}{2} \sum_j \left(\sum_n r_{nk} y_{nj}^2 - 2\mathbf{w}_{kj}^T \sum_n r_{nk} \mathbf{x}_n y_{nj} + \mathbf{w}_{kj}^T \left(\sum_n r_{nk} \mathbf{x}_n \mathbf{x}_n^T \right) \mathbf{w}_{kj} \right),
\end{aligned} \tag{7.27}$$

where $r_{nk} \equiv \mathbb{E}_Z(z_{nk})$ is the *responsibility* of classifier k for observation n, and y_{nj} is the jth element of $\mathbf{y}_n$. The constant represents the terms that are independent of $\mathbf{W}_k$ and τ_k.

$\mathbb{E}_\alpha(\ln p(\mathbf{W}_k, \tau_k|\alpha_k))$ is expanded by the use of (7.8) and results in

$$\begin{aligned}
&\mathbb{E}_\alpha(\ln p(\mathbf{W}_k, \tau_k|\alpha_k)) \\
&\quad= \sum_j \mathbb{E}_\alpha \left(\ln \mathcal{N}(\mathbf{w}_{kj}|\mathbf{0}, (\alpha_k \tau_k)^{-1}\mathbf{I}) + \ln \text{Gam}(\tau_k|a_\tau, b_\tau) \right) \\
&\quad= \sum_j \left(\frac{D_\mathcal{X}}{2} \ln \tau_k - \frac{\tau_k}{2} \mathbb{E}_\alpha(\alpha_k) \mathbf{w}_{kj}^T \mathbf{w}_{kj} + (a_\tau - 1) \ln \tau_k - b_\tau \tau_k \right) + \text{const.} \\
&\quad= \left(D_\mathcal{Y} a_\tau - D_\mathcal{Y} + \frac{D_\mathcal{X} D_\mathcal{Y}}{2} \right) \ln \tau_k \\
&\qquad - \frac{\tau_k}{2} \left(2 D_\mathcal{Y} b_\tau + \mathbb{E}_\alpha(\alpha_k) \sum_j \mathbf{w}_{kj}^T \mathbf{w}_{kj} \right) + \text{const.}
\end{aligned} \tag{7.28}$$

Thus, evaluating (7.26) gives

$$\ln q^*_{W,\alpha}(\mathbf{W}_k, \tau_k) = \left(D_{\mathcal{Y}} a_\tau - D_{\mathcal{Y}} + \frac{D_{\mathcal{X}} D_{\mathcal{Y}}}{2} + \frac{D_{\mathcal{Y}}}{2} \sum_n r_{nk} \right) \ln \tau_k$$
$$- \frac{\tau_k}{2} \left(2 D_{\mathcal{Y}} b_\tau + \sum_j \left(\sum_n r_{nk} y_{nj}^2 - 2 \mathbf{w}_{kj}^T \sum_n r_{nk} \mathbf{x}_n y_{nj} \right.\right.$$
$$\left.\left. + \mathbf{w}_{kj}^T \left(\mathbb{E}_\alpha(\alpha_k) \mathbf{I} + \sum_n r_{nk} \mathbf{x}_n \mathbf{x}_n^T \right) \mathbf{w}_{kj} \right) \right) + \text{const.}$$
$$= \ln \prod_j \left(\mathcal{N}(\mathbf{w}_{kj} | \mathbf{w}^*_{kj}, (\tau_k \mathbf{\Lambda}^*_k)^{-1}) \text{Gam}(\tau_k | a^*_{\tau_k}, b^*_{\tau_k}) \right), \quad (7.29)$$

with the distribution parameters

$$\mathbf{\Lambda}^*_k = \mathbb{E}_\alpha(\alpha_k) \mathbf{I} + \sum_n r_{nk} \mathbf{x}_n \mathbf{x}_n^T, \quad (7.30)$$

$$\mathbf{w}^*_{kj} = \mathbf{\Lambda}^{*-1}_k \sum_n r_{nk} \mathbf{x}_n y_{nj}, \quad (7.31)$$

$$a^*_{\tau_k} = a_\tau + \frac{1}{2} \sum_n r_{nk}, \quad (7.32)$$

$$b^*_{\tau_k} = b_\tau + \frac{1}{2 D_{\mathcal{Y}}} \left(\sum_j \left(\sum_n r_{nk} y_{nj}^2 - \mathbf{w}^{*T}_{kj} \mathbf{\Lambda}^*_k \mathbf{w}^*_{kj} \right) \right). \quad (7.33)$$

The second equality in (7.29) can be derived by expanding the final result and replacing all terms that are independent of $\mathbf{W}_k$ and τ_k by a constant. The distribution parameter update equations are that of a standard Bayesian weighted linear regression (for example, [19, 15, 72]).

Note that due to the use of conjugate priors, the variational posterior $q^*_{W,\alpha}$ $(\mathbf{W}_k, \tau_k)$ (7.29) has the same distribution form as the prior $p(\mathbf{W}_k, \tau_k | \alpha_k)$ (7.8). The resulting weight vector $\mathbf{w}_{kj}$, that models the relation between the inputs and the jth component of the outputs, is given by a Gaussian with mean $\mathbf{w}^*_{kj}$ and precision $\tau_k \mathbf{\Lambda}^*_k$. The same posterior weight mean can be found by minimising

$$\| \mathbf{X} \mathbf{w}_{kj} - \mathbf{y}_j \|^2_{R_k} + \mathbb{E}_\alpha(\alpha_k) \| \mathbf{w}_{kj} \|^2, \quad (7.34)$$

with respect to $\mathbf{w}_{kj}$, where $\mathbf{R}_k$ is the diagonal matrix $\mathbf{R}_k = \text{diag}(r_{1k}, \dots, r_{Nk})$, and $\mathbf{y}_j$ is the vector of jth output elements, $\mathbf{y}_j = (y_{1j}, \dots, y_{Nj})^T$, that is, the jth column of $\mathbf{Y}$. This shows that we are performing a responsibility-weighted ridge regression with ridge complexity $\mathbb{E}_\alpha(\alpha_k)$. Thus, the shrinkage is determined by the prior on α_k, as can be expected from the specification of the weight vector prior (7.8).

The noise precision posterior is the Gamma distribution $\text{Gam}(\tau_k | a^*_{\tau_k}, b^*_{\tau_k})$. Using the relation $\frac{\nu\lambda}{\chi^2_\nu} \sim \text{Gam}(\nu/2, \nu\lambda/2)$, where $\frac{\nu\lambda}{\chi^2_\nu}$ is the scaled inverse χ^2

distribution with ν degrees of freedom, (7.32) can be interpreted as incrementing the degrees of freedom from an initial $2a_\tau$ by $\sum_n r_{nk}$. Thus, while the prior has the weight of $2a_\tau$ observations, each added observation is weighted according to the responsibility that classifier k has for it. By using (7.30) and the relation

$$\sum_n r_{nk}(y_{nj} - \mathbf{w}_{kj}^{*\,T}\mathbf{x}_n)^2$$
$$= \sum_n r_{nk}y_{nj}^2 - 2\mathbf{w}_{kj}^{*\,T}\sum_n r_{nk}\mathbf{x}_n y_{nj} + \mathbf{w}_{kj}^{*\,T}\left(\sum_n r_{nk}\mathbf{x}_n\mathbf{x}_n^T\right)\mathbf{w}_{kj}^*,$$

Equation (7.33) can be reformulated to give

$$b_{\tau_k}^* = b_\tau + \frac{1}{2D_\mathcal{Y}}\left(\sum_n r_{nk}\|\mathbf{y}_n - \mathbf{W}_k^*\mathbf{x}_n\|^2 + \mathbb{E}_\alpha(\alpha_k)\sum_j \|\mathbf{w}_{kj}^*\|^2\right). \tag{7.35}$$

This shows that b_τ is updated by the responsibility-weighted sum of squared prediction errors, averaged over the different elements of the output vector, and the average size of the $\mathbf{w}_{kj}$'s, weighted by the expectation of the weight precision prior. Considering that $\mathbb{E}(\mathrm{Gam}(a,b)) = a/b$ [19], the mean of the noise variance posterior is therefore strongly influenced by the responsibility-weighted averaged squared prediction error, given a sufficiently uninformative prior.

7.3.3 Classifier Weight Priors $q_\alpha^*(\boldsymbol{\alpha})$

As by (7.17), $p(\boldsymbol{\alpha})$ factorises with respect to k, we can treat the variational posterior q_α^* for each classifier separately. For classifier k, this posterior is according to (7.15), (7.16), (7.17) and (7.24) given by

$$\ln q_\alpha^*(\alpha_k) = \mathbb{E}_{W,\tau}(\ln p(\mathbf{W}_k, \tau_k|\alpha_k)) + \ln p(\alpha_k) + \text{const.} \tag{7.36}$$

Using (7.8), the expectation of weights and noise precision evaluates to

$$\mathbb{E}_{W,\tau}(\ln p(\mathbf{W}_k, \tau_k|\alpha_k))$$
$$= \sum_j \mathbb{E}_{W,\tau}\left(\ln \mathcal{N}(\mathbf{w}_{kj}|\mathbf{0}, (\alpha_k\tau_k)^{-1}\mathbf{I}) + \ln \mathrm{Gam}(\tau_k|a_\tau, b_\tau)\right)$$
$$= \sum_j \left(\frac{D_\mathcal{X}}{2}\ln\alpha_k - \frac{\alpha_k}{2}\mathbb{E}_{W,\tau}(\tau_k\mathbf{w}_{kj}^T\mathbf{w}_{kj})\right) + \text{const.} \tag{7.37}$$

Also, by (7.9),

$$\ln p(\alpha_k) = (a_\alpha - 1)\ln\alpha_k - b_\alpha\alpha_k + \text{const.} \tag{7.38}$$

Together, that gives the variational posterior

$$\ln q_\alpha^*(\alpha_k) = \left(\frac{D_\mathcal{X}D_\mathcal{Y}}{2} + a_\alpha - 1\right)\ln\alpha_k$$
$$- \left(b_\alpha + \frac{1}{2}\sum_j \mathbb{E}_{W,\tau}(\tau_k\mathbf{w}_{kj}^T\mathbf{w}_{kj})\right)\alpha_k + \text{const.}$$
$$= \ln \mathrm{Gam}(\alpha_k|a_{\alpha_k}^*, b_{\alpha_k}^*), \tag{7.39}$$

with

$$a^*_{\alpha_k} = a_\alpha + \frac{D_\mathcal{X} D_\mathcal{Y}}{2}, \tag{7.40}$$

$$b^*_{\alpha_k} = b_\alpha + \frac{1}{2} \sum_j \mathbb{E}_{W,\tau}(\tau_k \mathbf{w}_{kj}^T \mathbf{w}_{kj}). \tag{7.41}$$

Utilising again the relation between the gamma distribution and the scaled inverse χ^2 distribution, (7.40) increments the initial $2a_\alpha$ degrees of freedom by $D_\mathcal{X} D_\mathcal{Y}$, which is the number of elements in $\mathbf{W}_k$.

The posterior mean of α_k is $\mathbb{E}(\alpha_k) = a^*_{\alpha_k}/b^*_{\alpha_k}$ and thus is inversely proportional to the size of the weight vectors $\|\mathbf{w}_{kj}\|^2 = \mathbf{w}_{kj}^T \mathbf{w}_{kj}$ and the noise precision τ_k. As the element-wise variance in the weight vector prior (7.8) is given by $(\alpha_k \tau_k)^{-1}$, the effect of τ_k on that prior is diminished. Thus, the weight vector prior variance is proportional to the expected size of the weight vectors, which has the effect of spreading the weight vector prior if the weight vector is expected to be large, effectively reducing the shrinkage. Intuitively, this is a sensible thing to do, as one should refrain from using an overly strong shrinkage prior if the weight vector is expected to have large elements.

7.3.4 Mixing Model $q^*_V(V)$

We get the variational posterior $q^*_V(\mathbf{V})$ on the mixing model parameters by solving (7.24) with (7.15), that is

$$\ln q^*_V(\mathbf{V}) = \mathbb{E}_Z(\ln p(\mathbf{Z}|\mathbf{V})) + \mathbb{E}_\beta(\ln p(\mathbf{V}|\boldsymbol{\beta})) + \text{const.}. \tag{7.42}$$

Even though q^*_V factorises with respect to k, we will solve it for all classifiers simultaneously due to the Laplace approximation that is applied thereafter.

Evaluating the expectations by using (7.12), (7.13) and (7.19) we get

$$\mathbb{E}_Z(\ln p(\mathbf{Z}|\mathbf{V})) = \sum_n \sum_k r_{nk} g_k(\mathbf{x}_n), \tag{7.43}$$

$$\mathbb{E}_\beta(\ln p(\mathbf{V}|\boldsymbol{\beta})) = \sum_k \mathbb{E}_\beta(\ln \mathcal{N}(\mathbf{v}_k|\mathbf{0}, \beta_k^{-1}\mathbf{I}))$$

$$= \sum_k \left(-\frac{\mathbb{E}_\beta(\beta_k)}{2} \mathbf{v}_k^T \mathbf{v}_k \right) + \text{const.}, \tag{7.44}$$

where $r_{nk} \equiv \mathbb{E}_Z(z_{nk})$ was used. Thus, the variational log-posterior evaluates to

$$\ln q^*_V(\mathbf{V}) = \sum_k \left(-\frac{\mathbb{E}_\beta(\beta_k)}{2} \mathbf{v}_k^T \mathbf{v}_k + \sum_n r_{nk} g_k(\mathbf{x}_n) \right) + \text{const.} \tag{7.45}$$

Note that the distribution form of this posterior differs from its prior (7.13), which would cause problems in further derivations. Thus, we proceed the same way as Waterhouse et al. [227, 226] by performing a Laplace approximation of the posterior.

The Laplace approximation aims at finding a Gaussian approximation to the posterior density, by centring the Gaussian on the mode of the density and deriving its covariance by a second-order Taylor expansion of the posterior [19]. The mode of the posterior is found by solving

$$\frac{\partial \ln q_V^*(\mathbf{V})}{\partial \mathbf{V}} = 0, \tag{7.46}$$

which, by using the posterior (7.45) and the definition of g_k (7.10), results in

$$\sum_n (r_{nk} - g_k(\mathbf{x}_n))\phi(\mathbf{x}) - \mathbb{E}_\beta(\beta_k)\mathbf{v}_k = 0, \qquad k = 1, \dots, K. \tag{7.47}$$

Note that, besides the addition of the $\mathbb{E}_\beta(\beta_k)\mathbf{v}_k$ term due to the shrinkage prior on $\mathbf{v}_k$, the minimum we seek is equivalent to the one of the prior-less generalised softmax function, given by (6.11). Therefore, we can find this minimum by applying the IRLS algorithm (6.5) with error function $E(\mathbf{V}) = -\ln q_V^*(\mathbf{V})$, where the required gradient vector and the $D_V \times D_V$ blocks $\mathbf{H}_{kj}$ of the Hessian matrix (6.9) are given by

$$\nabla_V E(\mathbf{V}) = \begin{pmatrix} \nabla_{v_1} E(\mathbf{V}) \\ \vdots \\ \nabla_{v_K} E(\mathbf{V}) \end{pmatrix}, \quad \nabla_{v_j} E(\mathbf{V}) = \sum_n (g_j(\mathbf{x}_n) - r_{nj})\phi(\mathbf{x}_n) + \mathbb{E}_\beta(\beta_j)\mathbf{v}_j, \tag{7.48}$$

and

$$\mathbf{H}_{kj} = \mathbf{H}_{jk} = \sum_n g_k(\mathbf{x}_n)(\mathbf{I}_{kj} - g_j(\mathbf{x}_n))\phi(\mathbf{x}_n)\phi(\mathbf{x}_n)^T + \mathbf{I}_{kj}\mathbb{E}_\beta(\beta_k)\mathbf{I}. \tag{7.49}$$

$\mathbf{I}_{kj}$ is the kjth element of the identity matrix, and the second $\mathbf{I}$ in the above expression is an identity matrix of size $D_V \times D_V$. As the resulting Hessian is positive definite [173], the posterior density is concave and has a unique maximum. More details on how to implement the IRLS algorithm are given in the next chapter.

Let $\mathbf{V}^*$ with components $\mathbf{v}_k^*$ denote the parameters that maximise (7.45). $\mathbf{V}^*$ gives the mode of the posterior density, and thus the mean vector of its Gaussian approximation. As the logarithm of a Gaussian distribution is a quadratic function of the variables, this quadratic form can be recovered by a second-order Taylor expansion of $\ln q_V^*(\mathbf{V})$ [19], which results in the precision matrix

$$\mathbf{\Lambda}_V^* = -\nabla\nabla \ln q_V^*(\mathbf{V}^*) = \nabla\nabla E(\mathbf{V}^*) = \mathbf{H}|_{V=V^*}, \tag{7.50}$$

where $\mathbf{H}$ is the Hessian matrix of $E(\mathbf{V})$ as used in the IRLS algorithm. Overall, the Laplace approximation to the posterior $q_V^*(\mathbf{V})$ is given by the multivariate Gaussian

$$q_V^*(\mathbf{V}) \approx \mathcal{N}(\mathbf{V}|\mathbf{V}^*, \mathbf{\Lambda}_V^{*\,-1}), \tag{7.51}$$

where $\mathbf{V}^*$ is the solution to (7.47), and $\mathbf{\Lambda}_V^*$ is the Hessian matrix evaluated at $\mathbf{V}^*$.

7.3.5 Mixing Weight Priors $q_\beta^*(\boldsymbol{\beta})$

By (7.19), $p(\boldsymbol{\beta})$ factorises with respect to k, and thus allows us to find $q_\beta^*(\boldsymbol{\beta})$ for each classifier separately, which, by (7.15), (7.18) and (7.24), requires the evaluation of

$$\ln q_\beta^*(\beta_k) = \mathbb{E}_V(\ln p(\mathbf{v}_k|\beta_k)) + \ln p(\beta_k). \tag{7.52}$$

Using (7.13) and (7.14), the expectation and log-density are given by

$$\mathbb{E}_V(\ln p(\mathbf{v}_k|\beta_k)) = \frac{D_V}{2}\ln \beta_k - \frac{\beta_k}{2}\mathbb{E}_V(\mathbf{v}_k^T\mathbf{v}_k) + \text{const.}, \tag{7.53}$$

$$\ln p(\beta_k) = (a_\beta - 1)\ln \beta_k - \beta_k b_\beta + \text{const.} \tag{7.54}$$

Combining the above, we get the variational posterior

$$\begin{aligned}\ln q_\beta^*(\beta_k) &= \left(a_\beta - 1 + \frac{D_V}{2}\right)\ln \beta_k - \left(b_\beta + \frac{1}{2}\mathbb{E}_V(\mathbf{v}_k^T\mathbf{v}_k)\right) b_\beta + \text{const.} \\ &= \ln \text{Gam}(\beta_k | a_{\beta_k}^*, b_{\beta_k}^*),\end{aligned} \tag{7.55}$$

with the distribution parameters

$$a_{\beta_k}^* = a_\beta + \frac{D_V}{2}, \tag{7.56}$$

$$b_{\beta_k}^* = b_\beta + \frac{1}{2}\mathbb{E}_V(\mathbf{v}_k^T\mathbf{v}_k). \tag{7.57}$$

As the priors on $\mathbf{v}_k$ are similar to the ones on $\mathbf{w}_k$, they cause the same effect: as $b_{\beta_k}^*$ increases proportionally to the expected size $\|\mathbf{v}_k\|^2$, the expectation of the posterior $\mathbb{E}_\beta(\beta_k) = a_{\beta_k}^*/b_{\beta_k}^*$ decreases in proportion to it. This expectation determines the shrinkage on $\mathbf{v}_k$ (see (7.47)), and thus, the strength of the shrinkage prior is reduced if $\mathbf{v}_k$ is expected to have large elements, which is an intuitively sensible procedure.

7.3.6 Latent Variables $q_Z^*(Z)$

To get the variational posterior over the latent variables $\mathbf{Z}$ we need to evaluate (7.24) by the use of (7.15), that is,

$$\ln q_Z^*(\mathbf{Z}) = \mathbb{E}_{W,\tau}(\ln p(\mathbf{Y}|\mathbf{W}, \boldsymbol{\tau}, \mathbf{Z})) + \mathbb{E}_V(\ln p(\mathbf{Z}|\mathbf{V})) + \text{const.} \tag{7.58}$$

The first expression can be evaluated by combining (7.6) and (7.7) to get

$$
\begin{aligned}
&\mathbb{E}_{W,\tau}(\ln p(\mathbf{Y}|\mathbf{W},\boldsymbol{\tau},\mathbf{Z})) \\
&= \sum_n \sum_k z_{nk} \sum_j \mathbb{E}_{W,\tau}(\ln \mathcal{N}(y_{nj}|\mathbf{w}_{kj}^T \mathbf{x}_n, \tau_k^{-1})) \\
&= \sum_n \sum_k z_{nk} \sum_j \left(-\frac{1}{2}\ln 2\pi\right) + \sum_n \sum_k z_{nk} \sum_j \frac{1}{2}\mathbb{E}_\tau(\ln \tau_k) \\
&\quad -\frac{1}{2}\sum_n \sum_k z_{nk} \sum_j \mathbb{E}_{W,\tau}\left(\tau_k (y_{nj} - \mathbf{w}_{kj}^T \mathbf{x}_n)^2\right) \\
&= \frac{D_{\mathcal{Y}}}{2}\sum_n \sum_k z_{nk}\mathbb{E}_\tau(\ln \tau_k) \\
&\quad -\frac{1}{2}\sum_n \sum_k z_{nk} \sum_j \mathbb{E}_{W,\tau}\left(\tau_k (y_{nj} - \mathbf{w}_{kj}^T \mathbf{x}_n)^2\right) + \text{const.},
\end{aligned}
\tag{7.59}
$$

where $\sum_k z_{nk} = 1$ was used. Using (7.12) and (7.11), the second expectation results in

$$
\begin{aligned}
\mathbb{E}_V(\ln p(\mathbf{Z}|\mathbf{V})) &= \sum_n \sum_k z_{nk}\mathbb{E}_V(\ln g_k(\mathbf{x}_n)) \\
&\approx \sum_n \sum_k z_{nk} \ln g_k(\mathbf{x})|_{v_k = v_k^*},
\end{aligned}
\tag{7.60}
$$

where the expectation of $\ln g_k(\mathbf{x}_n)$ was approximated by the logarithm of its maximum a-posteriori estimate, that is, $\ln g_k(\mathbf{x}_n)$ evaluated at $\mathbf{v}_k = \mathbf{v}_k^*$. This approximation was applied as a direct evaluation of the expectation does not yield a closed-form solution. The same approximation was applied by Waterhouse et al. [227, 226] for the MoE model.

Combining the above expectations results in the posterior

$$
\ln q_Z^*(\mathbf{Z}) = \sum_n \sum_k z_{nk} \ln \rho_{nk} + \text{const.},
\tag{7.61}
$$

with

$$
\ln \rho_{nk} = \ln g_k(\mathbf{x}_n)|_{v_k = v_k^*} + \frac{D_{\mathcal{Y}}}{2}\mathbb{E}_\tau(\ln \tau_k) - \frac{1}{2}\sum_j \mathbb{E}_{W,\tau}\left(\tau_k (y_{nj} - \mathbf{w}_{kj}^T \mathbf{x}_n)^2\right).
\tag{7.62}
$$

Without the logarithm, the posterior becomes $q_Z^*(\mathbf{Z}) \propto \prod_n \prod_k \rho_{nk}^{z_{nk}}$, and thus, under the constraint $\sum_k z_{nk} = 1$, we get

$$
q_Z^*(\mathbf{Z}) = \prod_n \prod_k r_{nk}^{z_{nk}}, \qquad \text{with} \quad r_{nk} = \frac{\rho_{nk}}{\sum_j \rho_{nj}} = \mathbb{E}_Z(z_{nk}).
\tag{7.63}
$$

As for all posteriors, the variational posterior for the latent variables has the same distribution form as its prior (7.12).

Note that r_{nk} gives the responsibility that is assigned to classifier k for modelling observation n, and is proportional to ρ_{nk} (7.62). Thus, the responsibilities are on one hand proportional to the current mixing weights $g_k(\mathbf{x})$, and on the other hand are higher for low-variance classifiers (note that τ_k is the inverse variance of classifier k) that feature a low expected squared prediction error $(y_{nj} - \mathbf{w}_{kj}^T \mathbf{x}_n)^2$ for the associated observation. Overall, the responsibilities are distributed such that the observations are modelled by the classifiers that are best at modelling them.

7.3.7 Required Moments of the Variational Posterior

Some of the variational distribution parameters require evaluation of the moments of one or the other random variable in our probabilistic model. In this section, these moments and the ones required at a later stage are evaluated. Throughout this section we use $\mathbb{E}_x(\mathbf{x}) = \mathbf{x}^*$ and $\text{cov}_x(\mathbf{x}, \mathbf{x}) = \mathbf{\Lambda}^{-1}$, where $\mathbf{x} \sim \mathcal{N}(\mathbf{x}^*, \mathbf{\Lambda}^{-1})$ is a random vector that is distributed according to a multivariate Gaussian with mean $\mathbf{x}^*$ and covariance matrix $\mathbf{\Lambda}^{-1}$.

Given that we have a random variable $X \sim \text{Gam}(a, b)$, then its expectation is $\mathbb{E}_X(X) = a/b$, and the expectation of its logarithm is $\mathbb{E}_X(\ln X) = \psi(a) - \ln b$, where $\psi(x) = \frac{x}{\text{d}x} \ln \Gamma(x)$ is the digamma function [19]. Thus the following are the posterior moments for $q_\alpha^*(\alpha_k)$, $q_\beta^*(\beta_k)$, and $q_\tau^*(\tau_k)$:

$$\mathbb{E}_\alpha(\alpha_k) = \frac{a_{\alpha_k}^*}{b_{\alpha_k}^*}, \tag{7.64}$$

$$\mathbb{E}_\alpha(\ln \alpha_k) = \psi(a_{\alpha_k}^*) - \ln b_{\alpha_k}, \tag{7.65}$$

$$\mathbb{E}_\beta(\beta_k) = \frac{a_{\beta_k}^*}{b_{\beta_k}^*}, \tag{7.66}$$

$$\mathbb{E}_\beta(\ln \beta_k) = \psi(a_{\beta_k}^*) - \ln b_{\beta_k}^*, \tag{7.67}$$

$$\mathbb{E}_\tau(\tau_k) = \frac{a_{\tau_k}^*}{b_{\tau_k}^*}, \tag{7.68}$$

$$\mathbb{E}_\tau(\ln \tau_k) = \psi(a_{\tau_k}^*) - \ln b_{\tau_k}^*. \tag{7.69}$$

To get the moments of $q_{W,\tau}^*(\mathbf{W}_k, \tau_k)$ and $q_V^*(\mathbf{v}_k)$, we can use $\text{var}(X) = \mathbb{E}(X^2) - \mathbb{E}(X)^2$, and thus, $\mathbb{E}(X^2) = \text{var}(X) + \mathbb{E}(X)^2$, to get

$$\begin{aligned}
\mathbb{E}(\mathbf{x}^T \mathbf{x}) &= \sum_i \mathbb{E}(x_i^2) \\
&= \sum_i \text{var}(x_i) + \sum_i \mathbb{E}(x_i)^2 \\
&= \text{Tr}(\text{cov}(\mathbf{x}, \mathbf{x})) + \mathbb{E}(\mathbf{x})^T \mathbb{E}(\mathbf{x}),
\end{aligned}$$

and similarly,

$$\mathbb{E}(\mathbf{x}\mathbf{x}^T) = \text{cov}(\mathbf{x}, \mathbf{x}) + \mathbb{E}(\mathbf{x})\mathbb{E}(\mathbf{x})^T,$$

where X is a random variable, and $\mathbf{x} = (x_i)^T$ is a random vector. Hence, as by (7.51), $q_V^*(\mathbf{V})$ is a multivariate Gaussian with covariance matrix $\mathbf{\Lambda}_V^{*\,-1}$, we get

$$\mathbb{E}_V(\mathbf{v}_k^T \mathbf{v}_k) = \mathrm{Tr}\left((\mathbf{\Lambda}_V^{*\,-1})_{kk} \right) + \mathbf{v}_k^{*T} \mathbf{v}_k^*, \tag{7.70}$$

where $(\mathbf{\Lambda}_V^{*\,-1})_{kk}$ denotes the kth $D_V \times D_V$ block element along the diagonal of $\mathbf{\Lambda}_V^{*\,-1}$.

Getting the moments of $q_{W,\tau}^*(\mathbf{W}_k, \tau_k)$ requires a bit more work. Let us first consider $\mathbb{E}_{W,\tau}(\tau_k \mathbf{w}_{kj})$, which by (7.29) and the previously evaluated moments gives

$$\begin{aligned}
&\mathbb{E}_{W,\tau}(\tau_k \mathbf{w}_{kj}) \\
&\quad = \int \tau_k \mathrm{Gam}(\tau_k | a_{\tau_k}^*, b_{\tau_k}^*) \left(\int \mathbf{w}_{kj} \mathcal{N}(\mathbf{w}_{kj} | \mathbf{w}_{kj}^*, (\tau_k \mathbf{\Lambda}_k^*)^{-1}) \mathrm{d}\mathbf{w}_{kj} \right) \mathrm{d}\tau_k \\
&\quad = \mathbf{w}_{kj}^* \int \tau_k \mathrm{Gam}(\tau_k | a_{\tau_k}^*, b_{\tau_k}^*) \mathrm{d}\tau_k \\
&\quad = \frac{a_{\tau_k}^*}{b_{\tau_k}^*} \mathbf{w}_{kj}^*.
\end{aligned} \tag{7.71}$$

For $\mathbb{E}_{W,\tau}(\tau_k \mathbf{w}_{kj}^T \mathbf{w}_{kj})$ we get

$$\begin{aligned}
&\mathbb{E}_{W,\tau}(\tau_k \mathbf{w}_{kj}^T \mathbf{w}_{kj}) \\
&\quad = \int \tau_k \mathrm{Gam}(\tau_k | a_{\tau_k}^*, b_{\tau_k}^*) \left(\int \mathbf{w}_{kj}^T \mathbf{w}_{kj} \mathcal{N}(\mathbf{w}_{kj} | \mathbf{w}_{kj}^*, (\tau_k \mathbf{\Lambda}_k^*)^{-1}) \mathrm{d}\mathbf{w}_{kj} \right) \mathrm{d}\tau_k \\
&\quad = \int \tau_k \mathrm{Gam}(\tau_k | a_{\tau_k}^*, b_{\tau_k}^*) \mathbb{E}_W(\mathbf{w}_{kj}^T \mathbf{w}_{kj}) \mathrm{d}\tau_k \\
&\quad = \mathbf{w}_{kj}^{*\,T} \mathbf{w}_{kj}^* \mathbb{E}_\tau(\tau_k) + \mathrm{Tr}(\mathbf{\Lambda}_k^{*\,-1}) \\
&\quad = \frac{a_{\tau_k}^*}{b_{\tau_k}^*} \mathbf{w}_{kj}^{*\,T} \mathbf{w}_{kj}^* + \mathrm{Tr}(\mathbf{\Lambda}_k^{*\,-1}).
\end{aligned} \tag{7.72}$$

$\mathbb{E}_{W,\tau}(\tau_k \mathbf{w}_{kj} \mathbf{w}_{kj}^T)$ can be derived in a similar way, and results in

$$\mathbb{E}_{W,\tau}(\tau_k \mathbf{w}_{kj} \mathbf{w}_{kj}^T) = \frac{a_{\tau_k}^*}{b_{\tau_k}^*} \mathbf{w}_{kj}^* \mathbf{w}_{kj}^{*\,T} + \mathbf{\Lambda}_k^{*\,-1}. \tag{7.73}$$

The last required moment is $\mathbb{E}_{W,\tau}(\tau_k (y_{nj} - \mathbf{w}_{kj}^T \mathbf{x}_n)^2)$, which we get by binomial expansion and substituting the previously evaluated moments, to get

$$\begin{aligned}
&\mathbb{E}_{W,\tau}(\tau_k (y_{nj} - \mathbf{w}_{kj}^T \mathbf{x}_n)^2) \\
&\quad = \mathbb{E}_\tau(\tau_k) y_{nj}^2 - 2\mathbb{E}_{W,\tau}(\tau_k \mathbf{w}_{kj})^T \mathbf{x}_n y_{nj} + \mathbf{x}_n^T \mathbb{E}_{W,\tau}(\tau_k \mathbf{w}_{kj} \mathbf{w}_{kj}^T) \mathbf{x}_n \\
&\quad = \frac{a_{\tau_k}^*}{b_{\tau_k}^*} (y_{nj} - \mathbf{w}_{kj}^{*\,T} \mathbf{x}_n)^2 + \mathbf{x}_n^T \mathbf{\Lambda}_k^{*\,-1} \mathbf{x}_n.
\end{aligned} \tag{7.74}$$

Now we have all the required expressions to compute the parameters of the variational posterior density.

7.3.8 The Variational Bound $\mathcal{L}(q)$

We are most interested in finding the value for $\mathcal{L}(q)$ by (7.21), as it provides us with an approximated lower bound on the logarithm of the model evidence $\ln p(\mathbf{Y})$, and is the actual expression that is to be maximised. Evaluating (7.21) by using the distribution decomposition according to (7.15), the variational bound is given by

$$\begin{aligned}\mathcal{L}(q) &= \int q(\mathbf{U}) \ln \frac{p(\mathbf{Y},\mathbf{U})}{q(\mathbf{U})} \mathrm{d}\mathbf{U} \\ &= \mathbb{E}_{W,\tau,\alpha,Z,V,\beta}(\ln p(\mathbf{Y},\mathbf{W},\boldsymbol{\tau},\mathbf{Z},\mathbf{V},\boldsymbol{\beta})) \\ &\qquad -\mathbb{E}_{W,\tau,\alpha,Z,V,\beta}(\ln q(\mathbf{W},\boldsymbol{\tau},\boldsymbol{\alpha},\mathbf{Z},\mathbf{V},\boldsymbol{\beta})) \\ &= \mathbb{E}_{W,\tau,Z}(\ln p(\mathbf{Y}|\mathbf{W},\boldsymbol{\tau},\mathbf{Z})) + \mathbb{E}_{W,\tau,\alpha}(\ln p(\mathbf{W},\boldsymbol{\tau}|\boldsymbol{\alpha})) + \mathbb{E}_{\alpha}(\ln p(\boldsymbol{\alpha})) \\ &\quad +\mathbb{E}_{Z,V}(\ln p(\mathbf{Z}|\mathbf{V})) + \mathbb{E}_{V,\beta}(\ln p(\mathbf{V}|\boldsymbol{\beta})) + \mathbb{E}_{\beta}(\ln p(\boldsymbol{\beta})) \\ &\quad -\mathbb{E}_{W,\tau}(\ln q(\mathbf{W},\boldsymbol{\tau})) - \mathbb{E}_{\alpha}(\ln q(\boldsymbol{\alpha})) - \mathbb{E}_{Z}(\ln q(\mathbf{Z})) \\ &\quad -\mathbb{E}_{V}(\ln q(\mathbf{V})) - \mathbb{E}_{\beta}(\ln q(\boldsymbol{\beta})),\end{aligned} \tag{7.75}$$

where all expectations are taken with respect to the variational distribution q. These are evaluated one by one, using the previously derived moments of the variational posteriors.

To derive $\mathbb{E}_{W,\tau,Z}(\ln p(\mathbf{Y}|\mathbf{W},\boldsymbol{\tau},\mathbf{Z}))$, we use (7.6) and (7.7) to get

$$\begin{aligned}&\mathbb{E}_{W,\tau,Z}(\ln p(\mathbf{Y}|\mathbf{W},\boldsymbol{\tau})) \\ &= \sum_n \sum_k \mathbb{E}_Z(z_{nk}) \sum_j \mathbb{E}_{W,\tau}(\ln \mathcal{N}(y_{nj}|\mathbf{w}_{kj}^T\mathbf{x}_n, \tau_k^{-1})) \\ &= \sum_n \sum_k r_{nk} \sum_j \left(\frac{1}{2}\mathbb{E}_\tau(\ln \tau_k) - \frac{1}{2}\ln 2\pi - \frac{1}{2}\mathbb{E}_{W,\tau}(\tau_k(y_{nj} - \mathbf{w}_{kj}^T\mathbf{x}_n)^2)\right) \\ &= \sum_k \left(\frac{D_{\mathcal{Y}}}{2}\left(\psi(a_{\tau_k}^*) - \ln b_{\tau_k}^* - \ln 2\pi\right)\sum_n r_{nk}\right. \\ &\qquad \left. -\frac{1}{2}\sum_n r_{nk}\sum_j\left(\frac{a_{\tau_k}^*}{b_{\tau_k}^*}(y_{nj} - \mathbf{w}_{kj}^{*T}\mathbf{x}_n)^2 + \mathbf{x}_n^T\mathbf{\Lambda}_k^{*-1}\mathbf{x}_n\right)\right) \\ &= \sum_k \left(\frac{D_{\mathcal{Y}}}{2}\left(\psi(a_{\tau_k}^*) - \ln b_{\tau_k}^* - \ln 2\pi\right)\sum_n r_{nk}\right. \\ &\qquad \left. -\frac{1}{2}\sum_n r_{nk}\left(\frac{a_{\tau_k}^*}{b_{\tau_k}^*}\|\mathbf{y}_n - \mathbf{W}_k^*\mathbf{x}_n\|^2 + D_{\mathcal{Y}}\mathbf{x}_n^T\mathbf{\Lambda}_k^{*-1}\mathbf{x}_n\right)\right).\end{aligned} \tag{7.76}$$

The classifier model parameters expectation $\mathbb{E}_{W,\tau,\alpha}(\ln p(\mathbf{W},\boldsymbol{\tau}|\boldsymbol{\alpha}))$ can be derived by using (7.7) and (7.16), and is given by

$$\begin{aligned}&\mathbb{E}_{W,\tau,\alpha}(\ln p(\mathbf{W},\boldsymbol{\tau}|\boldsymbol{\alpha})) \\ &= \sum_k \sum_j \left(\mathbb{E}_{W,\tau,\alpha}(\ln \mathcal{N}(\mathbf{w}_{kj}|\mathbf{0},(\alpha_k\tau_k)^{-1}\mathbf{I})) + \mathbb{E}_\tau(\ln \mathrm{Gam}(\tau_k|a_\tau,b_\tau))\right).\end{aligned} \tag{7.77}$$

Expanding for the densities and substituting the variational moments results in

$$\begin{aligned}
&\mathbb{E}_{W,\tau,\alpha}(\ln p(\mathbf{W},\boldsymbol{\tau}|\boldsymbol{\alpha})) \\
&= \sum_k \left(\frac{D_{\mathcal{X}} D_{\mathcal{Y}}}{2} \left(\psi(a^*_{\alpha_k}) - \ln b^*_{\alpha_k} + \psi(a^*_{\tau_k}) - \ln b^*_{\tau_k} - \ln 2\pi\right) \right. \\
&\quad -\frac{1}{2}\frac{a^*_{\alpha_k}}{b^*_{\alpha_k}} \left(\frac{a^*_{\tau_k}}{b^*_{\tau_k}} \sum_j {\mathbf{w}^*_{kj}}^T \mathbf{w}^*_{kj} + D_{\mathcal{Y}} \mathrm{Tr}({\boldsymbol{\Lambda}^*_k}^{-1}) \right) \qquad (7.78) \\
&\quad \left. +D_{\mathcal{Y}} \left(-\ln\Gamma(a_\tau) + a_\tau \ln b_\tau + (a_\tau - 1)(\psi(a^*_{\tau_k}) - \ln b^*_{\tau_k}) - b_\tau \frac{a^*_{\tau_k}}{b^*_{\tau_k}} \right) \right).
\end{aligned}$$

The negative entropy $\mathbb{E}_{W,\tau}(\ln q(\mathbf{W},\boldsymbol{\tau}))$ of $\{\mathbf{W},\boldsymbol{\tau}\}$ is based on (7.29) and results in

$$\begin{aligned}
&\mathbb{E}_{W,\tau}(\ln q(\mathbf{W},\boldsymbol{\tau})) \\
&= \mathbb{E}_{W,\tau}\left(\sum_k \sum_j \ln \mathcal{N}(\mathbf{w}_{kj}|\mathbf{w}^*_{kj}, (\tau_k \boldsymbol{\Lambda}^*_k)^{-1}) \mathrm{Gam}(\tau_k | a^*_{\tau_k}, b^*_{\tau_k}) \right) \\
&= \sum_k \sum_j \left(\frac{D_{\mathcal{X}}}{2} \mathbb{E}_\tau(\ln \tau_k) - \frac{D_{\mathcal{X}}}{2} \ln 2\pi + \frac{1}{2} \ln |\boldsymbol{\Lambda}^*_k| \right. \\
&\quad + \mathbb{E}_{W,\tau}\left(-\frac{\tau}{2} (\mathbf{w}_{kj} - \mathbf{w}_{kj})^2 \boldsymbol{\Lambda}^*_k (\mathbf{w}_{kj} - \mathbf{w}_{kj}) \right) - \ln\Gamma(a^*_{\tau_k}) \\
&\quad \left. + a^*_{\tau_k} \ln b^*_{\tau_k} + (a^*_{\tau_k} - 1)\mathbb{E}_\tau(\ln \tau_k) - b^*_{\tau_k} \mathbb{E}_\tau(\tau_k) \right) \\
&= D_{\mathcal{Y}} \sum_k \left(\left(a^*_{\tau_k} - 1 + \frac{D_{\mathcal{X}}}{2} \right) (\psi(a^*_{\tau_k}) - \ln b^*_{\tau_k}) - \frac{D_{\mathcal{X}}}{2} (\ln 2\pi + 1) \right. \\
&\quad \left. + \frac{1}{2} \ln |\boldsymbol{\Lambda}^*_k| - \ln\Gamma(a^*_{\tau_k}) + a^*_{\tau_k} \ln b^*_{\tau_k} - a^*_{\tau_k} \right), \qquad (7.79)
\end{aligned}$$

where the previously evaluated variational moments and

$$\mathbb{E}_{W,\tau}\left(-\frac{\tau}{2} (\mathbf{w}_{kj} - \mathbf{w}_{kj})^2 \boldsymbol{\Lambda}^*_k (\mathbf{w}_{kj} - \mathbf{w}_{kj}) \right) = -\frac{1}{2} D_{\mathcal{X}} \qquad (7.80)$$

was used.

We derive the expression $\mathbb{E}_\alpha(\ln p(\boldsymbol{\alpha})) - \mathbb{E}_\alpha(\ln q(\boldsymbol{\alpha}))$ in combination, as that allows for some simplification. Starting with $\mathbb{E}_\alpha(\ln p(\boldsymbol{\alpha}))$, we get from (7.17) and (7.9), by expanding the densities and substituting the variational moments,

$$\begin{aligned}
&\mathbb{E}_\alpha(\ln p(\boldsymbol{\alpha})) \qquad (7.81) \\
&= \sum_k \left(-\ln\Gamma(a_\alpha) + a_\alpha \ln b_\alpha + (a_\alpha - 1)(\psi(a^*_{\alpha_k}) - \ln b^*_{\alpha_k}) - b_\alpha \frac{a^*_{\alpha_k}}{b^*_{\alpha_k}} \right).
\end{aligned}$$

The expression for $\mathbb{E}_\alpha(\ln q(\boldsymbol{\alpha}))$ can be derived by observing that $-\mathbb{E}_\alpha(\ln q(\alpha_k))$ is the entropy of $q^*_\alpha(\alpha_k)$. Thus, using $q^*_\alpha(\boldsymbol{\alpha}) = \prod_k q^*_\alpha(\alpha_k)$, substituting (7.39) for $q^*_\alpha(\alpha_k)$, and applying the entropy of the Gamma distribution [19], we get

$$\mathbb{E}_\alpha(\ln q(\boldsymbol{\alpha})) = -\sum_k \left(\ln \Gamma(a^*_{\alpha_k}) - (a^*_{\alpha_k} - 1)\psi(a^*_{\alpha_k}) - \ln b^*_{\alpha_k} + a^*_{\alpha_k}\right) \tag{7.82}$$

Combining the above expressions and removing the terms that cancel out results in

$$\mathbb{E}_\alpha(\ln p(\boldsymbol{\alpha})) - \mathbb{E}_\alpha(\ln q(\boldsymbol{\alpha})) = \sum_k \Big(-\ln\Gamma(a_\alpha) + a_\alpha \ln b_\alpha + (a_\alpha - a^*_{\alpha_k})\psi(a^*_{\alpha_k}) \\ -a_\alpha \ln b^*_{\alpha_k} - b_\alpha \frac{a^*_{\alpha_k}}{b^*_{\alpha_k}} + \ln\Gamma(a^*_{\alpha_k}) + a^*_{\alpha_k}\Big). \tag{7.83}$$

The expression $\mathbb{E}_{Z,V}(\ln p(\mathbf{Z}|\mathbf{V})) - \mathbb{E}_Z(\ln q(\mathbf{Z}))$ is also derived in combination by using (7.12), (7.11) and (7.63), from which we get

$$\mathbb{E}_{Z,V}(\ln p(\mathbf{Z}|\mathbf{V})) - \mathbb{E}_Z(\ln q(\mathbf{Z})) = \sum_n \sum_k r_{nk} \ln \frac{g_k(\mathbf{x})|_{v_k = v^*_k}}{r_{nk}}, \tag{7.84}$$

where we have, as previously, approximated $\mathbb{E}_V(\ln g_k(\mathbf{x}_n))$ by $\ln g_k(\mathbf{x}_n)|_{v_k = v^*_k}$.

The derivation to get $\mathbb{E}_{V,\beta}(\ln p(\mathbf{V}|\boldsymbol{\beta}))$ is again based on simple expansion of the distribution given by (7.18) and (7.13), and substituting the variational moments, which results in

$$\mathbb{E}_{V,\beta}(\ln p(\mathbf{V}|\boldsymbol{\beta})) \tag{7.85}$$
$$= \sum_k \left(\frac{D_V}{2}\left(\psi(a^*_{\beta_k}) - \ln b^*_{\beta_k} - \ln 2\pi\right) - \frac{1}{2}\frac{a^*_{\beta_k}}{b^*_{\beta_k}}\left(\mathbf{v}^{*T}_k \mathbf{v}^*_k + \mathrm{Tr}((\boldsymbol{\Lambda}^{*\,-1}_V)_{kk})\right)\right).$$

We get $\mathbb{E}_V(\ln q(\mathbf{V}))$ by observing that it is the negative entropy of the Gaussian (7.51), and thus evaluates to [19]

$$\mathbb{E}_V(\ln q(\mathbf{V})) = -\left(\frac{1}{2}\ln|\boldsymbol{\Lambda}^{*\,-1}_V| + \frac{KD_V}{2}(1 + \ln 2\pi)\right). \tag{7.86}$$

As the priors on β_k are of the same distribution form as the ones on α_k, the expectations of their log-density results in a similar expression as (7.65) and is given by

$$\mathbb{E}_\beta(\ln p(\boldsymbol{\beta})) - \mathbb{E}_\beta(\ln q(\boldsymbol{\beta})) = \sum_k \Big(-\ln\Gamma(a_\beta) + a_\beta \ln b_\beta + (a_\beta - a^*_{\beta_k})\psi(a^*_{\beta_k}) \\ -a_\beta \ln b^*_{\beta_k} - b_\beta \frac{a^*_{\beta_k}}{b^*_{\beta_k}} + \ln\Gamma(a^*_{\beta_k}) + a^*_{\beta_k}\Big). \tag{7.87}$$

This completes the evaluation of the expectations required to compute the variational bound (7.75).

To simplify the computation of the variational bound, we define

$$\mathcal{L}_k(q) = \mathbb{E}_{W,\tau,Z}(\ln p(\mathbf{Y}|\mathbf{W}_k, \tau_k, \mathbf{z}_k)) + \mathbb{E}_{W,\tau,\alpha}(\ln p(\mathbf{W}_k, \tau_k|\alpha_k)) \\ +\mathbb{E}_{\alpha}(\ln p(\alpha_k)) - \mathbb{E}_{W,\tau}(\ln q(\mathbf{W}_k, \tau_k)) - \mathbb{E}_{\alpha}(\ln q(\alpha_k)), \quad (7.88)$$

which can be evaluated separately for each classifier by observing that all expectations except for $\mathbb{E}_V(\ln q(\mathbf{V}))$ are sums whose components can be evaluated independently for each classifier. Furthermore, $\mathcal{L}_k(q)$ can be simplified by using the relations

$$\frac{D_{\mathcal{X}} D_{\mathcal{Y}}}{2} = a^*_{\alpha_k} - a_\alpha, \quad (7.89)$$

$$\frac{1}{2}\left(\frac{a^*_{\tau_k}}{b^*_{\tau_k}} \sum_j \mathbf{w}^{*\,T}_{kj} \mathbf{w}^*_{kj} + D_{\mathcal{Y}} \mathrm{Tr}(\mathbf{\Lambda}^{*\,-1}_k)\right) = b^*_{\alpha_k} - b_\alpha, \quad (7.90)$$

which results from (7.40) and (7.41). Thus, the final, simplified expression for $\mathcal{L}_k(q)$ becomes

$$\mathcal{L}_k(q) = \frac{D_{\mathcal{Y}}}{2}\left(\psi(a^*_{\tau_k}) - \ln b^*_{\tau_k} - \ln 2\pi\right) \sum_n r_{nk} + \frac{D_{\mathcal{X}} D_{\mathcal{Y}}}{2} \\ -\frac{1}{2}\sum_n r_{nk}\left(\frac{a^*_{\tau_k}}{b^*_{\tau_k}} \|\mathbf{y}_n - \mathbf{W}^*_k \mathbf{x}_n\|^2 + D_{\mathcal{Y}} \mathbf{x}_n^T \mathbf{\Lambda}^{*\,-1}_k \mathbf{x}_n\right) \\ -\ln\Gamma(a_\alpha) + a_\alpha \ln b_\alpha + \ln\Gamma(a^*_{\alpha_k}) - a^*_{\alpha_k} \ln b^*_{\alpha_k} + \frac{D_{\mathcal{Y}}}{2} \ln|\mathbf{\Lambda}^{*\,-1}_k| \\ +D_{\mathcal{Y}}\Big(-\ln\Gamma(a_\tau) + a_\tau \ln b_\tau + (a_\tau - a^*_{\tau_k})\psi(a^*_{\tau_k}) - a_\tau \ln b^*_{\tau_k} - b_\tau \frac{a^*_{\tau_k}}{b^*_{\tau_k}} \\ + \ln\Gamma(a^*_{\tau_k}) + a^*_{\tau_k}\Big). \quad (7.91)$$

All leftover terms from (7.75) are assigned to the mixing model, which results in

$$\mathcal{L}_M(q) = \mathbb{E}_{Z,V}(\ln p(\mathbf{Z}|\mathbf{V})) + \mathbb{E}_{V,\beta}(\ln p(\mathbf{V}|\boldsymbol{\beta})) + \mathbb{E}_{\beta}(\ln p(\boldsymbol{\beta})) \\ -\mathbb{E}_Z(\ln q(\mathbf{Z})) - \mathbb{E}_V(\ln q(\mathbf{V})) - \mathbb{E}_\beta(\ln q(\boldsymbol{\beta})). \quad (7.92)$$

We can again derive a simplified expression for $\mathcal{L}_M(q)$ by using the relations

$$\frac{D_V}{2} = a^*_{\beta_k} - a_\beta, \quad (7.93)$$

$$\frac{1}{2}\left(\mathrm{Tr}\left((\mathbf{\Lambda}^{*\,-1}_V)_{kk}\right) + \mathbf{v}^{*\,T}_k \mathbf{v}^*_k\right) = b^*_{\beta_k} - b_\beta, \quad (7.94)$$

which result from (7.56) and (7.57). Overall, this leads to the final simplified expression

$$\mathcal{L}_M(q) = \sum_k \left(-\ln\Gamma(a_\beta) + a_\beta \ln b_\beta + \ln\Gamma(a^*_{\beta_k}) - a^*_{\beta_k} \ln b^*_{\beta_k}\right) \quad (7.95) \\ +\sum_n \sum_k r_{nk}\left(\ln g_k(\mathbf{x}_n)|_{v_k = v^*_k} - \ln r_{nk}\right) + \frac{1}{2}\ln|\mathbf{\Lambda}^{*\,-1}_V| + \frac{K D_V}{2}.$$

The get the variational bound of the whole model structure, and with it the lower bound on the logarithm of the model evidence $\ln p(\mathbf{Y})$, we need to compute

$$\mathcal{L}(q) = \mathcal{L}_M(q) + \sum_k \mathcal{L}_k(q), \tag{7.96}$$

where $\mathcal{L}_k(q)$ and $\mathcal{L}_M(q)$ are given by (7.91) and (7.95) respectively.

Training the model means maximising $\mathcal{L}(q)$ (7.96) with respect to its parameters $\{\mathbf{W}_k^*, \mathbf{\Lambda}_k^*, a_{\tau_k}^*, b_{\tau_k}^*, a_{\alpha_k}^*, b_{\alpha_k}^*, \mathbf{V}^*, \mathbf{\Lambda}_V^*, a_{\beta_k}^*, b_{\beta_k}^*\}$. In fact, deriving the maximum of $\mathcal{L}(q)$ with respect to each of these parameters separately while keeping the others constant results in the variational update equations that were derived in the previous sections [19].

7.3.9 Independent Classifier Training

As we can see from (7.91), we need to know the responsibilities $\{r_{nk}\}$ to train each of the classifiers. The mixing model, on the other hand, relies on the goodness-of-fit of the classifiers, as embedded in g_k in (7.95). Therefore, classifiers and mixing model need to be trained in combination to maximise (7.96). Taking this approach, however, introduces local optima in the training process, as already discussed for the non-Bayesian MoE model in Sect. 4.1.5. Such local optima make evaluating the model evidence for a single model structure too costly to perform efficient model structure search, and so the training process needs to be modified to remove these local optima. Following the same approach as in Sect. 4.4, we train the classifiers independently of the mixing model.

More specifically, the classifiers are fully trained on all observations that they match, independently of other classifiers, and then combined by the mixing model. Formally, this is achieved by replacing the responsibilities r_{nk} by the matching functions $m_k(\mathbf{x}_n)$.

The only required modification to the variational update equations is to change the classifier model updates from (7.30) – (7.33) to

$$\mathbf{\Lambda}_k^* = \mathbb{E}_\alpha(\alpha_k)\mathbf{I} + \sum_n m_k(\mathbf{x}_n)\mathbf{x}_n\mathbf{x}_n^T, \tag{7.97}$$

$$\mathbf{w}_{kj}^* = \mathbf{\Lambda}_k^{*-1} \sum_n m_k(\mathbf{x}_n)\mathbf{x}_n y_{nj}, \tag{7.98}$$

$$a_{\tau_k}^* = a_\tau + \frac{1}{2}\sum_n m_k(\mathbf{x}_n), \tag{7.99}$$

$$b_{\tau_k}^* = b_\tau + \frac{1}{2D_\mathcal{Y}}\left(\sum_j \left(\sum_n m_k(\mathbf{x}_n) y_{nj}^2 - \mathbf{w}_{kj}^{*\,T}\mathbf{\Lambda}_k^*\mathbf{w}_{kj}^*\right)\right). \tag{7.100}$$

Thus, we are now effectively finding a $\mathbf{w}_{kj}$ that minimises

$$\|\mathbf{X}\mathbf{w}_{kj} - \mathbf{y}_j\|_{M_k}^2 + \mathbb{E}_\alpha(\alpha_k)\|\mathbf{w}_{kj}\|^2, \tag{7.101}$$

as we have already discussed extensively in Sect. 5.3.5. The weight prior update (7.40) and (7.41), as well as all mixing model update equations remain unchanged.

Even though all r_{nk}'s in the classifier update equations are replaced with $m_k(\mathbf{x}_n)$'s, the classifier-specific component $\mathcal{L}_k(q)$ (7.91) remains unchanged. This is because the responsibilities enter $\mathcal{L}_k(q)$ through the expectation $\mathbb{E}_{W,\tau,Z}(\ln p(\mathbf{Y}|\mathbf{W},\boldsymbol{\tau},\mathbf{Z}))$, which is based on (7.6) and (7.7). Note that (7.6) combines the classifier models to form a global model, and is thus conceptually part of the mixing model rather than the classifier model. Thus, the r_{nk}'s in $\mathcal{L}_k(q)$ specify how classifier k contributes to the global model and remain unchanged.

Consequently, the variational posteriors for the classifiers only maximise the variational bound $\mathcal{L}(q)$ if we have $r_{nk} = m_k(\mathbf{x}_n)$ for all n, k. In all other cases, the variational bound remains below the one that we could achieve by training the classifiers according to their responsibilities. This effect is analogous to the reduced likelihood as discussed in Sect. 4.4.5. In cases where we only have one classifier per observation, we automatically have $r_{nk} = m_k(\mathbf{x}_n)$, and thus making classifier training independent only affects areas where several classifiers match the same input. Nonetheless, the model structure selection criterion is proportional to the value of the variational bound and therefore most likely prefers model structures that do not assign multiple classifiers to a single observation.

7.3.10 How to Get $p(\mathcal{M}|\mathcal{D})$ for Some $\mathcal{M}$

Recall that rather than finding the model parameters $\boldsymbol{\theta}$ for a fixed model structure, the aim is to find the model structure $\mathcal{M}$ that maximises $p(\mathcal{M}|\mathcal{D})$. This, however, cannot be done without also training the model.

Variational Bayesian inference yields a lower bound on $\ln p(\mathcal{D}|\mathcal{M})$ that is given by maximising the variational bound $\mathcal{L}(q)$. As $p(\mathcal{M}|\mathcal{D})$ results from $p(\mathcal{D}|\mathcal{M})$ by (7.3), $p(\mathcal{M}|\mathcal{D})$ can be approximated for a given model structure $\mathcal{M}$ by maximising $\mathcal{L}(q)$. Using the assumptions of factorial distributions, $\mathcal{L}(q)$ is maximised with respect to a group of hidden variables while keeping the other ones fixed by computing (7.24). Therefore, by iteratively updating the distribution parameters of $q^*_{W,\tau}(\mathbf{W},\boldsymbol{\tau})$, $q^*_\alpha(\boldsymbol{\alpha})$, $q^*_V(\mathbf{V})$, $q^*_\beta(\boldsymbol{\beta})$, and $q^*_Z(\mathbf{Z})$ in a sequential fashion, the variational bound increases monotonically until it reaches a maximum [26]. Independent classifier training simplifies this procedure by making the update of $q^*_{W,\tau}(\mathbf{W},\boldsymbol{\tau})$ and $q^*_\alpha(\boldsymbol{\alpha})$ independent of the update of the other variational densities. Firstly, the classifier are trained independently of each other and the mixing model, and secondly, the mixing model parameters are updated accordingly.

To summarise, finding $p(\mathcal{M}|\mathcal{D})$ for a given model structure can be done with the following steps:

1. Train the classifiers by iteratively updating the distribution parameters of $q^*_{W,\tau}(\mathbf{W},\tau)$ and $q^*_\alpha(\boldsymbol{\alpha})$ until convergence, for each classifier separately.
2. Train the mixing model by iteratively updating the distribution parameters of $q^*_V(\mathbf{V})$, $q^*_\beta(\boldsymbol{\beta})$, and $q^*_Z(\mathbf{Z})$ until convergence.
3. Compute the variational bound $\mathcal{L}(q)$ by (7.96).

4. $p(\mathcal{M}|\mathcal{D})$ is then given by (7.3), where $\ln p(\mathcal{D}|\mathcal{M})$ is replaced by its approximation $\mathcal{L}(q)$.

Appropriate convergence criteria are introduced in the next chapter.

7.4 Predictive Distribution

An additional bonus of a probabilistic basis for LCS is that it provides predictive distributions rather than simple point estimates. This gives additional information about the certainty of the prediction and the specification of confidence interval. Here, the predictive density for the Bayesian LCS model for regression is derived.

The question we are answering is: in the light of all available data, how likely are certain output values for a new input? This question is approached formally by providing the predictive density $p(\mathbf{y}'|\mathbf{x}', \mathcal{D}) \equiv p(\mathbf{y}'|\mathbf{x}', \mathbf{X}, \mathbf{Y})$, where $\mathbf{x}'$ is the new known input vector, and $\mathbf{y}'$ its associated unknown output vector, and all densities are, as before, implicitly conditional on the current model structure $\mathcal{M}$.

7.4.1 Deriving $p(y'|x', \mathcal{D})$

We get an expression for $p(\mathbf{y}'|\mathbf{x}', \mathcal{D})$ by using the relation

$$\begin{aligned}
&p(\mathbf{y}'|\mathbf{x}', \mathbf{X}, \mathbf{Y}) \qquad (7.102)\\
&= \sum_{z'} \iiint p(\mathbf{y}', \mathbf{z}', \mathbf{W}, \boldsymbol{\tau}, \mathbf{V}|\mathbf{x}', \mathbf{X}, \mathbf{Y}) \mathrm{d}\mathbf{W} \mathrm{d}\boldsymbol{\tau} \mathrm{d}\mathbf{V}\\
&= \sum_{z'} \iiint p(\mathbf{y}'|\mathbf{x}', \mathbf{z}', \mathbf{W}, \boldsymbol{\tau}) p(\mathbf{z}'|\mathbf{x}', \mathbf{V}) p(\mathbf{W}, \boldsymbol{\tau}, \mathbf{V}|\mathbf{X}, \mathbf{Y}) \mathrm{d}\mathbf{W} \mathrm{d}\boldsymbol{\tau} \mathrm{d}\mathbf{V}\\
&= \sum_{z'} \iiint \left(\prod_k \mathcal{N}(\mathbf{y}'|\mathbf{W}_k \mathbf{x}', \tau_k^{-1}\mathbf{I})^{z'_k} g_k(\mathbf{x}')^{z'_k} \right)\\
&\qquad \times p(\mathbf{W}, \boldsymbol{\tau}, \mathbf{V}|\mathbf{X}, \mathbf{Y}) \mathrm{d}\mathbf{W} \mathrm{d}\boldsymbol{\tau} \mathrm{d}\mathbf{V},
\end{aligned}$$

where $\mathbf{z}'$ is the latent variable associated with the observation $(\mathbf{x}', \mathbf{y}')$, and $p(\mathbf{y}'|\mathbf{x}', \mathbf{z}', \mathbf{W}, \boldsymbol{\tau})$ is replaced by (7.6), and $p(\mathbf{z}'|\mathbf{x}', \mathbf{V})$ by (7.11). As the real posterior $p(\mathbf{W}, \boldsymbol{\tau}, \mathbf{V}|\mathbf{X}, \mathbf{Y})$ is not known, it is approximated by the variational posterior, that is, $p(\mathbf{W}, \boldsymbol{\tau}, \mathbf{V}|\mathbf{X}, \mathbf{Y}) \approx q^*_{W,\tau}(\mathbf{W}, \boldsymbol{\tau}) q^*_V(\mathbf{V})$. Together with summing over all $\mathbf{z}'$, this results in

$$\begin{aligned}
&p(\mathbf{y}'|\mathbf{x}', \mathbf{X}, \mathbf{Y}) \qquad (7.103)\\
&= \sum_k \left(\int g_k(\mathbf{x}') q^*_V(\mathbf{v}_k) \mathrm{d}\mathbf{v}_k \right) \iint q^*_{W,\tau}(\mathbf{W}_k, \tau_k) \mathcal{N}(\mathbf{y}'|\mathbf{W}_k \mathbf{x}', \tau_k^{-1}\mathbf{I}) \mathrm{d}\mathbf{W}_k \mathrm{d}\tau_k,
\end{aligned}$$

where the factorisation of $q^*_V(\mathbf{V})$ and $q^*_{W,\tau}(\mathbf{W}, \boldsymbol{\tau})$ with respect to k and the independence of the two variational densities was utilised.

The first integral $\int g_k(\mathbf{x}')q_V^*(\mathbf{v}_k)\mathrm{d}\mathbf{v}_k$ is the expectation $\mathbb{E}_V(g_k(\mathbf{x}'))$ which does not have an analytical solution. Thus, following Ueda and Ghahramani [216], it is approximated by its maximum a-posteriori estimate

$$\int g_k(\mathbf{x}')q_V^*(\mathbf{v}_k)\mathrm{d}\mathbf{v}_k \approx g_k(\mathbf{x}')|_{v_k=v_k^*}. \tag{7.104}$$

The second integral $\iint q_{W,\tau}^*(\mathbf{W}_k,\tau_k)\mathcal{N}(\mathbf{y}'|\mathbf{W}_k\mathbf{x}',\tau_k^{-1}\mathbf{I})\mathrm{d}\mathbf{W}_k\mathrm{d}\tau_k$ is the expectation $\mathbb{E}_{W,\tau}(\mathcal{N}(\mathbf{y}'|\mathbf{W}_k\mathbf{x}',\tau_k^{-1}\mathbf{I}))$, that, by using (7.7) and (7.29), evaluates to

$$\begin{aligned}
&\mathbb{E}_{W,\tau}(\mathcal{N}(\mathbf{y}'|\mathbf{W}_k\mathbf{x}',\tau_k^{-1}\mathbf{I})\mathrm{d}\mathbf{W}_k\mathrm{d}\tau_k \\
&= \iint \mathcal{N}(\mathbf{y}'|\mathbf{W}_k\mathbf{x}',\tau_k^{-1}\mathbf{I})q_{W|\tau}^*(\mathbf{W}_k|\tau_k)q_\tau^*(\tau_k)\mathrm{d}\mathbf{W}_k\mathrm{d}\tau_k \\
&= \int \left(\prod_j \int \mathcal{N}(y_j'|\mathbf{w}_{kj}^T\mathbf{x}',\tau_k^{-1})\mathcal{N}(\mathbf{w}_{kj}|\mathbf{w}_{kj}^*,(\tau_k\mathbf{\Lambda}_k^*)^{-1})\mathrm{d}\mathbf{w}_{kj}\right) q_\tau^*(\tau_k)\mathrm{d}\tau_k \\
&= \prod_j \int \mathcal{N}(y_j'|{\mathbf{w}_{kj}^*}^T\mathbf{x}',\tau_k^{-1}(1+\mathbf{x}'^T{\mathbf{\Lambda}_k^*}^{-1}\mathbf{x}'))\mathrm{Gam}(\tau_k|a_{\tau_k}^*,b_{\tau_k}^*)\mathrm{d}\tau_k \\
&= \prod_j \mathrm{St}\left(y_j'|{\mathbf{w}_{kj}^*}^T\mathbf{x}',(1+\mathbf{x}'^T{\mathbf{\Lambda}_k^*}^{-1}\mathbf{x}')^{-1}\frac{a_{\tau_k}^*}{b_{\tau_k}^*},2a_{\tau_k}^*\right),
\end{aligned} \tag{7.105}$$

where $\mathrm{St}(y_j'|{\mathbf{w}_{kj}^*}^T\mathbf{x}',(1+\mathbf{x}'^T{\mathbf{\Lambda}_k^*}^{-1}\mathbf{x}')^{-1}a_{\tau_k}^*/b_{\tau_k}^*,2a_{\tau_k}^*)$ is the Student's t distribution with mean ${\mathbf{w}_{kj}^*}^T\mathbf{x}'$, precision $(1+\mathbf{x}'^T{\mathbf{\Lambda}_k^*}^{-1}\mathbf{x}')^{-1}a_{\tau_k}^*/b_{\tau_k}^*$, and $2a_{\tau_k}^*$ degrees of freedom. To derive the above we have used the convolution of two Gaussians [19], given by

$$\begin{aligned}
&\int \mathcal{N}(y_j'|\mathbf{w}_{kj}^T\mathbf{x}',\tau_k^{-1})\mathcal{N}(\mathbf{w}_{kj}|\mathbf{w}_{kj}^*,(\tau_k\mathbf{\Lambda}_k^*)^{-1})\mathrm{d}\mathbf{w}_{kj} \\
&\qquad = \mathcal{N}(y_j'|{\mathbf{w}_{kj}^*}^T\mathbf{x}',\tau_k^{-1}(1+\mathbf{x}'^T{\mathbf{\Lambda}_k^*}^{-1}\mathbf{x}')),
\end{aligned} \tag{7.106}$$

and the convolution of a Gaussian with a Gamma distribution [19],

$$\begin{aligned}
&\int \mathcal{N}(y_j'|{\mathbf{w}_{kj}^*}^T\mathbf{x}',\tau_k^{-1}(1+\mathbf{x}'^T{\mathbf{\Lambda}_k^*}^{-1}\mathbf{x}'))\mathrm{Gam}(\tau_k|a_{\tau_k}^*,b_{\tau_k}^*)\mathrm{d}\tau_k \\
&\qquad = \mathrm{St}\left(y_j'|{\mathbf{w}_{kj}^*}^T\mathbf{x}',(1+\mathbf{x}'^T{\mathbf{\Lambda}_k^*}^{-1}\mathbf{x}')^{-1}\frac{a_{\tau_k}^*}{b_{\tau_k}^*},2a_{\tau_k}^*\right).
\end{aligned} \tag{7.107}$$

Combining (7.103), (7.104), and (7.105) gives the final predictive density

$$\begin{aligned}
&p(\mathbf{y}'|\mathbf{x}',\mathbf{X},\mathbf{Y}) \\
&= \sum_k g_k(\mathbf{x}')|_{v_k=v_k^*}\prod_j \mathrm{St}\left(y_j'|{\mathbf{w}_{kj}^*}^T\mathbf{x}',(1+\mathbf{x}'^T{\mathbf{\Lambda}_k^*}^{-1}\mathbf{x}')^{-1}\frac{a_{\tau_k}^*}{b_{\tau_k}^*},2a_{\tau_k}^*\right),
\end{aligned} \tag{7.108}$$

which is a mixture of Student's t distributions.

7.4.2 Mean and Variance

Given the predictive density, point estimates and information about the prediction confidence are given by its mean and variance, respectively. As the mixture of Student's t distributions might be multi-modal, there exists no clear definition for the 95% confidence intervals, but a mixture density-related study that deals with this problem was performed by Hyndman [117]. Here, the variance is taken as a sufficient indicator of the prediction's confidence.

Let us first consider the mean and variance for arbitrary mixture densities, and subsequently apply it to (7.108). Let $\{X_k\}$ be a set of random variables that are mixed with mixing coefficients $\{g_k\}$ to give $X = \sum_k g_k X_k$. As shown by Waterhouse [227], the mean and variance of X are given by

$$\mathbb{E}(X) = \sum_k g_k \mathbb{E}(X_k), \quad \mathrm{var}(X) = \sum_k g_k(\mathrm{var}(X_k) + \mathbb{E}(X_k)^2) - \mathbb{E}(X)^2. \tag{7.109}$$

The Student's t distributions in (7.108) have mean $\mathbf{w}_{kj}^{*}{}^T \mathbf{x}'$ and variance $(1 + \mathbf{x}'^T \mathbf{\Lambda}_k^{*}{}^{-1} \mathbf{x}') 2b_{\tau_k}^* / (a_{\tau_k}^* - 1)$. Therefore, the mean vector of the predictive density is

$$\mathbb{E}(\mathbf{y}'|\mathbf{x}', \mathbf{X}, \mathbf{Y}) = \left(\sum_k g_k(\mathbf{x}')|_{v_k = v_k^*} \mathbf{W}_k^* \right) \mathbf{x}', \tag{7.110}$$

and each element y_j' of $\mathbf{y}'$ has variance

$$\begin{aligned} &\mathrm{var}(y_j'|\mathbf{x}', \mathbf{X}, \mathbf{Y}) \\ &= \sum_k g_k(\mathbf{x}')|_{v_k = v_k^*} \left(2 \frac{b_{\tau_k}^*}{a_{\tau_k}^* - 1} (1 + \mathbf{x}'^T \mathbf{\Lambda}_k^{*}{}^{-1} \mathbf{x}') + (\mathbf{w}_{kj}^{*}{}^T \mathbf{x}')^2 \right) \\ &\quad - \mathbb{E}(\mathbf{y}'|\mathbf{x}', \mathbf{X}, \mathbf{Y})_j^2, \end{aligned} \tag{7.111}$$

where $\mathbb{E}(\mathbf{y}'|\mathbf{x}', \mathbf{X}, \mathbf{Y})_j$ denotes the jth element of $\mathbb{E}(\mathbf{y}'|\mathbf{x}', \mathbf{X}, \mathbf{Y})$.

These expressions are used in the following chapter to plot the mean predictions of the LCS model, and to derive confidence intervals on these predictions.

7.5 Model Modifications to Perform Classification

In order to adjust the Bayesian LCS model to perform classification rather than regression, the input space will, as before, be assumed to be given by $\mathcal{X} = \mathbb{R}^{D_\mathcal{X}}$. The output space, on the other hand, is $\mathcal{Y} = \{0, 1\}^{D_\mathcal{Y}}$, where $D_\mathcal{Y}$ is the number of classes of the problem. For any observation $(\mathbf{x}, \mathbf{y})$, the output vector $\mathbf{y}$ defines the class j associated with input $\mathbf{x}$ by $y_j = 1$ and all other elements being 0. The task of the LCS model for a fixed model structure $\mathcal{M}$ is to model the probability $p(\mathbf{y}|\mathbf{x}, \mathcal{M})$ of any class being associated with a given input. A good model structure is one that assigns high probabilities to a single class, dependent on the input, without modelling the noise.

7.5.1 Local Classification Models and Their Priors

Taking the generative point-of-view, it is assumed that a single classifier k generates each of the classes with a fixed probability, independent of the input. Thus, its model is, as already introduced in Sect. 4.2.2, given by

$$p(\mathbf{y}|\mathbf{w}_k) = \prod w_{kj}^{y_j}, \qquad \text{with } \sum_j w_j = 1. \tag{7.112}$$

$\mathbf{w}_k \in \mathbb{R}^{D_\mathcal{Y}}$ is the parameter vector of that classifier, with each of its elements w_{kj} modelling the generative probability for its associated class j. As a consequence, its elements have to be non-negative and sum up to 1.

The conjugate prior $p(\mathbf{w}_k)$ on a classifier's parameters is given by the Dirichlet distribution

$$p(\mathbf{w}_k) = \text{Dir}(\mathbf{w}_k|\boldsymbol{\alpha}) = C(\boldsymbol{\alpha}) \prod_j w_{kj}^{\alpha_j - 1}, \tag{7.113}$$

parametrised by the vector $\boldsymbol{\alpha} \in \mathbb{R}^{D_\mathcal{Y}}$, that is equivalent for all classifiers, due to the lack of better knowledge. Its normalising constant $C(\boldsymbol{\alpha})$ is given by

$$C(\boldsymbol{\alpha}) = \frac{\Gamma(\tilde{\alpha})}{\prod_j \Gamma(\alpha_j))}, \tag{7.114}$$

where $\tilde{\alpha}$ denotes the sum of all elements of $\boldsymbol{\alpha}$, that is

$$\tilde{\alpha} = \sum_j \alpha_j. \tag{7.115}$$

Given this prior, we have $\mathbb{E}(\mathbf{w}_k) = \boldsymbol{\alpha}/\tilde{\alpha}$, and thus the elements of $\boldsymbol{\alpha}$ allow us to specify a prior bias towards one or the other class. Usually, nothing is known about the class distribution for different areas of the input space, and so all elements of $\boldsymbol{\alpha}$ should be set to the same value.

In contrast to the relation of the different elements of $\boldsymbol{\alpha}$ to each other, their absolute magnitude specifies the strength of the prior, that is, how strongly the prior affects the posterior in the light of further evidence. Intuitively speaking, a change of 1 to an element of $\boldsymbol{\alpha}$ represents one observation of the associated class. Thus, to keep the prior non-informative it should be set to small positive values, such as, for example, $\boldsymbol{\alpha} = (10^{-2}, \dots, 10^{-2})^T$.

Besides a different classifier model, no further modifications are required to the Bayesian LCS model. Its hidden variables are now $\mathbf{U} = \{\mathbf{W}, \mathbf{Z}, \mathbf{V}, \boldsymbol{\beta}\}$, where $\mathbf{W} = \{\mathbf{w}_k\}$ is the set of the classifier's parameters, whose distribution factorises with respect to k, that is

$$p(\mathbf{W}) = \prod_k p(\mathbf{w}_k). \tag{7.116}$$

Assuming knowledge of $\mathbf{X}$ and $\mathcal{M}$, the joint distribution of data and hidden variables is given by

$$p(\mathbf{Y}, \mathbf{U}|\mathbf{X}) = p(\mathbf{Y}|\mathbf{X}, \mathbf{W}, \mathbf{Z})p(\mathbf{W})p(\mathbf{Z}|\mathbf{X}, \mathbf{V})p(\mathbf{V}|\boldsymbol{\beta})p(\boldsymbol{\beta}). \tag{7.117}$$

The data likelihood is, similarly to (7.6), given by

$$p(\mathbf{Y}|\mathbf{X}, \mathbf{W}, \mathbf{Z}) = \prod_n \prod_k p(\mathbf{y}_n|\mathbf{w}_k)^{z_{nk}}. \tag{7.118}$$

The mixing model is equivalent to that of the Bayesian LCS model for regression (see Table 7.1).

7.5.2 Variational Posteriors and Moments

The posteriors are again evaluated by variational Bayesian inference. Starting with the individual classifiers, their variational posterior is found by applying (7.24) to (7.112), (7.113), (7.117) and (7.118), and for classifier k results in

$$q_w^*(\mathbf{w}_k) = \text{Dir}(\mathbf{w}_k|\boldsymbol{\alpha}_k^*), \tag{7.119}$$

with

$$\boldsymbol{\alpha}_k^* = \boldsymbol{\alpha} + \sum_n r_{nk}\mathbf{y}_n. \tag{7.120}$$

Assuming $\boldsymbol{\alpha} = \mathbf{0}$, $\mathbb{E}(\mathbf{w}_k|\boldsymbol{\alpha}_k^*) = \sum_n r_{nk}\mathbf{y}_n / \sum_n r_{nk}$ results in the same frequentist probability estimate as the maximum likelihood procedure described in Sect. 5.5.2. The prior $\boldsymbol{\alpha}$ acts like additional observations of particular classes.

The variational posterior of $\mathbf{Z}$ is the other posterior that is influenced by the classifier model. Solving (7.24) by combining (7.12), (7.112), (7.117) and (7.118) gives

$$q_Z^*(\mathbf{Z}) = \prod_n \prod_k r_{nk}^{z_{nk}}, \qquad \text{with } r_{nk} = \frac{\rho_{nk}}{\sum_j \rho_{nj}} = \mathbb{E}_Z(z_{nk}), \tag{7.121}$$

where ρ_{nk} satisfies

$$\begin{aligned}\ln \rho_{nk} &= \ln g_k(\mathbf{x}_n)|_{v_k=v_k^*} + \sum_j y_{nj}\mathbb{E}_W(\ln w_{kj}) \\ &= \ln g_k(\mathbf{x}_n)|_{v_k=v_k^*} + \sum_j y_{nj}\psi(\alpha_{kj}^*) - \psi(\tilde{\alpha}_k^*). \end{aligned} \tag{7.122}$$

$\tilde{\alpha}_k^*$, is, as before, the sum of the elements of $\boldsymbol{\alpha}_k^*$.

The variational posteriors of $\mathbf{V}$ and $\boldsymbol{\beta}$ remain unchanged, and are thus given by (7.51) and (7.55).

7.5.3 Variational Bound

For the classification model, the variational bound $\mathcal{L}(q)$ is given by

$$\begin{aligned}\mathcal{L}(q) = {} & \mathbb{E}_{W,Z}(\ln p(\mathbf{Y}|\mathbf{X}, \mathbf{W}, \mathbf{Z})) + \mathbb{E}_W(\ln p(\mathbf{W})) \\ & + \mathbb{E}_V(\ln p(\mathbf{Z}|\mathbf{X}, \mathbf{V})) + \mathbb{E}_{V,\beta}(\ln p(\mathbf{V}|\boldsymbol{\beta})) + \mathbb{E}_\beta(\ln p(\boldsymbol{\beta})) \\ & - \mathbb{E}_W(\ln q(\mathbf{W})) - \mathbb{E}_V(\ln q(\mathbf{V})) - \mathbb{E}_\beta(\ln q(\boldsymbol{\beta})) - \mathbb{E}_Z(\ln q(\mathbf{Z})). \end{aligned} \tag{7.123}$$

The only terms that differ from the ones evaluated in Sect. 7.3.8 are the ones that contain $\mathbf{W}$ and are for the classification model given by

$$\mathbb{E}_{W,Z}(\ln p(\mathbf{Y}|\mathbf{X},\mathbf{W},\mathbf{Z})) = \sum_n \sum_k r_{nk} \left(\sum_j y_{nj} (\psi(\alpha^*_{kj}) - \psi(\tilde{\alpha}^*_k)) \right), \tag{7.124}$$

$$\mathbb{E}_W(\ln p(\mathbf{W})) = \sum_k \left(\ln C(\boldsymbol{\alpha}) + \sum_j (\alpha_j - 1)(\psi(\alpha^*_{kj}) - \psi(\tilde{\alpha}^*_k)) \right), \tag{7.125}$$

$$\mathbb{E}_W(\ln q(\mathbf{W})) = \sum_k \left(\ln C(\boldsymbol{\alpha}^*_k) + \sum_j (\alpha^*_{kj} - 1)(\psi(\alpha^*_{kj}) - \psi(\tilde{\alpha}^*_k)) \right) \tag{7.126}$$

Splitting the variational bound again into $\mathcal{L}_k$'s and $\mathcal{L}_M$, $\mathcal{L}_k(q)$ for classifier k is defined as

$$\mathcal{L}_k(q) = \mathbb{E}_{W,Z}(\ln p(\mathbf{Y}|\mathbf{X},\mathbf{W},\mathbf{Z})) + \mathbb{E}_W(\ln p(\mathbf{W})) - \mathbb{E}_W(\ln q(\mathbf{W})), \tag{7.127}$$

and evaluates to

$$\mathcal{L}_k(q) = \ln C(\boldsymbol{\alpha}) - \ln C(\boldsymbol{\alpha}^*_k), \tag{7.128}$$

where (7.120) was used to simplify the expression. $\mathcal{L}_M(q)$ remains unchanged and is thus given by (7.95). As before, $\mathcal{L}(q)$ is given by (7.96).

7.5.4 Independent Classifier Training

As before, the classifiers can be trained independently by replacing r_{nk} by $m_k(\mathbf{x}_n)$. This only influences the classifier weight vector update (7.120) that becomes

$$\boldsymbol{\alpha}^*_k = \boldsymbol{\alpha} + \sum_n m_k(\mathbf{x}_n)\mathbf{y}_n. \tag{7.129}$$

This change invalidates the simplifications performed to get $\mathcal{L}_k(q)$ by (7.128). Instead,

$$\begin{aligned}\mathcal{L}_k(q) = {} & \ln C(\boldsymbol{\alpha}) - \ln C(\boldsymbol{\alpha}^*_k) \\ & + \sum_j \left(\sum_n r_{nk}\mathbf{y}_{nj} + \alpha_j - \alpha^*_{kj} \right) (\psi(\alpha^*_{kj}) - \psi(\tilde{\alpha}^*_k))\end{aligned} \tag{7.130}$$

has to be used.

If classifiers are trained independently, then they can be trained in a single pass by (7.129), as no hyperpriors are used. How the mixing model is trained and the variational bound is evaluated remains unchanged and is described in Sect. 7.3.10.

7.5.5 Predictive Density

Given a new observation $(\mathbf{y}', \mathbf{x}')$, its predictive density is given by $p(\mathbf{y}'|\mathbf{x}', \mathcal{D})$. The density's mixing-model component is essentially the same as in Sect. 7.4.

What remains to evaluate is the marginalised classifier prediction

$$\int q_W^*(\mathbf{w}_k) \prod_j w_{kj}^{y'_j} \mathrm{d}\mathbf{w}_k = \frac{C(\boldsymbol{\alpha}_k^*)}{C(\boldsymbol{\alpha}'_k)}, \tag{7.131}$$

where $\boldsymbol{\alpha}'_k = \boldsymbol{\alpha}_k^* + \mathbf{y}'$. Thus, the predictive density is given by

$$p(\mathbf{y}'|\mathbf{x}', \mathbf{X}, \mathbf{Y}) = \sum_k g_k(\mathbf{x}')|_{v_k = v_k^*} \frac{C(\boldsymbol{\alpha}_k^*)}{C(\boldsymbol{\alpha}'_k)}. \tag{7.132}$$

Due to the 1-of-$D_{\mathcal{Y}}$ structure of $\mathbf{y}'$, only a single element y'_j, associated with class j, is 1. Thus, using the definition of $C(\cdot)$ by (7.114) and $\Gamma(x+1) = x\Gamma(x)$ allows us to simplify the above expression to

$$p(y'_j = 1|\mathbf{x}', \mathbf{X}, \mathbf{Y}) = \sum_k g_k(\mathbf{x}')|_{v_k = v_k^*} \frac{\alpha_{kj}^*}{\sum_{\bar{j}} \alpha_{k\bar{j}}^*}. \tag{7.133}$$

The predicted class j is the one that is considered as being the most likely to have generated the observation, and is the one that maximises the above expression. This completes the Bayesian LCS model for classification.

7.6 Alternative Model Selection Methods

Bayesian model selection is not the only model selection criterion that might be applicable to LCS. In this section a set of alternatives and their relation to LCS are reviewed.

As described in Sect. 7.1.2, model selection criteria might differ in their philosophical background, but they all result in the principle of minimising a combination of model error and model complexity. Their main difference lies in how they define the model complexity. Very crude approaches, like the two-part MDL, only consider the coarse model structure, whereas more refined criteria, like the refined MDL, SRM, and BYY, are based on the functional form of the model. However, they usually do not take the distribution of the training data into consideration when evaluating the model complexity. Recent research has shown that consider this distribution, like cross-validation, Bayesian model selection, or Rademacher complexity, are usually better in approximating the target function [125].

7.6.1 Minimum Description Length

The principle of Minimum Description Length (MDL) [188, 189, 190] is based on the idea of Occam's Razor, that amongst models that explain the data equally well, the simplest one is the one to prefer. MDL uses Kolmogorov complexity as a baseline to describe the complexity of the model, but as that is uncomputable, coding theory is used as an approximation to find minimum coding lengths that then represent the model complexity [101].

In its crudest form, the two-part MDL requires a binary representation of both the model error and the model itself, where the combined representation is to be minimised [188, 189]. Using such an approach for LCS makes its performance highly dependent on the representation used for the matching functions and the model parameters, and is therefore rather arbitrary. Its dependence on the chosen representation and the lack of guidelines on how to decide upon a particular representation are generally considered the biggest weakness of the two-part MDL [101].

A more refined approach is to use the Bayesian MDL [101] that — despite a different philosophical background — is mathematically identical to Bayesian model selection as applied here. In that sense, the approach presented in this chapter can be said to be using the Bayesian MDL model selection criterion.

The latest MDL approach is theoretically optimal in the sense that it minimises the worst-case coding length of the model. Mathematically, it is expressed as the maximum likelihood normalised by the model complexity, where the model complexity is its coding length summed over all possible model parameter values [191]. Therefore, given continuous model parameters, as used here, the complexity is infinite, which makes model comparison impossible. In addition, the LCS structure makes computing the model complexity even for a finite set of parameters extremely complicated, which makes it unlikely that, in its pure form, the latest MDL measure will be of any use for LCS.

7.6.2 Structural Risk Minimisation

Structural Risk Minimisation (SRM) is based on minimising an upper bound on the expected risk (3.1), given the sum of the empirical risk (3.2) and a model complexity metric based on the functional form of the model [218]. The functional form of the model complexity enters SRM in the form of the model's Vapnik-Chervonenkis (VC) dimensions. Having the empirical risk and the VC dimensions of the model, we can find a model that minimises the expected risk.

The difficulty of the SRM approach when applied to LCS is to find the VC dimensions of the LCS model. For linear regression classifiers, the VC dimensions are simply the dimensionality of the input space $D_{\mathcal{X}}$. Mixing these models, however, introduces non-linearity that makes evaluation of the VC dimensions difficult. An additional weakness of SRM is that it deals with worst-case bounds that do apply to any distribution of the data, which causes the bound on the expected risk to be quite loose and reduces its usefulness for model selection [19].

A more powerful approach that provides us with a tighter bound to the expected risk is to use data-dependent SRM. Such an approach has been applied to the Mixtures-of-Expert model by Azran et al. [5, 4]. It still remains to be seen if this approach can be generalised to the LCS model, such as was done here with the Bayesian MoE model to provide the Bayesian LCS model. If this is possible, data-dependent SRM might be a viable alternative for defining the optimal set of classifiers.

7.6.3 Bayesian Ying-Yang

Bayesian Ying Yang (BYY) defines a unified framework that lets one derive many statistics-based machine learning methods [243]. It describes the probability distribution given by the data, and the one described by the model, and aims at finding models that are closest in distribution to the data. Using the Kullback-Leibler divergence as a distribution comparison metric results in maximum likelihood learning, and therefore will cause overfitting of the model. An alternative is Harmony Learning which is based on minimising the cross entropy between the data distribution and the model distribution, and prefers statistically simple distributions, that is, distributions of low entropy.

Even though it is very likely applicable to LCS as it has already been applied to the Mixtures-of-Expert model [242], there is no clear philosophical background that justifies the use of the cross entropy. Therefore, the Bayesian approach that was introduced in this chapter seems to be a better alternative.

7.6.4 Training Data-Based Approaches

It has been shown that penalising the model complexity based on some structural properties of the model alone cannot compete on all scales with data-based methods like cross validation [125]. Furthermore, using the training data rather than an independent test set gives even better results in minimising the expected risk [13]. Two examples of such complexity measures are the Rademacher complexity and the Gaussian complexity [14]. Both of them are defined as the expected error of the model when trying to fit the data perturbed by a sequence of either Rademacher random variables (uniform over $\{\pm 1\}$) or Gaussian $\mathcal{N}(0,1)$ random variables. Hence, they measure the model complexity by the model's ability to match a noisy sequence.

Using such methods in LCS would require training two models for the same model structure, where one is trained with the normal training data, and the other with the perturbed data. It is questionable if such additional space and computational effort justifies the application of the methods. Furthermore, using sampling of random variables to find the model complexity makes it impossible to find an analytical expression for the utility of the model and thus provides little insight in how a particular model structure is selected. Nonetheless, it might still be of use as a benchmark method.

7.7 Discussion and Summary

This chapter tackled the core question of LCS: what is the best set of classifiers that explains the given data? Rather than relying on intuition, this question was approached formally by aiming to find the best model structure $\mathcal{M}$ that explains the given data $\mathcal{D}$. More specifically, the principles of Bayesian model selection were applied to define the best set of classifiers to be the most likely one given the data, that is, the one that maximises $p(\mathcal{M}|\mathcal{D})$.

Computing this probability density requires a Bayesian LCS model that was introduced by adding priors to the probabilistic model from Chap. 4. Additionally, the flexibility of the regression classifier model was increased from univariate to multivariate regression. The requirement of specifying prior parameters is not a weakness of this approach, but rather a strength, as the priors make explicit the commonly implicit assumptions made about the data-generating process.

Variational Bayesian inference was employed to find a closed-form solution to $p(\mathcal{M}|\mathcal{D})$, in combination with various approximation to handle the generalised softmax function that is used to combine the local classifier models to a global model. Whilst variational Bayesian inference usually provides a lower bound $\mathcal{L}(q)$ on $\ln p(\mathcal{D}|\mathcal{M})$ that is directly related to $p(\mathcal{M}|\mathcal{D})$, these approximations invalidate the lower bound nature of $\mathcal{L}(q)$. Even without these approximations, the use of $\mathcal{L}(q)$ for selecting the best set of classifiers depends very much on the tightness of the bound, and if this tightness is consistent for different model structures $\mathcal{M}$. Variational Bayesian inference has been shown to perform well in practice [216, 19], and the same approximations that were used here were successfully used for the Mixtures-of-Experts model [226, 227]. Thus, the presented method can be expected to feature good performance when applied to LCS, but more definite statements require further empirical investigation.

What was introduced in this chapter is the first formal and general definition of what if means for a set of classifiers to be optimal, using the best applicable of the currently common model selection approaches. The definition is general as i) it is independent of the representation of the matching function, ii) it can be used for both discrete and continuous input spaces, iii) it can handle matching by degree, and iv) it is not restricted to the LCS model that is introduced in this book but is applicable to all LCS model types that can be described probabilistically, including the linear LCS model. The reader is reminded that the definition itself is independent of the variational inference, and thus is not affected by the issues that are introduced through approximating the posterior. A further significant advancement that comes with the definition of optimality is a Bayesian model for LCS that goes beyond the probabilistic model as it makes the prior assumptions about the data-generating process explicit. Additionally, the use of multivariate regression is also a novelty in the LCS context.

Defining the best set of classifiers as a maximisation problem also promotes its theoretical investigation: depending on the LCS model type, one could, for example, ask the question if the optimal set of classifiers is ever overlapping. In other words, does the optimal set of classifiers include classifiers that are responsible for the same input and thus have overlapping matching? Given that the removal of overlaps increases $p(\mathcal{M}|\mathcal{D})$ in all cases, then this is not the case. Such knowledge can guide model structure search itself, as it can avoid classifier constellations that are very likely suboptimal. Thus, further research in this area is not only of theoretical value but can guide the design of other LCS components.

After this rather abstract introduction of the definition of the optimal classifier set and a method of computing the model probability, a more concrete description of how it can be implemented will be provided. Also, a set of simple experiments demonstrates that Bayesian model selection is indeed able to identify good sets of classifiers.

8 An Algorithmic Description

In the previous chapter, the optimal set of classifiers given some data $\mathcal{D}$ was defined as the one given by the model structure $\mathcal{M}$ that maximises $p(\mathcal{M}|\mathcal{D})$. In addition, a Bayesian LCS model for both regression and classification was introduced, and it was shown how to apply variational Bayesian inference to compute a lower bound on $\ln p(\mathcal{M}|\mathcal{D})$ for some given $\mathcal{M}$ and $\mathcal{D}$.

To demonstrate that the definition of the optimal classifier set leads to useful results, a set of simple algorithms are introduced that demonstrate its use on a set of regression tasks. This includes two possible approaches to search the model structure space in order to maximise $p(\mathcal{M}|\mathcal{D})$, one based on a basic genetic algorithm to create a simple Pittsburgh-style LCS, and the other on sampling from the model posterior $p(\mathcal{M}|\mathcal{D})$ by Markov Chain Monte Carlo (MCMC) methods. These approaches are by no means supposed to act as viable competitors to current LCS, but rather as prototype implementations to demonstrate the correctness and usefulness of the optimality definition. Additionally, the presented formulation of the algorithm seeks for readability rather than performance. Thus, there might still be plenty of room for optimisation.

The core of both approaches is the evaluation of $p(\mathcal{M}|\mathcal{D})$ and its comparison for different classifier sets in order to find the best set. The evaluation of $p(\mathcal{M}|\mathcal{D})$ is approached by variational Bayesian inference, as introduced in the previous chapter. Thus, the algorithmic description of how to find $p(\mathcal{M}|\mathcal{D})$ also provides a summary of the variational approach for regression classifier models and a better understanding of how it can be implemented. Even though not described here, the algorithm can easily be modified to handle classification rather than regression. A general drawback of the algorithm as it is presented here is that it does not scale well with the number of classifiers, and that it can currently only operate in batch mode. The reader is reminded, however, that the algorithmic description is only meant to show that the definition of the optimal set of classifiers is a viable one. Possible extensions to this work, as described later in this chapter, allude on how this definition can be incorporated into current LCS or can kindle the development of new LCS.

Firstly, a set of functions are introduced, that in combination compute a measure of the quality of a classifier set given the data. As this measure can

J. Drugowitsch: Des. & Anal. of Learn. Class. Sys.: A Prob. Approach, SCI 139, pp. 165–201, 2008.
springerlink.com

subsequently by used by any global search algorithm that is able to find its maximum in the space of possible model structures, its algorithmic description is kept separate from the model structure search. For the structure search, two simple alternatives are provided in a later section, one based on genetic algorithms, and another based on sampling the model posterior $p(\mathcal{M}|\mathcal{D})$ by MCMC methods. Finally, both approaches are applied to simple regression tasks to demonstrate the usefulness of the classifier set optimality criterion.

8.1 Computing $p(\mathcal{M}|\mathcal{D})$

Let us start with a set of functions that allow the computation of an approximation to $p(\mathcal{M}|\mathcal{D})$ for a given data set $\mathcal{D}$ and model structure $\mathcal{M}$. These functions rely on a small set of global system parameters and constants that are given in Table 8.1. The functions are presented in a top-down order, starting with a function that returns $p(\mathcal{M}|\mathcal{D})$, and continuing with the sub-functions that it calls. The functions use a small set of non-standard operators and global functions that are described in Table 8.2.

The data is assumed to be given by the $N \times D_{\mathcal{X}}$ input matrix $\mathbf{X}$ and the $N \times D_{\mathcal{Y}}$ output matrix $\mathbf{Y}$, as described in Sect. 7.2.1. The model structure is fully defined by the $N \times K$ matching matrix $\mathbf{M}$, that is given by

Table 8.1. Description of the system parameters and constants. These include the distribution parameters of the priors and hyperpriors, and constants that parametrise the stopping criteria of parameter update iterations. The recommended values specify rather uninformative priors and hyperpriors, such that the introduced bias due to these priors is negligible.

Symbol	*Recom.*	*Description*
a_α	10^{-2}	Scale parameter of weight vector variance prior
b_α	10^{-4}	Shape parameter of weight vector variance prior
a_β	10^{-2}	Scale parameter of mixing weight vector variance prior
b_β	10^{-4}	Shape parameter of mixing weight vector variance prior
a_τ	10^{-2}	Scale parameter of noise variance prior
b_τ	10^{-4}	Shape parameter of noise variance prior
$\Delta_s \mathcal{L}_k(q)$	10^{-4}	Stopping criterion for classifier update
$\Delta_s \mathcal{L}_M(q)$	10^{-2}	Stopping criterion for mixing model update
$\Delta_s \mathrm{KL}(\mathbf{R}\|\mathbf{G})$	10^{-8}	Stopping criterion for mixing weight update
$\exp_{\min}$	–	lowest real number x on system such that $\exp(x) > 0$
$\ln_{\max}$	–	$\ln(x)$, where x is the highest real number on system

Table 8.2. Operators and global functions used in the algorithmic descriptions

Fn. / Op.	*Description*
$\mathbf{A} \otimes \mathbf{B}$	given an $a \times b$ matrix or vector $\mathbf{A}$, and $c \times d$ matrix or vector $\mathbf{B}$, and $a = c, b = d$, $\mathbf{A} \otimes \mathbf{B}$ returns an $a \times b$ matrix that is the result of an element-wise multiplication of $\mathbf{A}$ and $\mathbf{B}$. If $a = c, d = 1$, that is, if $\mathbf{B}$ is a column vector with c elements, then every column of $\mathbf{A}$ is multiplied element-wise by $\mathbf{B}$, and the result is returned. Analogously, if $\mathbf{B}$ is a row vector with b elements, then each row of $\mathbf{A}$ is multiplied element-wise by $\mathbf{B}$, and the result is returned.
$\mathbf{A} \oslash \mathbf{B}$	the same as $\mathbf{A} \otimes \mathbf{B}$, only performing division rather than multiplication.
`Sum`($\mathbf{A}$)	returns the sum over all elements of matrix or vector $\mathbf{A}$.
`RowSum`($\mathbf{A}$)	given an $a \times b$ matrix $\mathbf{A}$, returns a column vector of size a, where its ith element is the sum of the b elements of the ith row of $\mathbf{A}$.
`FixNaN`($\mathbf{A}, b$)	replaces all `NaN` elements in matrix or vector $\mathbf{A}$ by b.

$$\mathbf{M} = \begin{pmatrix} m_1(\mathbf{x}_1) & \cdots & m_K(\mathbf{x}_1) \\ \vdots & \ddots & \vdots \\ m_1(\mathbf{x}_N) & \cdots & m_K(\mathbf{x}_N) \end{pmatrix}. \tag{8.1}$$

Thus, column k of this matrix specifies the degree of matching of classifier k for all available observations. Note that the definition of $\mathbf{M}$ differs from the one in Chap. 5, where $\mathbf{M}$ was a diagonal matrix that specified the matching for a single classifier.

In addition to the matching matrix, we also need to define the $N \times D_V$ mixing feature matrix $\mathbf{\Phi}$, that is given by

$$\mathbf{\Phi} = \begin{pmatrix} -\phi(\mathbf{x}_1)^T- \\ \vdots \\ -\phi(\mathbf{x}_N)^T- \end{pmatrix}, \tag{8.2}$$

and thus specifies the feature vector $\phi(\mathbf{x})$ for each observation. In LCS, we usually have $\phi(\mathbf{x}) = 1$ for all $\mathbf{x}$, and thus also $\mathbf{\Phi} = (1, \dots 1)^T$, but the algorithm presented here also works for other definitions of ϕ.

8.1.1 Model Probability and Evidence

The Function `ModelProbability` takes the model structure and the data as arguments and returns $\mathcal{L}(q) + \ln p(\mathcal{M})$ as an approximation to the unnormalised

Function. `ModelProbability(` $\mathbf{M}, \mathbf{X}, \mathbf{Y}, \mathbf{\Phi}$`)`

Input: matching matrix $\mathbf{M}$, input matrix $\mathbf{X}$, output matrix $\mathbf{Y}$, mixing feature matrix $\mathbf{\Phi}$

Output: approximate model probability $\mathcal{L}(q) + \ln p(\mathcal{M})$

1 get K from shape of $\mathbf{M}$
2 **for** $k \leftarrow 1$ **to** K **do**
3 $\quad \mathbf{m}_k \leftarrow k$th column of $\mathbf{M}$
4 $\quad \mathbf{W}_k^*, \mathbf{\Lambda}_k^{*-1}, a_{\tau_k}^*, b_{\tau_k}^*, a_{\alpha_k}^*, b_{\alpha_k}^* \leftarrow$ `TrainClassifier(`$\mathbf{m}_k$, $\mathbf{X}$, $\mathbf{Y}$`)`
5 $\mathbf{W}, \mathbf{\Lambda}^{-1} \leftarrow \{\mathbf{W}_1, \dots, \mathbf{W}_K\}, \{\mathbf{\Lambda}_1^{-1}, \dots, \mathbf{\Lambda}_K^{-1}\}$
6 $\mathbf{a}_\tau, \mathbf{b}_\tau \leftarrow \{a_{\tau_1}, \dots, a_{\tau_K}\}, \{b_{\tau_1}, \dots, b_{\tau_K}\}$
7 $\mathbf{a}_\alpha, \mathbf{b}_\alpha \leftarrow \{a_{\alpha_1}, \dots, a_{\alpha_K}\}, \{b_{\alpha_1}, \dots, b_{\alpha_K}\}$
8 $\mathbf{V}, \mathbf{\Lambda}_V^{-1}\mathbf{a}_\beta, \mathbf{b}_\beta \leftarrow$ `TrainMixing(`$\mathbf{M}, \mathbf{X}, \mathbf{Y}, \mathbf{\Phi}, \mathbf{W}, \mathbf{\Lambda}^{-1}, \mathbf{a}_\tau, \mathbf{b}_\tau, \mathbf{a}_\alpha, \mathbf{b}_\alpha$`)`
9 $\boldsymbol{\theta} \leftarrow \{\mathbf{W}, \mathbf{\Lambda}^{-1}, \mathbf{a}_\tau, \mathbf{b}_\tau, \mathbf{a}_\alpha, \mathbf{b}_\alpha, \mathbf{V}, \mathbf{\Lambda}_V^{-1}\mathbf{a}_\beta, \mathbf{b}_\beta\}$
10 $\mathcal{L}(q) \leftarrow$ `VarBound(`$\mathbf{M}, \mathbf{X}, \mathbf{Y}, \mathbf{\Phi}, \boldsymbol{\theta}$`)`
11 **return** $\mathcal{L}(q) + \ln K!$

$\ln p(\mathcal{M}|\mathcal{D})$. Thus, it replaces the model evidence $p(\mathcal{D}|\mathcal{M})$ in (7.3) by its approximation $\mathcal{L}(q)$. The function assumes that the order of the classifiers can be arbitrarily permuted without changing the model structure and therefore uses the $p(\mathcal{M})$ given by (7.4). In approximating $\ln p(\mathcal{M}|\mathcal{D})$, the function does not add the normalisation constant. Hence, even though the return values are not proper probabilities, they can still be used for the comparison of different model structures, as the normalisation term is shared between all of them.

The computation of $\mathcal{L}(q) + \ln p(\mathcal{M})$ is straightforward: Lines 2 to 7 compute and assemble the parameters of the classifiers by calling `TrainClassifier` for each classifier k separately, and provide it with the data and the matching vector $\mathbf{m}_k$ for that classifier. After that, the mixing model parameters are computed in Line 8 by calling `TrainMixing`, based on the fully trained classifiers.

Having evaluated all classifiers, all parameters are collected in Line 9 to give $\boldsymbol{\theta}$ and are used in Line 10 to compute $\mathcal{L}(q)$ by calling `VarBound`. After that, the function returns $\mathcal{L}(q) + \ln K!$, based on (7.3) and (7.4).

8.1.2 Training the Classifiers

The Function `TrainClassifier` takes the data $\mathbf{X}, \mathbf{Y}$ and the matching vector $\mathbf{m}_k$ and returns all model parameters for the trained classifier k. The model parameters are found by iteratively updating the distribution parameters of the variational posteriors $q_{W,\tau}^*(\mathbf{W}_k, \tau_k)$ and $q_\alpha^*(\alpha_k)$ until the convergence criterion is satisfied. This criterion is given by the classifier-specific components $\mathcal{L}_k(q)$ of the variational bound $\mathcal{L}(q)$, as given by (7.91). However, rather than evaluating $\mathcal{L}_k(q)$ with the responsibilities r_{nk}, as done in (7.91), the matching function $m_k(\mathbf{x}_n)$ are used instead. The underlying idea is that – as each classifier is trained independently – the responsibilities are equivalent to the matching function values. This has the effect that by updating the classifier parameters according

Function. `TrainClassifier(` $\mathbf{m}_k$, $\mathbf{X}$, $\mathbf{Y}$`)`

Input: matching vector $\mathbf{m}_k$, input matrix $\mathbf{X}$, output matrix $\mathbf{Y}$
Output: $D_\mathcal{Y} \times D_\mathcal{X}$ weight matrix $\mathbf{W}_k$, $D_\mathcal{X} \times D_\mathcal{X}$ covariance matrix $\mathbf{\Lambda}_k^{-1}$, noise precision parameters a_{τ_k}, b_{τ_k}, weight vector prior parameters $a_{\alpha_k}, b_{\alpha_k}$

1 get $D_\mathcal{X}, D_\mathcal{Y}$ from shape of $\mathbf{X}, \mathbf{Y}$
2 $\mathbf{X}_k \leftarrow \mathbf{X} \otimes \sqrt{\mathbf{m}_k}$
3 $\mathbf{Y}_k \leftarrow \mathbf{Y} \otimes \sqrt{\mathbf{m}_k}$
4 $a_{\alpha_k}, b_{\alpha_k} \leftarrow a_\alpha, b_\alpha$
5 $a_{\tau_k}, b_{\tau_k} \leftarrow a_\tau, b_\tau$
6 $\mathcal{L}_k(q) \leftarrow -\infty$
7 $\Delta\mathcal{L}_k(q) \leftarrow \Delta_s\mathcal{L}_k(q) + 1$
8 **while** $\Delta\mathcal{L}_k(q) > \Delta_s\mathcal{L}_k(q)$ **do**
9     $\mathbb{E}_\alpha(\alpha_k) \leftarrow a_{\alpha_k}/b_{\alpha_k}$
10     $\mathbf{\Lambda}_k \leftarrow \mathbb{E}_\alpha(\alpha_k)\mathbf{I} + \mathbf{X}_k^T\mathbf{X}_k$
11     $\mathbf{\Lambda}_k^{-1} \leftarrow (\mathbf{\Lambda}_k)^{-1}$
12     $\mathbf{W}_k \leftarrow \mathbf{Y}_k^T\mathbf{X}_k\mathbf{\Lambda}_k^{-1}$
13     $a_{\tau_k} \leftarrow a_\tau + \frac{1}{2}$ `Sum(`$\mathbf{m}_k$`)`
14     $b_{\tau_k} \leftarrow b_\tau + \frac{1}{2D_\mathcal{Y}}\big($ `Sum(`$\mathbf{Y}_k \otimes \mathbf{Y}_k$`)` $-$ `Sum(`$\mathbf{W}_k \otimes \mathbf{W}_k\mathbf{\Lambda}_k$`)` $\big)$
15     $\mathbb{E}_\tau(\tau_k) \leftarrow a_{\tau_k}/b_{\tau_k}$
16     $a_{\alpha_k} \leftarrow a_\alpha + \frac{D_\mathcal{X}D_\mathcal{Y}}{2}$
17     $b_{\alpha_k} \leftarrow b_\alpha + \frac{1}{2}\big(\mathbb{E}_\tau(\tau_k)$ `Sum(`$\mathbf{W}_k \otimes \mathbf{W}_k$`)` $+ D_\mathcal{Y}\mathrm{Tr}(\mathbf{\Lambda}_k^{-1})\big)$
18     $\mathcal{L}_{k,prev}(q) \leftarrow \mathcal{L}_k(q)$
19     $\mathcal{L}_k(q) \leftarrow$ `VarClBound(`$\mathbf{X}, \mathbf{Y}, \mathbf{W}_k, \mathbf{\Lambda}_k^{-1}, a_{\tau_k}, b_{\tau_k}, a_{\alpha_k}, b_{\alpha_k}, \mathbf{m}_k$`)`
20     $\Delta\mathcal{L}_k(q) \leftarrow \mathcal{L}_k(q) - \mathcal{L}_{k,prev}(q)$
21   **assert** $\Delta\mathcal{L}_k(q) \geq 0$
22 **return** $\mathbf{W}_k, \mathbf{\Lambda}_k^{-1}, a_{\tau_k}, b_{\tau_k}, a_{\alpha_k}, b_{\alpha_k}$

to (7.97) – (7.100), $\mathcal{L}_k(q)$ is indeed maximised, which is not necessarily the case if $r_{nk} \neq m_k(\mathbf{x}_n)$, as discussed in Sect. 7.3.4. Therefore, every parameter update is guaranteed to increase $\mathcal{L}_k(q)$, until the algorithm converges.

In more detail, Lines 2 and 3 compute the matched input vector $\mathbf{X}_k$ and output vector $\mathbf{Y}_k$, based on $\sqrt{m_k(\mathbf{x})}\sqrt{m_k(\mathbf{x})} = m_k(\mathbf{x})$. Note that each column of $\mathbf{X}$ and $\mathbf{Y}$ is element-wise multiplied by $\sqrt{\mathbf{m}_k}$, where the square root is applied to each element of $\mathbf{m}_k$ separately. The prior and hyperprior parameters are initialised with their prior parameter values in Lines 4 and 5.

In the actual iteration, Lines 9 to 14 compute the parameters of the variational posterior $q^*_{W,\tau}(\mathbf{W}_k, \tau_k)$ by the use of (7.97) – (7.100) and (7.64). To get the weight vector covariance $\mathbf{\Lambda}_k^{-1}$ the equality $\mathbf{X}_k^T\mathbf{X}_k = \sum_n m_k(\mathbf{x_n})\mathbf{x}_n\mathbf{x}_n^T$ is used. The weight matrix $\mathbf{W}_k$ is evaluated by observing that the jth row of $\mathbf{Y}_k^T\mathbf{X}_k\mathbf{\Lambda}_k^{-1}$, giving $\mathbf{w}_{kj}$, is equivalent to $\mathbf{\Lambda}_k^{-1}\sum_n m_k(\mathbf{x}_n)\mathbf{x}_n y_{nj}$. The update of b_{τ_k} uses `Sum(`$\mathbf{Y}_k \otimes \mathbf{Y}_k$`)` that effectively squares each element of $\mathbf{Y}_k$ before returning the sum over all elements, that is $\sum_j\sum_n m_k(\mathbf{x}_n)y_{nj}^2$. $\sum_j \mathbf{w}_{kj}^T\mathbf{\Lambda}_k\mathbf{w}_{kj}$ in (7.100) is computed by observing that it can be reformulated to the sum over all elements of the element-wise multiplication of $\mathbf{W}_k$ and $\mathbf{W}_k\mathbf{\Lambda}_k$.

Lines 15 to 17 update the parameters of the variational posterior $q_\alpha^*(\alpha_k)$, as given by (7.40), (7.41), and (7.72). Here, the sum over all squared elements of $\mathbf{W}_k$ is used to evaluate $\sum_j \mathbf{w}_{kj}^T \mathbf{w}_{kj}$.

The function determines convergence of the parameter updates in Lines 18 to 21 by computing the change of $\mathcal{L}_k(q)$ over two successive iterations. If this change drops below the system parameter $\Delta_s \mathcal{L}_k(q)$, then the function returns. The value of $\mathcal{L}_k(q)$ is computed by Function `VarClBound`, which is described in Sect. 8.1.4. Its last argument is a vector of responsibilities for classifier k, which is substituted by the matching function values for reasons mentioned above. Each parameter update either increases $\mathcal{L}_k(q)$ or leaves it unchanged, which is specified in Line 21. If this is not the case, then the implementation is faulty and/or suffers from numerical instabilities. In the experiments that were performed, convergence was usually reached after 3–4 iterations.

8.1.3 Training the Mixing Model

Training the mixing model is more complex than training the classifiers, as the IRLS algorithm is used to find the parameters of $q_V^*(\mathbf{V})$. The function `TrainMixing` takes the model structure, data, and the parameters of the fully trained classifiers, and returns the parameters of the mixing model.

As with training the classifiers, the parameters of the mixing model are found incrementally, by sequentially updating the parameters of the variational posteriors $q_V^*(\mathbf{V})$, $q_\beta^*(\boldsymbol{\beta})$ and $q_Z^*(\mathbf{Z})$. Convergence of the updates is determined by

---

Function. `TrainMixing`$(\mathbf{M}, \mathbf{X}, \mathbf{Y}, \mathbf{\Phi}, \mathbf{W}, \mathbf{\Lambda}^{-1}, \mathbf{a}_\tau, \mathbf{b}_\tau, \mathbf{a}_\alpha, \mathbf{b}_\alpha)$

Input: matching matrix $\mathbf{M}$, input matrix $\mathbf{X}$, output matrix $\mathbf{Y}$, mixing feature matrix $\mathbf{\Phi}$, classifier parameters $\mathbf{W}, \mathbf{\Lambda}^{-1}, \mathbf{a}_\tau, \mathbf{b}_\tau, \mathbf{a}_\alpha, \mathbf{b}_\alpha$

Output: $D_V \times K$ mixing weight matrix $\mathbf{V}$, $(KD_V) \times (KD_V)$ mixing weight covariance matrix, mixing weight vector prior parameters $\mathbf{a}_\beta, \mathbf{b}_\beta$

get $D_\mathcal{X}, D_\mathcal{Y}, D_V, K$ from shape of $\mathbf{X}, \mathbf{Y}, \mathbf{\Phi}, \mathbf{W}$
$\mathbf{V} \leftarrow D_V \times K$ matrix with elements sampled from $\mathcal{N}\left(0, \left(\frac{a_\beta}{b_\beta}\right)\right)$
$\mathbf{a}_\beta \leftarrow \{a_{\beta_1}, \dots, a_{\beta_K}\}$, all initialised to $a_{\beta_k} = a_\beta$
$\mathbf{b}_\beta \leftarrow \{b_{\beta_1}, \dots, b_{\beta_K}\}$, all initialised to $b_{\beta_k} = b_\beta$
$\mathcal{L}_M(q) \leftarrow -\infty$
$\Delta \mathcal{L}_M(q) \leftarrow \Delta_s \mathcal{L}_M(q) + 1$
while $\Delta \mathcal{L}_M(q) > \Delta_s \mathcal{L}_M(q)$ **do**
  $\mathbf{V}, \mathbf{\Lambda}_V^{-1} \leftarrow$ `TrainMixWeights`$(\mathbf{M}, \mathbf{X}, \mathbf{Y}, \mathbf{\Phi}, \mathbf{W}, \mathbf{\Lambda}^{-1}, \mathbf{a}_\tau, \mathbf{b}_\tau, \mathbf{V}, \mathbf{a}_\beta, \mathbf{b}_\beta)$
  $\mathbf{a}_\beta, \mathbf{b}_\beta \leftarrow$ `TrainMixPriors`$(\mathbf{V}, \mathbf{\Lambda}_V^{-1})$
  $\mathbf{G} \leftarrow$ `Mixing`$(\mathbf{M}, \mathbf{\Phi}, \mathbf{V})$
  $\mathbf{R} \leftarrow$ `Responsibilities`$(\mathbf{X}, \mathbf{Y}, \mathbf{G}, \mathbf{W}, \mathbf{\Lambda}^{-1}, \mathbf{a}_\tau, \mathbf{b}_\tau)$
  $\mathcal{L}_{M,prev}(q) \leftarrow \mathcal{L}_M(q)$
  $\mathcal{L}_M(q) \leftarrow$ `VarMixBound`$(\mathbf{G}, \mathbf{R}, \mathbf{V}, \mathbf{\Lambda}_V^{-1}, \mathbf{a}_\beta, \mathbf{b}_\beta)$
  $\Delta \mathcal{L}_M(q) \leftarrow |\mathcal{L}_M(q) - \mathcal{L}_{M,prev}(q)|$
return $\mathbf{V}, \mathbf{\Lambda}_V^{-1}, \mathbf{a}_\beta, \mathbf{b}_\beta$

---

monitoring the change of the mixing model-related components $\mathcal{L}_M(q)$ of the variational bound $\mathcal{L}(q)$, as given by (7.95). If the magnitude of change of $\mathcal{L}_M(q)$ between two successive iterations is lower than the system parameter $\Delta_s\mathcal{L}_M(q)$, then the algorithm assumes convergence and returns.

The parameters are initialised in Lines 2 to 4 of `TrainMixing`. The $D_V \times K$ mixing matrix $\mathbf{V}$ holds the vector $\mathbf{v}_k$ that corresponds to classifier k in its kth column. As by (7.13) the prior on each element of $\mathbf{v}_k$ is given by a zero-mean Gaussian with variance β_k^{-1}, each element of $\mathbf{V}$ is initialised by sampling from $\mathcal{N}(0, b_\beta/a_\beta)$ where the value of the random variable β_k is approximated by its prior expectation. The distribution parameters for $q_\beta(\beta_k)$ are initialised by setting them to the prior parameters.

An iteration starts by calling `TrainMixWeights` in Line 8 to get the parameters of the variational posterior $q_V^*(\mathbf{V})$. These are subsequently used in Line 9 to update the parameters of $q_\beta^*(\beta_k)$ for each k by calling `TrainMixPriors`. Lines 10 to 14 determine the magnitude of change of $\mathcal{L}_M(q)$ when compared to the last iteration. This is achieved by computing the $N \times K$ mixing matrix $\mathbf{G} = (g_k(\mathbf{x}_n))$ by calling `Mixing`. Based on $\mathbf{G}$, the responsibility matrix $\mathbf{R} = (r_{nk})$ is evaluated by calling `Responsibilities` in Line 11. This allows for the evaluation of $\mathcal{L}_M(q)$ in Line 13 by calling `VarMixBound`, and determines the magnitude of change $\Delta\mathcal{L}_M(q)$ in the next Line, which is subsequently used to determine if the parameter updates converged. In the performed experiments, the function usually converged after 5–6 iterations.

Next, the Functions `TrainMixWeights`, `TrainMixPriors`, `Mixing` and `Responsibilities` will be introduced, as they are all used by `TrainMixing` to train the mixing model. `VarMixBound` is described in the later Sect. 8.1.4.

Function. `Mixing`$(\mathbf{M}, \mathbf{\Phi}, \mathbf{V})$

Input: matching matrix $\mathbf{M}$, mixing feature matrix $\mathbf{\Phi}$, mixing weight matrix $\mathbf{V}$
Output: $N \times K$ mixing matrix $\mathbf{G}$

1 get K from shape of $\mathbf{V}$
2 $\mathbf{G} \leftarrow \mathbf{\Phi V}$
3 limit all elements of $\mathbf{G}$ such that $\exp_{\min} \leq g_{nk} \leq \ln_{\max} - \ln K$
4 $\mathbf{G} \leftarrow \exp(\mathbf{G}) \otimes \mathbf{M}$
5 $\mathbf{G} \leftarrow \mathbf{G} \oslash$ `RowSum`$(\mathbf{G})$
6 `FixNaN`$(\mathbf{G}, 1/K)$
7 **return** $\mathbf{G}$

Starting with `Mixing`, this function is used to compute the mixing matrix $\mathbf{G}$ that contains the values for $g_k(\mathbf{x}_n)$ for each classifier/input combination. It takes the matching matrix $\mathbf{M}$, the mixing features $\mathbf{\Phi}$, and the mixing weight matrix $\mathbf{V}$ as arguments, and returns $\mathbf{G}$.

The mixing matrix $\mathbf{G}$ is evaluated by computing (7.10) in several steps: firstly, in Line 2, $\mathbf{v}_k^T\phi(\mathbf{x}_n)$ is computed for each combination of n and k. Before the exponential of these values is taken, it needs to be ensured that this does not cause any overflow/underflow. This is done by limiting the values in $\mathbf{G}$ in Line 3 to a certain range, with the following underlying idea [173]: they are limited from below by $\exp_{\min}$ to ensure that their exponential is positive, as their logarithm might be later taken. Additionally, they are limited from above by $\ln_{\max} - \ln K$ such that summing over K such elements does not cause an overflow. Once this is done, the element-wise exponential can be taken, and each element is multiplied by the corresponding matching function value, as done in Line 4. This essentially gives the nominator of (7.10) for all combinations of n and k. Normalisation over k is performed in the next line by dividing each element in a certain row by the element sum of this row. If rows in $\mathbf{G}$ were zero before normalisation, 0/0 was performed, which is fixed in Line 6 by assigning equal weights to all classifiers for inputs that are not matched by any classifier. Usually, this should never happen as only model structures are accepted where $\sum_k m_k(\mathbf{x}_n) > 0$ for all n. Nonetheless, this check was added to ensure that even these cases are handled gracefully.

Function. `Responsibilities`$(\mathbf{X}, \mathbf{Y}, \mathbf{G}, \mathbf{W}, \mathbf{\Lambda}^{-1}, \mathbf{a}_\tau, \mathbf{b}_\tau)$

Input: input matrix $\mathbf{X}$, output matrix $\mathbf{Y}$, gating matrix $\mathbf{G}$, classifier parameters $\mathbf{W}, \mathbf{\Lambda}^{-1}, \mathbf{a}_\tau, \mathbf{b}_\tau$

Output: $N \times K$ responsibility matrix $\mathbf{R}$

1 get $K, D_\mathcal{Y}$ from shape of $\mathbf{Y}, \mathbf{G}$

2 **for** $k = 1$ **to** K **do**

3 $\quad \mathbf{W}_k, \mathbf{\Lambda}_k^{-1}, a_{\tau_k}, b_{\tau_k} \leftarrow$ pick from $\mathbf{W}, \mathbf{\Lambda}^{-1}, \mathbf{a}_\tau, \mathbf{b}_\tau$

4 $\quad k$th column of $\mathbf{R} \leftarrow \exp\Big(\frac{D_\mathcal{Y}}{2}(\psi(a_{\tau_k}) - \ln b_{\tau_k})$

5 $\quad\quad -\frac{1}{2}\big(\frac{a_{\tau_k}}{b_{\tau_k}}$ `RowSum`$((\mathbf{Y} - \mathbf{X}\mathbf{W}_k{}^T)^2) + D_\mathcal{Y}$ `RowSum`$(\mathbf{X} \otimes \mathbf{X}\mathbf{\Lambda}_k^{-1})\big)\Big)$

6 $\mathbf{R} \leftarrow \mathbf{R} \otimes \mathbf{G}$

7 $\mathbf{R} \leftarrow \mathbf{R} \oslash$ `RowSum`$(\mathbf{R})$

8 `FixNaN`$(\mathbf{R}, 0)$

9 **return** $\mathbf{R}$

Based on the gating matrix $\mathbf{G}$ and the goodness-of-fit of the classifiers, the Function `Responsibilities` computes the $N \times K$ responsibility matrix, with r_{nk} as its nkth element. Its elements are evaluated by following (7.62), (7.63), (7.69) and (7.74).

The loop from Line 2 to 5 in `Responsibilities` iterates over all k to fill the columns of $\mathbf{R}$ with the values for ρ_{nk} according to (7.62), but without the term $g_k(\mathbf{x}_n)$[1]. This is simplified by observing that the term $\sum_j (y_{nj} - \mathbf{w}_{kj}^T \mathbf{x}_n)^2$, which is by (7.74) part of $\sum_j \mathbb{E}_{W,\tau}(\tau_k (y_{nj} - \mathbf{w}_{kj}^T \mathbf{x}_n)^2)$, is given for each observation separately in the vector that results from summing over the rows of $(\mathbf{Y} - \mathbf{X}\mathbf{W}_k^T)^2$, where the square is taken element-wise. Similarly, $\mathbf{x}_n^T \mathbf{\Lambda}_k^{-1} \mathbf{x}_n$ of the same expectation is given for each observation by the vector that results from summing over

[1] Note that we are operating on ρ_{nk} rather than $\ln \rho_{nk}$, as given by (7.62), as we certainly have $g_k(\mathbf{x}_n) = 0$ in cases where $m_k(\mathbf{x}_n) = 0$, which would lead to subsequent numerical problems when evaluating $\ln g_k(\mathbf{x}_n)$.

the rows of $\mathbf{X} \otimes \mathbf{X}\boldsymbol{\Lambda}_k^{-1}$, based on $\mathbf{x}_n^T \boldsymbol{\Lambda}_k^{-1} \mathbf{x}_n = \sum_i (\mathbf{x}_n)_i (\boldsymbol{\Lambda}_k^{-1} \mathbf{x}_n)_i$. The values of $g_k(\mathbf{x}_n)$ are added to ρ_{nk} in Line 6, and the normalisation step by (7.63) is performed in Line 7. For the same reason as in the `Mixing` function, all `NaN` values in $\mathbf{R}$ need to be subsequently replaced by 0 to not assign responsibility to any classifiers for inputs that are not matched.

Function. `TrainMixWeights`$(\mathbf{M}, \mathbf{X}, \mathbf{Y}, \boldsymbol{\Phi}, \mathbf{W}, \boldsymbol{\Lambda}^{-1}, \mathbf{a}_\tau, \mathbf{b}_\tau, \mathbf{V}, \mathbf{a}_\beta, \mathbf{b}_\beta)$

Input: matching matrix $\mathbf{M}$, input matrix $\mathbf{X}$, output matrix $\mathbf{Y}$, mixing feature matrix $\boldsymbol{\Phi}$, classifier parameters $\mathbf{W}, \boldsymbol{\Lambda}^{-1}, \mathbf{a}_\tau, \mathbf{b}_\tau$, mixing weight matrix $\mathbf{V}$, mixing weight prior parameters $\mathbf{a}_\beta, \mathbf{b}_\beta$

Output: $D_V \times K$ mixing weight matrix $\mathbf{V}$, $(KD_V) \times (KD_V)$ mixing weight covariance matrix $\boldsymbol{\Lambda}_V^{-1}$

$\mathbb{E}_\beta(\boldsymbol{\beta}) \leftarrow$row vector with elements $\left(\frac{a_{\beta_1}}{b_{\beta_1}}, \dots, \frac{a_{\beta_K}}{b_{\beta_K}}\right)$
$\mathbf{G} \leftarrow$ `Mixing`$(\mathbf{M}, \boldsymbol{\Phi}, \mathbf{V})$
$\mathbf{R} \leftarrow$ `Responsibilities`$(\mathbf{X}, \mathbf{Y}, \mathbf{G}, \mathbf{W}, \boldsymbol{\Lambda}^{-1}, \mathbf{a}_\tau, \mathbf{b}_\tau)$
$\mathrm{KL}(\mathbf{R}\|\mathbf{G}) \leftarrow \infty$
$\Delta\mathrm{KL}(\mathbf{R}\|\mathbf{G}) \leftarrow \Delta_s\mathrm{KL}(\mathbf{R}\|\mathbf{G}) + 1$
while $\Delta KL(\mathbf{R}\|\mathbf{G}) > \Delta_s KL(\mathbf{R}\|\mathbf{G})$ **do**
  $\mathbf{E} \leftarrow \boldsymbol{\Phi}^T(\mathbf{G} - \mathbf{R}) + \mathbf{V} \otimes \mathbb{E}_\beta(\boldsymbol{\beta})$
  $\mathbf{e} \leftarrow (\mathbf{E}_{11}, \dots, \mathbf{E}_{D_V 1}, \mathbf{E}_{12}, \dots, \mathbf{E}_{D_V 2}, \dots, \mathbf{E}_{1K}, \dots, \mathbf{E}_{D_V K})^T$
  $\mathbf{H} \leftarrow$ `Hessian`$(\boldsymbol{\Phi}, \mathbf{G}, \mathbf{a}_\beta, \mathbf{b}_\beta)$
  $\Delta\mathbf{v} \leftarrow -\mathbf{H}^{-1}\mathbf{e}$
  $\Delta\mathbf{V} \leftarrow D_V \times K$ matrix with jkth element
    given by $((k-1)K + j)$th element of $\Delta\mathbf{v}$
  $\mathbf{V} \leftarrow \mathbf{V} + \Delta\mathbf{V}$
  $\mathbf{G} \leftarrow$ `Mixing`$(\mathbf{M}, \boldsymbol{\Phi}, \mathbf{V})$
  $\mathbf{R} \leftarrow$ `Responsibilities`$(\mathbf{X}, \mathbf{Y}, \mathbf{G}, \mathbf{W}, \boldsymbol{\Lambda}^{-1}, \mathbf{a}_\tau, \mathbf{b}_\tau)$
  $\mathrm{KL}_{prev}(\mathbf{R}\|\mathbf{G}) \leftarrow \mathrm{KL}(\mathbf{R}\|\mathbf{G})$
  $\mathrm{KL}(\mathbf{R}\|\mathbf{G}) \leftarrow$ `Sum`$(\mathbf{R} \otimes$ `FixNaN`$(\ln(\mathbf{G} \oslash \mathbf{R}), 0))$
  $\Delta\mathrm{KL}(\mathbf{R}\|\mathbf{G}) = |\mathrm{KL}_{prev}(\mathbf{R}\|\mathbf{G}) - \mathrm{KL}(\mathbf{R}\|\mathbf{G})|$
$\mathbf{H} \leftarrow$ `Hessian`$(\boldsymbol{\Phi}, \mathbf{G}, \mathbf{a}_\beta, \mathbf{b}_\beta)$
$\boldsymbol{\Lambda}_V^{-1} \leftarrow \mathbf{H}^{-1}$
return $\mathbf{V}, \boldsymbol{\Lambda}_V^{-1}$

The Function `TrainMixWeights` approximates the mixing weights variational posterior $q_V^*(\mathbf{V})$ (7.51) by performing the IRLS algorithm. It takes the matching matrix, the data and mixing feature matrix, the trained classifier parameters, the mixing weight matrix, and the mixing weight prior parameters. As the IRLS algorithm performs incremental updates of the mixing weights $\mathbf{V}$ until convergence, $\mathbf{V}$ is not re-initialised every time `TrainMixWeights` is called, but rather the previous estimates are used as their initial values to reduce the number of iterations that are required until convergence.

As the aim is to model the responsibilities by finding mixing weights that make the mixing coefficients given by $g_k(\mathbf{x}_n)$ similar to r_{nk}, convergence is determined by the Kullback-Leibler divergence measure $\text{KL}(\mathbf{R}\|\mathbf{G})$ that measures the distance between the probability distributions given by $\mathbf{R}$ and $\mathbf{G}$. Formally, it is defined by $\text{KL}(\mathbf{R}\|\mathbf{G}) = \sum_n \sum_k r_{nk} \ln(g_k(\mathbf{x}_n)/r_{nk})$, and is represented in $\mathcal{L}_M(q)$ (7.95) by the terms $\mathbb{E}_{Z,V}(\ln p(\mathbf{Z}|\mathbf{V}) - \mathbb{E}_Z(\ln q(\mathbf{Z}))$, given by (7.84). As the Kullback-Leibler divergence is non-negative and zero if and only if $\mathbf{R} = \mathbf{G}$ [232], the algorithm assumes convergence of the IRLS algorithm if the change in $\text{KL}(\mathbf{R}\|\mathbf{G})$ between two successive iterations is below the system parameter $\Delta_s\text{KL}(\mathbf{R}\|\mathbf{G})$.

`TrainMixWeights` starts by computing the expectation $\mathbb{E}_\beta(\beta_k)$ for all k in Line 1. The IRLS iteration (6.5) requires the error gradient $\nabla E(\mathbf{V})$ and the Hessian $\mathbf{H}$, which are by (7.48) and (7.49) based on the values of $g_k(\mathbf{x}_n)$ and r_{nk}. Hence, `TrainMixWeights` continues by computing $\mathbf{G}$ and $\mathbf{R}$ in Lines 2 and 3.

The error gradient $\nabla E(\mathbf{V})$ by (7.48) is evaluated in Lines 7 and 8. Line 7 uses the fact that $\mathbf{\Phi}^T(\mathbf{G} - \mathbf{R})$ results in a $D_V \times K$ matrix that has the vector $\sum_n (g_j(\mathbf{x}_n) - r_{nj})\phi(\mathbf{x}_n)$ as its jth column. Similarly, $\mathbf{V} \otimes \mathbb{E}_\beta(\boldsymbol{\beta})$ results in a matrix of the same size, with $\mathbb{E}_\beta(\beta_j)\mathbf{v}_j$ as its jth column. Line 8 rearranges the matrix $\mathbf{E}$, which has $\nabla_{\mathbf{v}_j} E(\mathbf{V})$ as its jth column, to the gradient vector $\mathbf{e} = \nabla E(\mathbf{V})$. The Hessian $\mathbf{H}$ is assembled in Line 9 by calling the Function `Hessian`, and is used in the next line to compute the vector $\Delta\mathbf{v}$ by which the mixing weights need to be changed according to the IRLS algorithm (6.5). The mixing weight vector is updated by rearranging $\Delta\mathbf{v}$ to the shape of $\mathbf{V}$ in Line 12, and adding it to $\mathbf{V}$ in the next line.

As the mixing weights have changed, $\mathbf{G}$ and $\mathbf{R}$ are recomputed with the updated weights, to get $\text{KL}(\mathbf{R}\|\mathbf{G})$, and eventually to use it in the next iteration. The Kullback-Leibler divergence between the responsibilities $\mathbf{R}$ and their model $\mathbf{G}$ are evaluated in Line 17, and then compared to its value of the last iteration to determine convergence of the IRLS algorithm. Note that due to the use of matrix operations, the elements in $\mathbf{R}$ are not checked for being $r_{nk} = 0$ due to $g_k(\mathbf{x}) = 0$ when computing $\mathbf{G} \oslash \mathbf{R}$, which might cause `NaN` entries in the resulting matrix. Even though these entries are multiplied by $r_{nk} = 0$ thereafter, they firstly need to be replaced by zero, as otherwise we would still get $0 \times$ `NaN` = `NaN`.

The IRLS algorithm gives the mean of $q_V^*(\mathbf{V})$ by the mixing weights that minimise the error function $E(\mathbf{V})$. The covariance matrix $\mathbf{\Lambda}_V^{-1}$ still needs to be evaluated and is by (7.50) the inverse Hessian, as evaluated in Line 19. Due to its dependence on $\mathbf{G}$, the last Hessian in the IRLS iteration in Line 9 cannot be used for that purpose, as $\mathbf{G}$ has changed thereafter.

To complete `TrainMixWeights`, let us consider how the Function `Hessian` assembles the Hessian matrix $\mathbf{H}$: it first creates an empty $(KD_V) \times (KD_V)$ matrix that is thereafter filled by its block elements $\mathbf{H}_{kj} = \mathbf{H}_{jk}$, as given by (7.49). Here, the equality

$$\sum_n \phi(\mathbf{x}_n)\left(g_k(\mathbf{x}_n) g_j(\mathbf{x}_n) \phi(\mathbf{x}_n)^T\right) = \mathbf{\Phi}^T(\mathbf{\Phi} \otimes (\mathbf{g}_k \otimes \mathbf{g}_j)) \tag{8.3}$$

is used for the off-diagonal blocks of $\mathbf{H}$ where $\mathbf{I}_{kj} = 0$ in (7.49), and a similar relation is used to get the diagonal blocks of $\mathbf{H}$.

Function. `Hessian`$(\mathbf{\Phi}, \mathbf{G}, \mathbf{a}_\beta, \mathbf{b}_\beta)$

Input: mixing feature matrix $\mathbf{\Phi}$, mixing matrix $\mathbf{G}$, mixing weight prior parameters $\mathbf{a}_\beta, \mathbf{b}_\beta$
Output: $(KD_V) \times (KD_V)$ Hessian matrix $\mathbf{H}$

get D_V, K from shape of $\mathbf{V}$
$\mathbf{H} \leftarrow$ empty $(KD_V) \times (KD_V)$ matrix
for $k = 1$ **to** K **do**
  $\mathbf{g}_k \leftarrow k$th column of $\mathbf{G}$
  for $j = 1$ **to** $k - 1$ **do**
    $\mathbf{g}_j \leftarrow j$th column of $\mathbf{G}$
    $\mathbf{H}_{kj} \leftarrow -\mathbf{\Phi}^T (\mathbf{\Phi} \otimes (\mathbf{g}_k \otimes \mathbf{g}_j))$
    kjth $D_V \times D_V$ block of $\mathbf{H} \leftarrow \mathbf{H}_{kj}$
    jkth $D_V \times D_V$ block of $\mathbf{H} \leftarrow \mathbf{H}_{kj}$
  $a_{\beta_k}, b_{\beta_k} \leftarrow$ pick from $\mathbf{a}_\beta, \mathbf{b}_\beta$
  $\mathbf{H}_{kk} \leftarrow \mathbf{\Phi}^T (\mathbf{\Phi} \otimes (\mathbf{g}_k \otimes (1 - \mathbf{g}_k))) + \frac{a_{\beta_k}}{b_{\beta_k}} \mathbf{I}$
  kth $D_V \times D_V$ block along diagonal of $\mathbf{H} \leftarrow \mathbf{H}_{kk}$
return $\mathbf{H}$

Function. `TrainMixPriors`$(\mathbf{V}, \mathbf{\Lambda}_V^{-1})$

Input: mixing weight matrix $\mathbf{V}$, mixing weight covariance matrix $\mathbf{\Lambda}_V^{-1}$
Output: mixing weight vector prior parameters $\mathbf{a}_\beta, \mathbf{b}_\beta$

1 get D_V, K from shape of $\mathbf{V}$
2 **for** $k = 1$ **to** K **do**
3   $\mathbf{v} \leftarrow k$th column of $\mathbf{V}$
4   $(\mathbf{\Lambda}_V^{-1})_{kk} \leftarrow k$th $D_V \times D_V$ block along diagonal of $\mathbf{\Lambda}_V^{-1}$
5   $a_{\beta_k} \leftarrow a_\beta + \frac{D_V}{2}$
6   $b_{\beta_k} \leftarrow b_\beta + \frac{1}{2} \left(\mathrm{Tr}\left((\mathbf{\Lambda}_V^{-1})_{kk} \right) + \mathbf{v}_k^T \mathbf{v}_k \right)$
7 $\mathbf{a}_\beta, \mathbf{b}_\beta \leftarrow \{a_{\beta_1}, \dots, a_{\beta_K}\}, \{b_{\beta_1}, \dots, b_{\beta_K}\}$
8 **return** $\mathbf{a}_\beta, \mathbf{b}_\beta$

The posterior parameters of the prior on the mixing weights are evaluated according to (7.56), (7.57), and (7.70) in order to get $q_\beta^*(\beta_k)$ for all k. Function `TrainMixPriors` takes the parameters of $q_V^*(\mathbf{V})$ and returns the parameters for all $q_\beta^*(\beta_k)$. The posterior parameters are computed by iterating over all k, and in Lines 5 and 6 by performing a straightforward evaluation of (7.56) and (7.57), where in the latter, (7.70) replaces $\mathbb{E}_V(\mathbf{v}_k^T \mathbf{v}_k)$.

8.1.4 The Variational Bound

The variational bound $\mathcal{L}(q)$ is evaluated in Function `VarBound` according to (7.96). The function takes the model structure, the data, and the trained classifier and mixing model parameters, and returns the value for $\mathcal{L}(q)$. The classifier-specific

Function. `VarBound`$(\mathbf{M}, \mathbf{X}, \mathbf{Y}, \mathbf{\Phi}, \boldsymbol{\theta})$

Input: matching matrix $\mathbf{M}$, input matrix $\mathbf{X}$, output matrix $\mathbf{Y}$, mixing feature matrix $\mathbf{\Phi}$, trained model parameters $\boldsymbol{\theta}$
Output: variational bound $\mathcal{L}(q)$

1 get K from shape of $\mathbf{V}$
2 $\mathbf{G} \leftarrow$ `Mixing`$(\mathbf{M}, \mathbf{\Phi}, \mathbf{V})$
3 $\mathbf{R} \leftarrow$ `Responsibilities`$(\mathbf{X}, \mathbf{Y}, \mathbf{G}, \mathbf{W}, \mathbf{\Lambda}^{-1}, \mathbf{a}_\tau, \mathbf{b}_\tau)$
4 $\mathcal{L}_K(q) \leftarrow 0$
5 **for** $k = 1$ **to** K **do**
6 $\quad \mathbf{r}_k \leftarrow k$th column of $\mathbf{R}$
7 $\quad \mathcal{L}_K(q) \leftarrow \mathcal{L}_K(q)$
8 $\quad\quad +$ `VarClBound`$(\mathbf{X}, \mathbf{Y}, \mathbf{W}_k, \mathbf{\Lambda}_k^{-1}, a_{\tau_k}, b_{\tau_k}, a_{\alpha_k}, b_{\alpha_k}, \mathbf{r}_k)$
9 $\mathcal{L}_M(q) \leftarrow$ `VarMixBound`$(\mathbf{G}, \mathbf{R}, \mathbf{V}, \mathbf{\Lambda}_V^{-1}, \mathbf{a}_\beta, \mathbf{b}_\beta)$
10 **return** $\mathcal{L}_K(q) + \mathcal{L}_M(q)$

Function. `VarClBound`$(\mathbf{X}, \mathbf{Y}, \mathbf{W}_k, \mathbf{\Lambda}_k^{-1}, a_{\tau_k}, b_{\tau_k}, a_{\alpha_k}, b_{\alpha_k}, \mathbf{r}_k)$

Input: input matrix $\mathbf{X}$, output matrix $\mathbf{Y}$, classifier parameters $\mathbf{W}_k, \mathbf{\Lambda}_k^{-1}, a_{\tau_k}, b_{\tau_k}, a_{\alpha_k}, b_{\alpha_k}$, responsibility vector $\mathbf{r}_k$
Output: classifier component $\mathcal{L}_k(q)$ of variational bound

1 get $D_\mathcal{X}, D_\mathcal{Y}$ from shape of $\mathbf{X}, \mathbf{Y}$
2 $\mathbb{E}_\tau(\tau_k) \leftarrow a_{\tau_k} / b_{\tau_k}$
3 $\mathcal{L}_{k,1}(q) \leftarrow \frac{D_\mathcal{Y}}{2} (\psi(a_{\tau_k}) - \ln b_{\tau_k} - \ln 2\pi)$ `Sum`$(\mathbf{r}_k)$
4 $\mathcal{L}_{k,2}(q) \leftarrow -\frac{1}{2}\mathbf{r}_k^T (\mathbb{E}_\tau(\tau_k)$ `RowSum`$((\mathbf{Y} - \mathbf{X}\mathbf{W}_k^T)^2)$ $+ D_\mathcal{Y}$ `RowSum`$(\mathbf{X} \otimes \mathbf{X}\mathbf{\Lambda}_k^{-1})$ $)$
5 $\mathcal{L}_{k,3}(q) \leftarrow - \ln \Gamma(a_\alpha) + a_\alpha \ln b_\alpha + \ln \Gamma(a_{\alpha_k}) - a_{\alpha_k} \ln b_{\alpha_k} + \frac{D_\mathcal{X} D_\mathcal{Y}}{2} + \frac{D_\mathcal{Y}}{2} \ln |\mathbf{\Lambda}_k^{-1}|$
6 $\mathcal{L}_{k,4}(q) \leftarrow D_\mathcal{Y}(- \ln \Gamma(a_\tau) + a_\tau \ln b_\tau + (a_\tau - a_{\tau_k})\psi(a_{\tau_k}) - a_\tau \ln b_{\tau_k} - b_\tau \mathbb{E}_\tau(\tau_k)$
7 $\quad + \ln \Gamma(a_{\tau_k}) + a_{\tau_k})$
8 **return** $\mathcal{L}_{k,1}(q) + \mathcal{L}_{k,2}(q) + \mathcal{L}_{k,3}(q) + \mathcal{L}_{k,4}(q)$

components $\mathcal{L}_k(q)$ are computed separately for each classifier k in Line 8 by calling `VarClBound`. Note that in contrast to calling `VarClBound` with the matching function values of the classifiers, as done in Function `TrainClassifier`, we here conform to (7.91) and provide `VarClBound` with the previously evaluated responsibilities. The full variational bound is found by adding the mixing model-specific components $\mathcal{L}_M(q)$, that are computed in Line 8 by a call to `VarMixBound`, to the sum of all $\mathcal{L}_k(q)$'s.

By evaluating (7.91), the Function `VarClBound` returns the components of $\mathcal{L}(q)$ that are specific to classifier k. It takes the data, the trained classifier parameters, and the responsibilities with respect to that classifier, and returns the value for $\mathcal{L}_k(q)$. This values is computed by splitting (7.91) into the components $\mathcal{L}_{k,1}(q)$ to $\mathcal{L}_{k,4}(q)$, evaluating them one by one, and then returning their sum. To get $\mathcal{L}_{k,2}(q)$, the same matrix simplifications as in Line 5 of Function `Responsibilities` have been used to get $\|\mathbf{y}_n - \mathbf{W}_k \mathbf{x}_n\|^2$ and $\mathbf{x}_n^T \mathbf{\Lambda}_k^{-1} \mathbf{x}_n$.

Finally, Function `VarMixBound` takes mixing values and responsibilities, and the mixing model parameters, and returns the mixing model-specific components

Function. `VarMixBound`$(\mathbf{G}, \mathbf{R}, \mathbf{V}, \mathbf{\Lambda}_V^{-1}, \mathbf{a}_\beta, \mathbf{b}_\beta)$

Input: mixing matrix $\mathbf{G}$, responsibilities matrix $\mathbf{R}$, mixing weight matrix $\mathbf{V}$, mixing covariance matrix $\mathbf{\Lambda}_V^{-1}$ mixing weight prior parameters $\mathbf{a}_\beta, \mathbf{b}_\beta$
Output: mixing component $\mathcal{L}_M(q)$ of variational bound

1 get D_V, K from shape of $\mathbf{V}$
2 $\mathcal{L}_{M,1}(q) \leftarrow K\,(-\ln\Gamma(a_\beta) + a_\beta \ln b_\beta)$
3 **for** $k = 1$ **to** K **do**
4 $\quad a_{\beta_k}, b_{\beta_k} \leftarrow$ pick from $\mathbf{a}_\beta, \mathbf{b}_\beta$
5 $\quad \mathcal{L}_{M,1}(q) \leftarrow \mathcal{L}_{M,1}(q) + \ln\Gamma(a_{\beta_k}) - a_{\beta_k} \ln b_{\beta_k}$
6 $\mathcal{L}_{M,2}(q) \leftarrow$ `Sum`$(\mathbf{R} \otimes$ `FixNaN`$(\ln(\mathbf{G} \oslash \mathbf{R}), 0))$
7 $\mathcal{L}_{M,3}(q) \leftarrow \frac{1}{2}\ln|\mathbf{\Lambda}_V^{-1}| + \frac{KD_V}{2}$
8 **return** $\mathcal{L}_{M,1}(q) + \mathcal{L}_{M,2}(q) + \mathcal{L}_{M,3}(q)$

$\mathcal{L}_M(q)$ of $\mathcal{L}(q)$ by evaluating (7.95). As in `VarClBound`, the computation of $\mathcal{L}_M(q)$ is split into the components $\mathcal{L}_{M,1}(q)$, $\mathcal{L}_{M,2}(q)$, and $\mathcal{L}_{M,3}(q)$, whose sum is returned. $\mathcal{L}_{M,1}(q)$ contains the components of $\mathcal{L}_M(q)$ that depend on the parameters $q^*_\beta(\boldsymbol{\beta})$, and is computed in Lines 2 to 5 by iterating over all k. $\mathcal{L}_{M,2}(q)$ is the Kullback-Leibler divergence $\mathrm{KL}(\mathbf{R}\|\mathbf{G})$, as given by (7.84), which is computed in the same way as in Line 17 of Function `TrainMixWeights`.

8.1.5 Scaling Issues

Let us now consider how the presented algorithm scales with the dimensionality of the input space $D_\mathcal{X}$, output space $D_\mathcal{Y}$, the mixing feature space D_V, the number N of observations that are available, and the number K of classifiers. All $\mathcal{O}(\cdot)$ are based on the observation that the multiplication of an $a \times b$ matrix with a $b \times c$ matrix scales with $\mathcal{O}(abc)$, and the inversion and getting the determinant of an $a \times a$ matrix have complexity $\mathcal{O}(a^3)$ and $\mathcal{O}(a^2)$, respectively.

Table 8.1 gives an overview of how the different functions scale with N, K, $D_\mathcal{X}$, $D_\mathcal{Y}$ and D_V. Unfortunately, even though `ModelProbability` scales linearly with N and $D_\mathcal{Y}$, it neither scales well with $D_\mathcal{X}$, nor with K and D_V. In all three cases, the 3rd polynomial is caused by a matrix inversion.

Considering that $D_\mathcal{X}^3$ is due to inverting the precision matrix $\mathbf{\Lambda}_k$, it might be reducible to $D_\mathcal{X}^2$ by using the Sherman-Morrison formula, as shown in Sect. 5.3.5. $D_\mathcal{X}$ is the dimensionality of the input space with respect to the classifier model, and is given by $D_\mathcal{X} = 1$ for averaging classifiers, and by $D_\mathcal{X} = 2$ for classifiers that model straight lines. Thus, it is in general not too high and $D_\mathcal{X}^3$ will not be the most influential complexity component. In any case, as long as we are required to maintain a covariance matrix $\mathbf{\Lambda}_k^{-1}$ of size $D_\mathcal{X} \times D_\mathcal{X}$, the influence of $D_\mathcal{X}$ is unlikely to be reducible below $D_\mathcal{X}^2$.

The biggest weakness of the prototype algorithm that was presented here is that the number of operations required to find the parameters of the mixing model scale with $K^3 D_V^3$. This is due to the inversion of the $(KD_V) \times (KD_V)$ Hessian matrix that is required at each iteration of the IRLS algorithm. To apply

Function	$\mathcal{O}(\cdot)$	*Comments*
`ModelProbability`	$NK^3D_{\mathcal{X}}^3D_{\mathcal{Y}}D_V^3$	$K^3D_V^3$ from `TrainMixing`, $D_{\mathcal{X}}^3$ from `TrainClassifier`
`TrainClassifier`	$ND_{\mathcal{X}}^3D_{\mathcal{Y}}$	$D_{\mathcal{X}}^3$ due to $\mathbf{\Lambda}_k^{-1}$
`TrainMixing`	$NK^3D_{\mathcal{X}}^2D_{\mathcal{Y}}D_V^3$	$K^3D_{\mathcal{X}}^2D_V^3$ from `TrainMixWeights`
`Mixing`	NKD_V	–
`Responsibilities`	$NKD_{\mathcal{X}}^2D_{\mathcal{Y}}$	$D_{\mathcal{X}}^2$ due to $\mathbf{X\Lambda}_k^{-1}$
`TrainMixWeights`	$NK^3D_{\mathcal{X}}^2D_{\mathcal{Y}}D_V^3$	$(KD_V)^3$ due to $\mathbf{H}^{-1}$, $D_{\mathcal{X}}^2$ from `Responsibilities`
`Hessian`	$NK^2D_V^2$	K^2 due to nested iteration, D_V^2 due to $\mathbf{\Phi}^T(\mathbf{\Phi} \otimes (\mathbf{g}_k \otimes \mathbf{g}_j))$
`TrainMixPriors`	KD_V	–
`VarClBound`	$ND_{\mathcal{X}}^2D_{\mathcal{Y}}$	$D_{\mathcal{X}}^2$ due to $\mathbf{X\Lambda}_k^{-1}$ or $\lvert\mathbf{\Lambda}_k^{-1}\rvert$
`VarMixBound`	$NK^2D_V^2$	$(KD_V)^2$ due to $\lvert\mathbf{\Lambda}_V^{-1}\rvert$

Fig. 8.1. Complexity of the different functions with respect to the number of observations N, the number of classifiers K, the dimensionality of the input space $D_{\mathcal{X}}$, the dimensionality of the output space $D_{\mathcal{Y}}$, and the dimensionality of the mixing feature space D_V

variational inference to real-world problems, the algorithm would be required to scale linearly with the number of classifiers K. This is best achieved by approximating the optimal mixing weights by well-tuned heuristics, as was already done for the prior-free LCS model in Chap. 6. What remains to do is to find similar heuristics that honour the prior. The mixing feature space dimensionality, on the other hand, is usually $D_V = 1$, and its influence is therefore negligible.

In summary, the presented algorithm scales with $\mathcal{O}(NK^3D_{\mathcal{X}}^3D_{\mathcal{Y}}D_V^3)$. While it might be possible to reduce $D_{\mathcal{X}}^3$ to $D_{\mathcal{X}}^2$, it still scales super-linearly with the number of classifiers K. This is due to the use of the generalised softmax function that requires the application of the IRLS algorithm to find its parameters. To reduce the complexity, the softmax function needs to either be replaced by another model that is easier to train, or well-tuned heuristics that provide a good approximation to it.

8.2 Two Alternatives for Model Structure Search

Recall that the optimal set of classifiers $\mathcal{M}$ was defined as the set that maximises $p(\mathcal{M}|\mathcal{D})$. Therefore, in order to find this optimal set we need to search the space $\{\mathcal{M}\}$ for the $\mathcal{M}$ such that $p(\mathcal{M}|\mathcal{D}) \geq p(\bar{\mathcal{M}}|\mathcal{D})$ for all $\bar{\mathcal{M}}$. This can theoretically be approached by any method that is able to find some element in a set that maximises some function of the elements in that set, such as simulated annealing [217], or genetic algorithms [95, 167].

The two methods that will be described here are the ones that have been used to test the usefulness of the optimality definition. They are conceptually simple and not particularly intelligent, as neither of them uses any information embedded in the probabilistic LCS model besides the value proportional to $\ln p(\mathcal{M}|\mathcal{D})$ to form the search trajectory through the model structure space. Consequently, there is still plenty of room for improvement.

The reason why two alternatives are introduced is i) to emphasise the conceptual separation between evaluating the quality of a set of classifiers, and searching for better ones, and ii) to show that in theory any global optimiser can be used to perform the task of model structure search. As the aim is independent of the search procedure, reaching this aim only depends on the compatibility of the search procedure with the model structure space. After having introduced the two alternatives, a short discussion in Sect. 8.2.3 deals with their differences, and what might in general be good guidelines to improve the effectiveness of searching for good sets of classifiers.

Note that the optimal set of classifiers strongly depends on the chosen representation for the matching functions, as we can only find solutions that we are able to represent. Nonetheless, to keep the description of the methods representation-independent, the discussion of representation-dependent components of the methods are postponed until choosing some representation becomes inevitable; that is, in Sect. 8.3.

8.2.1 Model Structure Search by a Genetic Algorithm

Genetic algorithms (GA) are a family of global optimisers that are conceptually based on Darwinian evolution. The reader is expected to be familiar with their underlying idea and basic implementations, of which good overviews are available by Goldberg [95] and Mitchell [167].

An individual in the population that the GA operates on is defined by an LCS model structure $\mathcal{M}$, and its fitness is given by the value that `ModelProbability` returns for this model structure. As the genetic algorithm seeks to increase the fitness of the individuals in the population, its goal is to find the model structure that maximises $p(\mathcal{M}|\mathcal{D})$. An allele of an individual's genome is given by the representation of a single classifier's matching function, which makes the genome's length determined by the number of classifiers of the associated model structure. As this number is not fixed, the individuals in the population can be of variable length[2].

[2] Variable-length individuals might cause bloat, which is a common problem when using Evolutionary Computation algorithms with such individuals, as frequently observed in genetic programming [158]. It also plagues some Pittsburgh-style LCS that use variable-length individuals, such as LS-1 [198] and GAssist [6], and counteracting measures have to be taken to avoid its occurrence. Here, this is not an issue, as overly complex model structures will receive a lower fitness due to the preference of the applied model selection criterion for models of low complexity.

Function. `Crossover`($\mathcal{M}_a, \mathcal{M}_b$)

Input: two model structures $\mathcal{M}_a, \mathcal{M}_b$
Output: resulting two model structures $\mathcal{M}'_a, \mathcal{M}'_b$ after crossover

$K_a, K_b \leftarrow$ number of classifiers in $\mathcal{M}_a, \mathcal{M}_b$
$\mathbf{M}_a, \mathbf{M}_b$ matching function sets from $\mathcal{M}_a, \mathcal{M}_b$
$\mathbf{M}'_a \leftarrow \mathbf{M}_a \cup \mathbf{M}_b$
$K'_b \leftarrow$ random integer K such that $1 \leq K < K_a + K_b$
$\mathbf{M}'_b \leftarrow \emptyset$
for $k = 1$ **to** K'_b **do**
  $m_k \leftarrow$randomly selected matching function from $\mathbf{M}'_a$
  $\mathbf{M}'_b \leftarrow \mathbf{M}'_b \cup \{m_k\}$
  $\mathbf{M}'_a \leftarrow \mathbf{M}'_a \setminus m_k$
$\mathcal{M}'_a, \mathcal{M}'_b \leftarrow \{K_a + K_b - K'_b, \mathbf{M}'_a\}, \{K'_b, \mathbf{M}'_b\}$
return $\mathcal{M}'_a, \mathcal{M}'_b$

Starting with an initial population of P randomly generated individuals, a single iteration of the genetic algorithm is performed as follows: firstly, the matching matrix $\mathbf{M}$ is determined after (8.1) for each individual, based on its representation of the matching functions and the input matrix $\mathbf{X}$. This matching matrix is subsequently used to determine each individual's fitness by calling `ModelProbability`. After that, a new population is created by selecting two individuals from the current population and applying crossover with probability p_c and mutation with probability p_m. The last step is repeated until the new population again holds P individuals. Then, the new population replaces the current one, and the next iteration begins.

An individual is initially generated by randomly choosing the number of classifiers it represents, and then initialising the matching function of each of its classifiers, again randomly. How these matching functions are initialised depends on the representation and is thus discussed later. To avoid the influence of fitness scaling, the individuals from the current population are selected by deterministic tournament selection with tournament size t_s. Mutation is again dependent on the chosen representation, and will be discussed later.

As two selected individuals can be of different length, standard uniform crossover cannot be applied. Instead different means have to be used: the aim is to keep total number of classifiers constant, but as the location of the classifiers in the genome of an individual do not provide any information, their location is allowed to change. Thus, we proceed as shown in function `Crossover` by randomly choosing the new number K'_a and K'_b of classifiers in each of the new individuals $\mathcal{M}'_a$ and $\mathcal{M}'_b$ such that the sum of classifiers $K_a + K_b = K'_a + K'_b$ remains unchanged, and each new individual has at least one classifier. The matching functions of individual $\mathcal{M}'_b$ are determined by randomly picking K'_b matching functions from either of the old individuals. The other individual $\mathcal{M}'_a$ received all the remaining $K_a + K_b - K'_b$ matching functions. In summary, crossover is performed by collecting the matching functions of both individuals, and randomly redistributing them.

No particular criteria determine the convergence of the genetic algorithm when used in the following experiments. Rather, the number of iterations that it performs is pre-specified. Additionally, an elitist strategy is employed by separately maintaining the highest-fitness model structure $\mathcal{M}^*$ found so far. This model structure is not part of the normal population, but is replaced as soon as a fitter model structure is found.

This completes the description of the genetic algorithm that was used. It is kept deliberately simple to not distract from the task it has to solve, which is to find the model structure that maximises $p(\mathcal{M}|\mathcal{D})$. In the presented form, it *might* be considered as being a simple Pittsburgh-style LCS.

8.2.2 Model Structure Search by Markov Chain Monte Carlo

The given use of the MCMC algorithm provides a sample sequence $\mathcal{M}_1, \mathcal{M}_2, \dots$ from the model structure space that follows a Markov chain with steady state probabilities $p(\mathcal{M}|\mathcal{D})$, and thus allows sampling from $p(\mathcal{M}|\mathcal{D})$ [19]. As such a sampling process spends more time in high-probability ares of $p(\mathcal{M}|\mathcal{D})$, it takes more samples from high-probability model structures. Hence, the MCMC algorithm can be seen as a stochastic hill-climber that aims at finding the $\mathcal{M}$ that maximises $p(\mathcal{M}|\mathcal{D})$. The algorithm presented here is based on a similar algorithm developed for CART model search in [63].

The sample sequence is generated by the Metropolis-Hastings algorithm [103], which is give by the following procedure: given an initial model structure $\mathcal{M}_0$, a candidate model structure $\mathcal{M}'$ is created in step $t+1$, based on the current model structure $\mathcal{M}_t$. This candidate is accepted, that is, $\mathcal{M}_{t+1} = \mathcal{M}'$, with probability

$$\min\left(\frac{p(\mathcal{M}_t|\mathcal{M}')}{p(\mathcal{M}'|\mathcal{M}_t)}\frac{p(\mathcal{M}'|\mathcal{D})}{p(\mathcal{M}_t|\mathcal{D})}, 1\right), \tag{8.4}$$

and otherwise rejected, in which case the sequence continues with the previous model, that is, $\mathcal{M}_{t+1} = \mathcal{M}_t$. $p(\mathcal{M}_t|\mathcal{M}')$ and $p(\mathcal{M}'|\mathcal{M}_t)$ are the probability distributions that describes the process of generating the candidate model $\mathcal{M}'$. As the search procedure tends to prefer model structures that improve $p(\mathcal{M}|\mathcal{D})$, it is prone to spending many steps in areas of the model structure space where $p(\mathcal{M}|\mathcal{D})$ is locally optimal. To avoid being stuck in such areas, random restarts are performed after a certain number of steps, which are executed by randomly reinitialising the current model structure.

The initial model structure $\mathcal{M}_0$, as well as the model structure after a random restart, is generated by randomly initialising K classifiers, where K needs to be given. The matching function is assumed to be sampled from a probability distribution $p(m_k)$. Thus, $\mathcal{M}_0$ is generated by taking K samples from $p(m_k)$. The exact form of $p(m_k)$ depends on the chosen representation, and thus will be discussed later.

A new candidate model structure $\mathcal{M}'$ is created from the current model structure $\mathcal{M}_t$ with K_t classifiers similarly to the procedure used by Chipman, George and McCulloch [63], by choosing one of the following actions:

change. Picks one classifier of $\mathcal{M}_t$ at random, and re-initialises its matching function by taking a sample from $p(m_k)$.

add. Adds one classifier to $\mathcal{M}_t$, with a matching function sampled from $p(m_k)$, resulting in $K_t + 1$ classifiers.

remove. Removes one classifier from $\mathcal{M}_t$ at random, resulting in $K_t - 1$ classifiers.

The actions are chosen by taking samples from the discrete random variable $A \in \{\text{change}, \text{add}, \text{remove}\}$, where we assume $p(A = \text{add}) = p(A = \text{remove})$ and $p(A = \text{change}) = 1 - 2p(A = \text{add})$.

Let us now consider how to compute the acceptance probability (8.4) for each of these actions. We have $p(\mathcal{M}|\mathcal{D}) \propto p(\mathcal{D}|\mathcal{M})p(\mathcal{M}|K)p(K)$ by Bayes' Theorem, where, different to (7.3), we have separated the number of classifiers K from the model structure $\mathcal{M}$. As in (7.4), a uniform prior over unique models is assumed, resulting in $p(K) \propto 1/K!$. Additionally, every classifier in $\mathcal{M}$ is created independently by sampling from $p(m_k)$, which results in $p(\mathcal{M}|K) = p(m_k)^K$. Using variational inference, the model evidence is approximated by the variational bound $p(\mathcal{D}|\mathcal{M}) \propto \exp(\mathcal{L}_{\mathcal{M}}(q))$, where $\mathcal{L}_{\mathcal{M}}(q)$ denotes the variational bound of model $\mathcal{M}$. Thus, in combination we have

$$\frac{p(\mathcal{M}'|\mathcal{D})}{p(\mathcal{M}_t|\mathcal{D})} \approx \frac{\exp(\mathcal{L}_{\mathcal{M}'}(q))p(m_k)^{K'}(K'!)^{-1}}{\exp(\mathcal{L}_{\mathcal{M}_t}(q))p(m_k)^{K_t}(K_t!)^{-1}}, \tag{8.5}$$

where K' denotes the number of classifiers in $\mathcal{M}'$.

We get the model transition probability $p(\mathcal{M}'|\mathcal{M}_t)$ by marginalising over the actions A, to get

$$\begin{aligned} p(\mathcal{M}'|\mathcal{M}_t) = {} & p(\mathcal{M}'|\mathcal{M}_t, A = \text{change})p(A = \text{change}) \\ & + p(\mathcal{M}'|\mathcal{M}_t, A = \text{add})p(A = \text{add}) \\ & + p(\mathcal{M}'|\mathcal{M}_t, A = \text{remove})p(A = \text{remove}), \end{aligned} \tag{8.6}$$

and a similar expression for $p(\mathcal{M}_t|\mathcal{M}')$. When choosing action *add*, then $K' = K_t + 1$, and $p(\mathcal{M}'|\mathcal{M}_t, A = \text{change}) = p(\mathcal{M}'|\mathcal{M}_t, A = \text{remove}) = 0$, as neither the action *change* nor the action *remove* cause a classifier to be added. $\mathcal{M}_t$ and $\mathcal{M}'$ differ in a single classifier that is picked from $p(m_k)$, and therefore $p(\mathcal{M}_t|\mathcal{M}', A = \text{add}) = p(m_k)$. Similarly, when choosing the action *remove* for $\mathcal{M}_t$, an arbitrary classifier is picked with probability $1/K_t$, and therefore $p(\mathcal{M}'|\mathcal{M}_t, A = \text{remove}) = 1/K_t$. The action *change* requires choosing a classifier with probability $1/K_t$ and reinitialising it with probability $p(m_k)$, giving $p(\mathcal{M}'|\mathcal{M}_t, A = \text{change}) = p(m_k)/K_t$. The reverse transitions $p(\mathcal{M}_t|\mathcal{M}')$ can be evaluated by observing that the only possible action that causes the reverse transition from $\mathcal{M}'$ to $\mathcal{M}_t$ after the action *add* is the action *remove*, and vice versa. Equally, *change* causes the reverse transition after performing action *change*.

Overall, the candidate model $\mathcal{M}'$ that was created by *add* from $\mathcal{M}_t$ is accepted by (8.4) with probability

$$\begin{aligned} & \min\left(\frac{p(\mathcal{M}_t|\mathcal{M}', A = \text{remove})p(A = \text{remove})}{p(\mathcal{M}'|\mathcal{M}_t, A = \text{add})p(A = \text{add})} \frac{p(\mathcal{M}'|\mathcal{D})}{p(\mathcal{M}_t|\mathcal{D})}, 1\right) \\ & \approx \min\left(\exp\left(\mathcal{L}_{\mathcal{M}'}(q) - \mathcal{L}_{\mathcal{M}_t}(q) - 2\ln(K_t + 1)\right), 1\right), \end{aligned} \tag{8.7}$$

where we have used our previous assumption $p(A = \text{add}) = p(A = \text{remove})$, $K' = K_t + 1$, and (8.5). When choosing the action *remove*, on the other hand, the candidate model $\mathcal{M}'$ is accepted with probability

$$\min\left(\frac{p(\mathcal{M}_t|\mathcal{M}', A = \text{add})p(A = \text{add})}{p(\mathcal{M}'|\mathcal{M}_t, A = \text{remove})p(A = \text{remove})}\frac{p(\mathcal{M}'|\mathcal{D})}{p(\mathcal{M}_t|\mathcal{D})}, 1\right)$$
$$\approx \min\left(\exp\left(\mathcal{L}_{\mathcal{M}'}(q) - \mathcal{L}_{\mathcal{M}_t}(q) - 2\ln K_t\right), 1\right), \tag{8.8}$$

based on $K' = K_t - 1$, and (8.5). Note that in case of having $K' = 0$, the variational bound will be $\mathcal{L}_{\mathcal{M}'}(q) = -\infty$, and the candidate model will be always rejected, which confirms that a model without a single classifier is of no value. Finally, a candidate model $\mathcal{M}'$ where a single classifier from $\mathcal{M}_t$ has been changed by action *change* is accepted with probability

$$\min\left(\frac{p(\mathcal{M}_t|\mathcal{M}', A = \text{change})p(A = \text{change})}{p(\mathcal{M}'|\mathcal{M}_t, A = \text{change})p(A = \text{change})}\frac{p(\mathcal{M}'|\mathcal{D})}{p(\mathcal{M}_t|\mathcal{D})}, 1\right)$$
$$\approx \min\left(\exp\left(\mathcal{L}_{\mathcal{M}'}(q) - \mathcal{L}_{\mathcal{M}_t}(q)\right), 1\right). \tag{8.9}$$

To summarise, the MCMC algorithm starts with a randomly initialised model structure $\mathcal{M}_0$ with K_0 classifiers and at each step $t+1$ performs either *change*, *add*, or *remove* to create a candidate model structure $\mathcal{M}'$ from $\mathcal{M}_t$ that is either accepted ($\mathcal{M}_{t+1} = \mathcal{M}'$) with a probability that, dependent on the chosen action, is given by (8.7), (8.8) or (8.9), and otherwise rejected ($\mathcal{M}_{t+1} = \mathcal{M}_t$).

8.2.3 Building Blocks in Classifier Sets

As apparent from the above descriptions, the most pronounced difference between the GA and the MCMC search procedures is that the MCMC search only considers a single model structure at a time, while the GA operates on a population of them simultaneously. This parallelism allows the GA to maintain several competing model structure hypotheses that might contain valuable building blocks to form better model structures. In GA, *building blocks* refer to a group of alleles that in combination provide a part of the solution [95]. With respect to the model structure search, a building block is a subset of the classifiers in a model structure that *in combination* provides a good model for a subset of the data. A good model structure search maintains such building blocks and recombines them with other building blocks to form new model structure hypotheses.

Do such building blocks really exist in the given LCS model, and in LCS in general? Let us consider a simple example where the model structure contains a single classifier that matches all inputs with about equal probability. The only sensible action that MCMC search can perform is to add another classifier to see if it improves the model structure, which results in a classifier that matches all observations about equally, and a possibly more specific classifier that concentrates on a subset of the data. Only in rare cases will such a combination provide a better model for the data (see Sect. 8.3.3 for an example where it does). Rather, the globally matching classifier should be rearranged such that it does not directly compete with the specific classifier in modelling its part of the data. The resulting pair of classifiers would then cooperate to model a part of the data and can be seen as a building block of a potentially good model structure. Thus, while these building blocks exist, they are not exploited when using the MCMC algorithm for model structure search.

When using a GA for model structure search, on the other hand, the population of individuals can contain several potentially useful building blocks, and it is the responsibility of the crossover operator to identify and recombine them. As shown by Syswerda [210], uniform crossover generally yields better results that one-point and two-point crossover. The crossover operator that is used aims at uniform crossover for variable-length individuals. Further improvement in identifying building blocks can be made by using Estimation of Distribution Algorithms (EDAs) [183], but as there are currently no EDAs that directly apply to the problem structure at hand [150] this topic requires further investigation.

8.3 Empirical Demonstration

To demonstrate the usefulness of the optimality criterion that was introduced in the last chapter, the previously described algorithms are used to find a good set of classifiers for a set of simple regression tasks. These tasks are kept simple in the sense that the number of classifiers that are expected to be required are low, such that the $\mathcal{O}(K^3)$ complexity of `ModelProbability` does not cause any computational problems. Additionally, the crudeness of the model structure search procedures does not allow us to handle problems where the best solution is given by a complex agglomeration of classifiers. All regression tasks have $D_\mathcal{X} = 1$ and $D_\mathcal{Y} = 1$ such that the results can be visualised easily. The mixing features are given by $\phi(x) = 1$ for all x. Not all functions are standardised, but their domain is always within [-1:4] and their range is within [-1:1]. For all experiments, classifiers that model straight lines are used, together with uninformative priors and hyperpriors as given in Table 8.1.

Even though the prime problems that most new LCS are tested against are Multiplexer problems of various lengths [237], they are a challenge for the model structure search rather than the optimality criterion and thus are not part of the provided test set. Rather, a significant amount of noise is added to the data, as the aim is to provide a criterion that defines the minimal model, and can separate the underlying patterns from the noise, given that enough data is available.

Firstly, two different representations that are used for the matching functions are introduced. Then, the four regression tasks, their aim, and the found results are described, one by one.

8.3.1 Representations

The two representations that are going to be used are matching by radial-bases functions, and matching by soft intervals. Starting with matching by radial-basis functions, their matching functions as well as their initialisation and mutation is described.

Matching by Radial-Basis Functions

The matching function for matching by radial-basis functions is defined by

$$m_k(x) = \exp\left(\frac{1}{2\sigma_k^2}(x - \mu)^2\right), \tag{8.10}$$

which is an unnormalised Gaussian that is parametrised by a scalar μ_k and a positive spread σ_k. Thus, the probability of classifier k matching input x decreases with the distance from μ_k, where the strength of the decrease is determined by σ_k. If σ_k is small, then the matching probability decreases rapidly with the squared distance of x from μ_k. Note that, as $m_k(x) > 0$ for all $-\infty < x < \infty$, all classifiers match all inputs, even if only with a very low probability. Thus, we always guarantee that $\sum_k m_k(x_n) > 0$ for all n, that is, that all inputs are matched by at least one classifier, as required. Examples for

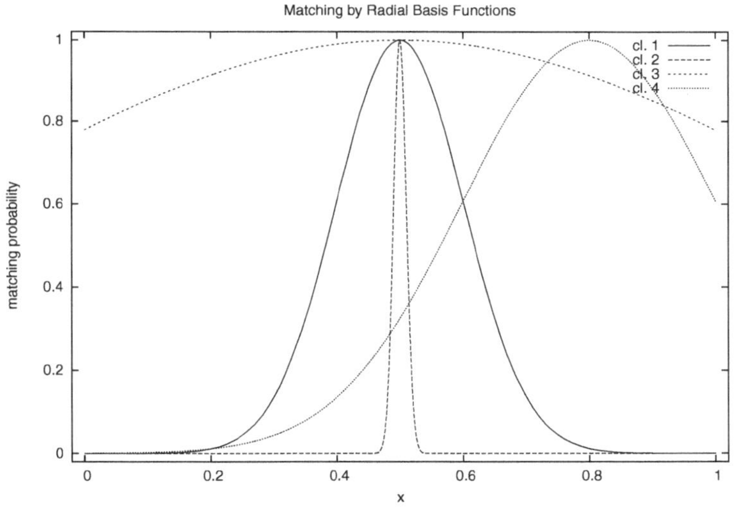

Fig. 8.2. Matching probability for matching by radial basis functions for different parameters. Classifiers 1, 2, and 3 all have their matching functions centred on $\mu_1 = \mu_2 = \mu_3 = 0.5$, but have different spreads $\sigma_1 = 0.1$, $\sigma_2 = 0.01$, $\sigma_3 = 1$. This visualises how a larger spread causes the classifier to match a larger area of the input space with higher probability. The matching function of classifier 4 is centred on $\mu_4 = 0.8$ and has spread $\sigma_4 = 0.2$, showing that μ controls the location x of the input space where the classifier matches with probability 1.

the shape of the radial-basis matching function are shown in Fig. 8.2. This form of matching function was chosen to demonstrate the possibility of matching by probability.

Rather than declaring μ_k and σ_k directly, the matching parameters $0 \leq a_k \leq 100$ and $0 \leq b_k \leq 50$ determine μ_k and σ_k by $\mu_k = l + (u - l)a_k/100$ and $\sigma_k^2 = 10^{-b_k/10}$, where $[l, u]$ is that range of the input x. Thus, a_k determines the centre of the classifier, where 0 and 100 specify the lower and higher end of x, respectively. σ_k is given by b_k such that $10^{-50} \leq \sigma_k^2 \leq 1$, and a low b_k gives a wide spread of the classifier matching function. A new classifier is initialised by randomly choosing a_k uniformly from $[0, 100)$, and b_k uniformly from $[0, 50)$. The two values are mutated by adding a sample from $\mathcal{N}(0, 10)$ to a_k, and a sample from $\mathcal{N}(0, 5)$ to b_k, but ensuring thereafter that they still conform to $0 \leq a_k \leq 100$ and $0 \leq b_k \leq 50$. The reason for operating on a_k, b_k rather than μ_k, σ_k is that it simplifies the mutation operation by making it independent of the range of x for μ_k and allows for non-linearity with respect to σ_k. Alternatively, one could simply acquire the mutation operator that was used by Butz, Lanzi and Wilson [52].

Matching by Soft Intervals

Matching by soft intervals is similar to the interval matching that was introduced in XCS by Wilson [239], with the difference that here, the intervals have soft boundaries. The reason for using soft rather than hard boundaries is to express the fact that we are never absolutely certain about the exact location of these boundaries, and to avoid the need to explicitly care about having each input matched by at least one classifier.

To avoid the representational bias of the centre/spread representation of Wilson [239], the lower/upper bound representation that was introduced and analysed by Stone and Bull [203] is used instead. The softness of the boundary is provided by an unnormalised Gaussian that is attached to both sides of the interval within which the classifier matches with probability 1. To avoid the boundaries from being too soft, they are partially included in the interval. More precisely, when specifying the interval for classifier k by its lower bound l_k and upper bound u_k, exactly one standard deviation of the Gaussian is to lie inside this interval, with the additional requirement of having 95% of the area underneath the matching function inside this interval. More formally, we need $0.95(b'_k + \sqrt{2\pi}\sigma_k) = b_k$ to hold to have the interval $b_k = u_k - l_k$ specify 95% of the area underneath the matching function, where b'_k gives the width of the interval where the classifier matches with probability 1, using the area $\sqrt{2\pi}\sigma$ underneath an unnormalised Gaussian with standard deviation σ. The requirement of the specified interval extending by one standard deviation to either side of the Gaussian is satisfied by $b'_k + 0.6827\sqrt{2\pi}\sigma_k = b_k$, based on the fact that the area underneath the unnormalised Gaussian within one standard deviation from its centre is $0.6827\sqrt{2\pi}\sigma$. Solving these equations with respect to b'_k and σ_k for a given b_k results in

$$\sigma_k = \frac{\frac{1}{0.95} - 1}{1 - 0.6827} \frac{1}{\sqrt{2\pi}} b_k \approx 0.0662 b_k, \tag{8.11}$$

$$b'_k = b_k - 0.6827\sqrt{2\pi}\sigma_k \approx 0.8868 b_k. \tag{8.12}$$

Thus, about 89% of the specified interval are matched with probability 1, and the leftover 5.5% to either side are matched according to one standard deviation of a Gaussian. Therefore, the matching function for soft interval matching is given by

$$m_k(x) = \begin{cases} \exp\left(-\frac{1}{2\sigma_k^2}(x - l'_k)^2\right) & \text{if } x < l'_k, \\ \exp\left(-\frac{1}{2\sigma_k^2}(x - u'_k)^2\right) & \text{if } x > u'_k \\ 1 & \text{otherwise,} \end{cases} \tag{8.13}$$

where l'_k and u'_k are the lower and upper bound of the interval that the classifier matches with probability 1, and are given by $l'_k \approx l_k + 0.0566 b_k$ and $u'_k \approx u_k - 0.0566 b_k$, such that $u'_k - l'_k = b'_k$. Fig. 8.3 shows examples for the shape of the matching function for soft interval matching.

Classifier k is initialised as by Stone and Bull, by sampling l_k and u_k from by a uniform distribution over $[l, u]$, which is the range of x. If $l_k > u_k$, then their values are swapped. While Stone and Bull [203] and Wilson [239] mutate the boundary values a uniform random variable, here the changes are sampled from a Gaussian to make small changes more likely than large changes. Thus, the boundaries after mutation are given by perturbing both bounds by $\mathcal{N}(0, (u - l)/10)$, that is, a sample from a zero-mean Gaussian with a standard deviation that is a 10th of the range of x. After that, it is again made sure that $l \leq l_k < u_k \leq u$ by swapping and bounding their values if required.

Even though both matching functions are only introduced for the case when $D_\mathcal{X} = 1$, they can be easily extended to higher-dimensional input spaces. In the case of radial-basis function matching, the matching function is specified by a multivariate Gaussian, analogous to the hyper-ellipsoidal conditions for XCS [41, 52]. Matching by a soft interval becomes slightly more complex due to the interval-specification of the matching function, but its computation can be simplified by defining the matching function as the product of one single-dimensional matching function per dimension of the input space.

8.3.2 Generated Function

To see if the optimality criterion is correct if the data conforms to the underlying assumptions of the model, it is firstly tested on a function that was generated to satisfy these assumptions. The data is generated by taking 300 samples from 3 linear classifiers with models $\mathcal{N}(y|0.05 + 0.5x, 0.1)$, $\mathcal{N}(y|2 - 4x, 0.1)$, and $\mathcal{N}(y| - 1.5 + 2.5x, 0.1)$ which use radial-basis function matching with (μ, σ^2) parameters $(0.2, 0.05)$, $(0.5, 0.01)$, $(0.8, 0.05)$ and mixing weights $v_1 = 0.5$, $v_2 = 1.0$, $v_3 = 0.4$,

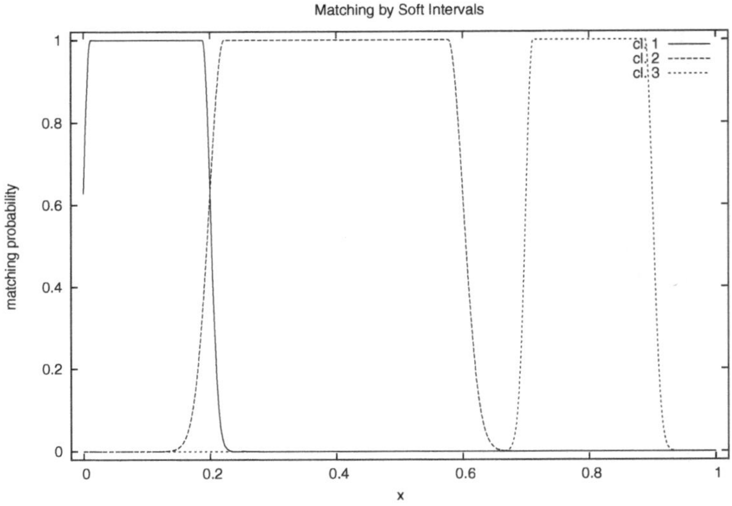

Fig. 8.3. Matching probability for matching by soft interval for different parameters. Classifiers 1 and 2 are adjacent as $l_1 = 0$, $u_1 = l_2 = 0.2$, and $u_2 = 0.5$. The area where these two classifiers overlap shows that the classifiers do not match their full interval with probability 1 due to the soft boundaries of the intervals. Nonetheless, 95% of the area beneath the matching function are within the specified interval. Classifier 3 matches the interval $l_3 = 0.7$, $u_3 = 0.9$. Comparing the boundary of classifier 2 and 3 shows that the spread of the boundary grows with the width of the interval that it matches.

respectively. A plot of the classifiers' means, their generated function mean, and the available data can be found in Fig. 8.4.

Both GA and MCMC model structure search were tested, where the GA is in this and all other experiments initialised with a population of size $P = 20$, crossover and mutation probability $p_c = p_m = 0.4$, and tournament size $t_s = 5$. The number of classifiers in each of the individuals is sampled from the binomial distribution $\mathcal{B}(8, 0.5)$, such that, on average, an individual has 4 classifiers. The performance of the GA model structure search is not sensitive to the initial size of the individuals and gives similar results for different initialisations of its population.

The result after a single run with 250 GA iterations are shown in Fig. 8.5. As can be seen, the model was not correctly identified as the number of classifiers of the best found individual is 2 rather than the desired 3, with $\mathcal{L}(q) - \ln K! \approx 118.81$. Nonetheless, the generated function mean is still within the first standard deviation of the predicted mean.

The MCMC model structure search was applied to the same data, using for this and all further experiments 10 restarts with 500 steps each, and $p(A = add) = p(A = remove) = 1/4$. Thus, MCMC search uses the same number

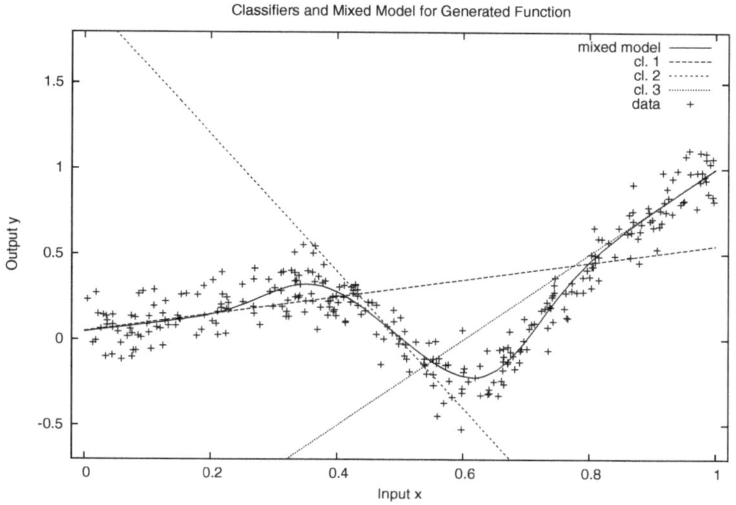

Fig. 8.4. Classifier models, mixed model and available data for the generated function

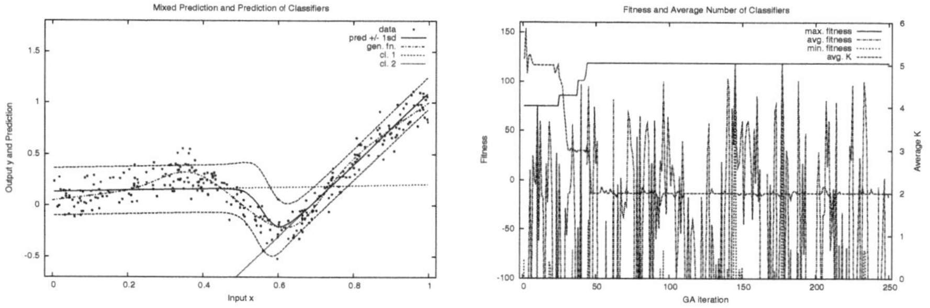

(a) (b)

Fig. 8.5. Plots showing the best found model structure for the generated function using GA model structure search, and fitness and average number of classifiers over the GA iterations. Plot (a) shows the available data, the model of the classifiers, and their mixed prediction with 1 standard deviation to either side, and additionally the mean of the generating function. The matching function parameters of the classifiers are $\mu_1 = 0.09, \sigma_1^2 = 0.063$ and $\mu_2 = 0.81, \sigma_2^2 = 0.006$. Plot (b) shows the maximum, average, and minimum fitness of the individuals in the population after each GA iteration. The minimum fitness is usually below the lower edge of the plot. The plot also shows the average number of classifiers for all individuals in the current population.

of model structure evaluations as the GA. The initial number of classifiers is after each restart sampled from the binomial distribution $\mathcal{B}(8, 0.5)$, resulting in 4 classifiers on average.

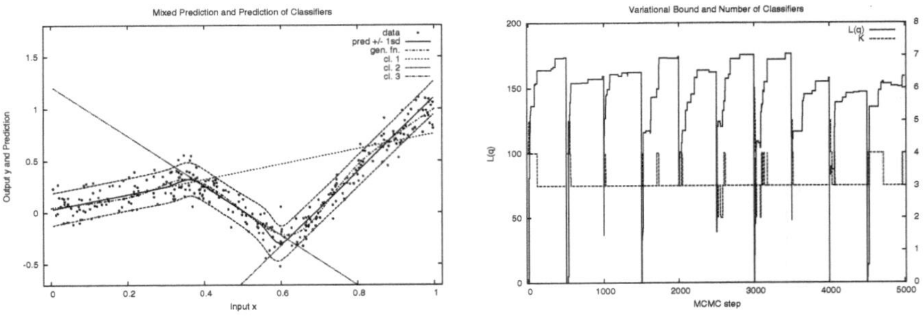

(a) (b)

Fig. 8.6. Plots showing the best discovered model structure for the generated function using MCMC model structure search, and variational bound and number of classifiers over the MCMC steps. Plot (a) shows the available data, the model of the classifiers, and their mixed prediction with 1 standard deviation to either side, and additionally the mean of the generating function. The matching function parameters of the classifiers are $\mu_1 = 0.16, \sigma_1^2 = 0.01$, $\mu_2 = 0.461, \sigma_2^2 = 0.025$, and $\mu_3 = 0.78, \sigma_3^2 = 0.006$. Plot (b) shows the variational bound $\mathcal{L}(q)$ for each step of the MCMC algorithm, and clearly visualises the random restarts after 500 steps. It also shows the number of classifiers K in the current model structure for each step of the MCMC search.

As can be seen in Fig. 8.6, MCMC model structure search performed better than the GA by correctly identifying all 3 classifiers with $\mathcal{L}(q) - \ln K! \approx 174.50$, indicating a higher $p(\mathcal{M}|\mathcal{D})$ than for the one found by the GA. While the discovered model structure is not exactly that of the data-generating process, it is intriguingly similar, given the rather crude search procedure. The reject rate of the MCMC algorithm was about 96.9%, which shows that the algorithm quickly finds a local optimum and remains there.

8.3.3 Sparse, Noisy Data

While the noise of the generated function is rather low and there is plenty of data available, the next experiment investigates if the optimality criterion can handle more noise and less data. For this purpose the test function from Waterhouse et al. [227] is taken, where it was used to test the performance of the Bayesian MoE model with a fixed model structure. The function is given by $f(x) = 4.25(\mathrm{e}^{-x} - 4\mathrm{e}^{-2x} + 3\mathrm{e}^{-3x}) + \mathcal{N}(0, 0.2)$ over $0 \leq x \leq 4$, and is shown in Fig. 8.7, together with the 200 sampled observations. Waterhouse et al. used additive noise with variance 0.44 which was here reduced to 0.2 as otherwise no pattern was apparent in the data. It is assumed that the Bayesian MoE model was only able to identify a good model despite the high noise due to its pre-determined model structure.

Again using radial-basis function matching, the GA and MCMC settings are the same as in the previous experiment, except for the initial number of

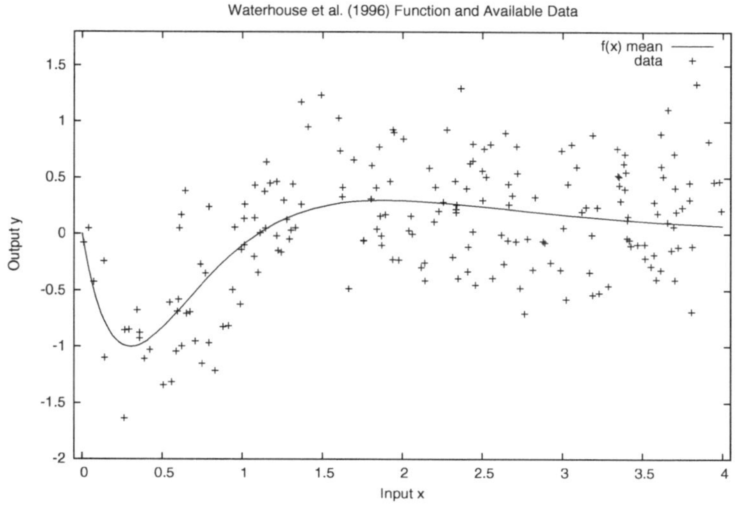

Fig. 8.7. Plot showing the test function used in [227], and the 200 available observations

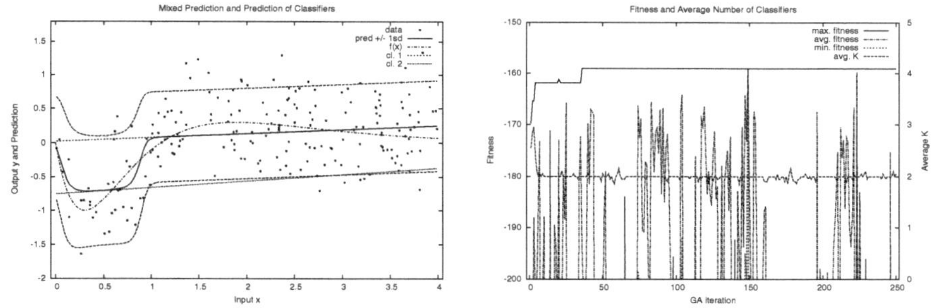

(a) (b)

Fig. 8.8. Plots similar to the ones in Fig. 8.5, when using a GA for model structure search applied to the function as used by Waterhouse et al. [227]. The best discovered model structure is given by $\mu_1 = 0.52, \sigma_1 = 0.016$ and $\mu_2 = 3.32, \sigma_2 = 1.000$.

classifiers, which is in both cases sampled from $\mathcal{B}(4, 0.5)$. As before, the result is insensitive to this number. The best discovered model structures are shown in Fig. 8.8 for the GA, with $\mathcal{L}(q) - \ln K! \approx -159.07$, and in Fig. 8.9 for the MCMC, with $\mathcal{L}(q) - \ln K! \approx -158.55$. The MCMC search had a reject rate of about 97.0% over its 5000 steps.

Both the GA and the MCMC search resulted in about the same model structure which at the first sight seems slightly surprising: looking at Fig. 8.7, one would initially expect the function to be modelled by a flat line over $1.5 < x < 4$, and 2 straight lines for the bump at around $x = 0.4$, requiring altogether 3

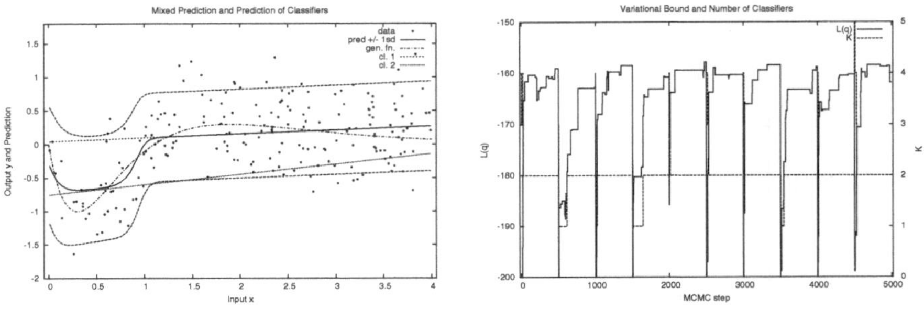

(a) (b)

Fig. 8.9. Plots similar to the ones in Fig. 8.6, when using MCMC model structure search applied to the function as given in [227]. The best discovered model structure is given by $\mu_1 = 0.56, \sigma_1 = 0.025$ and $\mu_2 = 2.40, \sigma_2 = 0.501$.

classifier. The model structure search, however, has identified a model that only requires 2 classifiers by having a global classifier that models the straight line, interleaved by a specific classifier that models the bump. This clearly shows that the applied model selection method prefers simpler models over more complex ones, in addition to the ability of handling rather noisy data.

8.3.4 Function with Variable Noise

One of the disadvantages of XCS, as discussed in Sect. 7.1.1, is that the desired mean absolute error of each classifier is globally specified by the system parameter ϵ_0. Therefore, XCS cannot properly handle data where the noise level varies significantly over the input space. The introduced LCS model assumes constant noise variance at the classifier level, but does not make such an assumption at the global level. Thus, it can handle cases where each classifier requires to accept a different level of noise, as is demonstrated by the following experiment.

Similar, but not equal to the study by Waterhouse et al. [227], the target function has two different noise levels. It is given for $-1 \leq x \leq 1$ by $f(x) = -1 - 2x + \mathcal{N}(0, 0.6)$ if $x < 0$, and $f(x) = -1 + 2x + \mathcal{N}(0, 0.1)$ otherwise. Thus, the V-shaped function has a noise variance of 0.6 below $x = 0$, and a noise variance of 0.1 above it. Its mean and 200 data points that are used as the data set are shown in Fig. 8.10. To assign each classifier to a clear interval of the input space, soft interval matching is used.

Both GA and MCMC search were applied with with the same settings as before, with the initial number of classifiers sampled from $\mathcal{B}(8, 0.5)$. The best discovered model structures are shown for the GA in Fig. 8.11, with $\mathcal{L}(q) + \ln K! \approx -63.12$, and for MCMC search in Fig. 8.12, with a slightly better $\mathcal{L}(q) + \ln K! \approx -58.59$. The reject rate of the MCMC search was about 96.6%.

In both cases, the model structure search was able to identify two classifiers with different noise variance. The difference in the modelled noise variance is

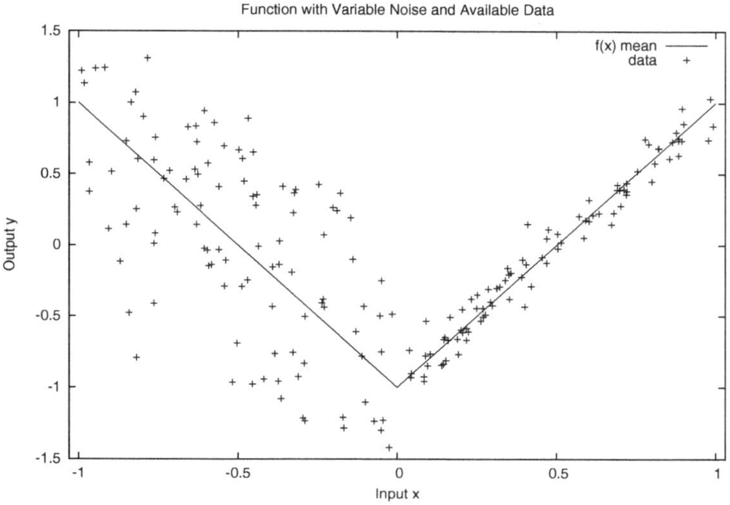

Fig. 8.10. Plot showing the mean of the function with variable noise, and the 200 observations that are available from this function

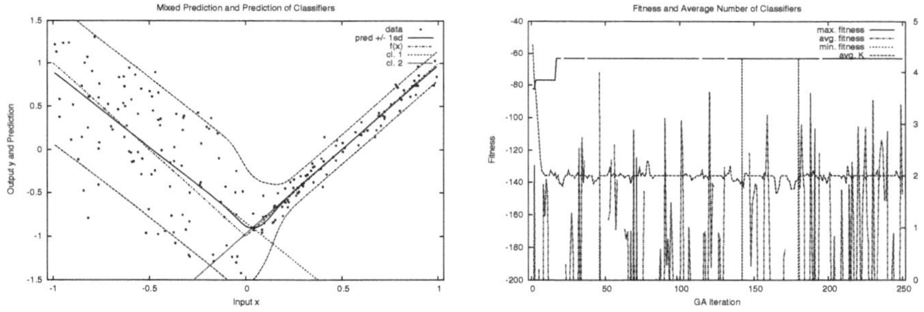

(a) (b)

Fig. 8.11. Plots similar to the ones in Fig. 8.5, where GA model structure search was applied to a function with variable noise. The best discovered model structure is given by $l_1 = -0.82, u_1 = 0.08$ and $l_2 = 0.04, u_2 = 1.00$.

clearly visible in both Fig. 8.11 and 8.12 by the plotted prediction standard deviation. This demonstrates that the LCS model is suitable for data where the level of noise differs for different areas of the input space.

8.3.5 A Slightly More Complex Function

To demonstrate the limitations of the rather naïve model structure search methods as introduced in this chapter, the last experiment is performed on a slightly

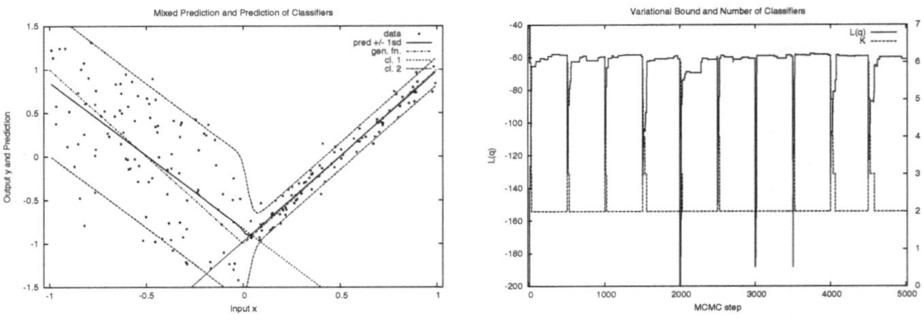

(a) (b)

Fig. 8.12. Plots similar to the ones in Fig. 8.6, where MCMC model structure search was applied to a function with variable noise. The best discovered model structure is given by $l_1 = -0.98, u_1 = -0.06$ and $l_2 = 0.08, u_2 = 0.80$.

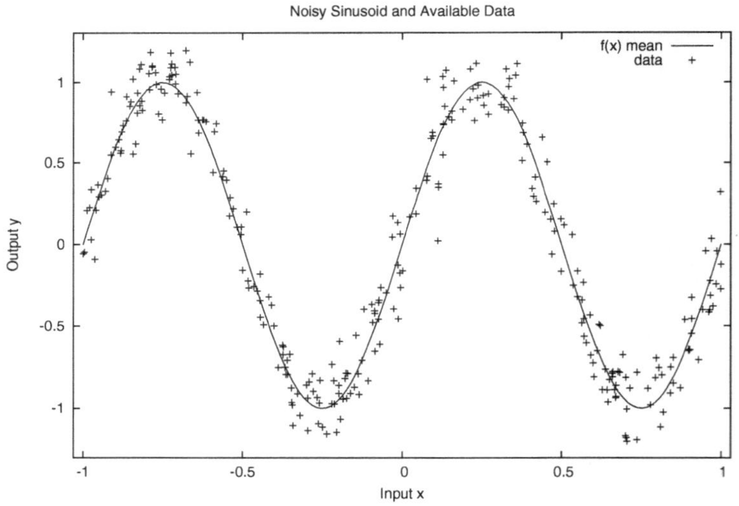

Fig. 8.13. Plot showing the mean of the noisy sinusoidal function, and the 300 observations that are available from this function

more complex function. The used function is the noisy sinusoid given over the range $-1 \leq x \leq 1$ by $f(x) = \sin(2\pi x) + \mathcal{N}(0, 0.15)$, as shown in Fig. 8.13. Soft interval matching is again used to clearly specify the area of the input space that a classifier models. The data set is given by 300 samples from $f(x)$.

Both GA and MCMC search are initialised as before, with the number of classifiers sampled from $\mathcal{B}(8, 0.5)$. The GA search identified 7 classifiers with $\mathcal{L}(q) + \ln K! \approx -155.68$, as shown in Fig. 8.14. It is apparent that the model can be improved by reducing the number of classifiers to 5 and moving them to

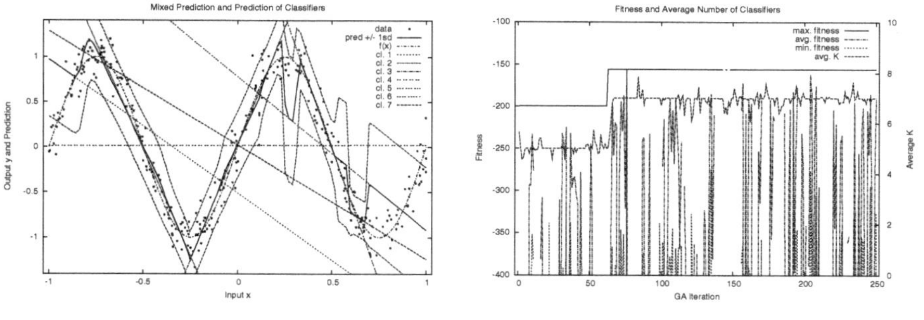

(a) (b)

Fig. 8.14. Plots similar to the ones in Fig. 8.5, using GA model structure search applied to the noisy sinusoidal function. The best discovered model structure is given by $l_1 = -0.98, u_1 = -0.40$, $l_2 = -0.78, u_2 = -0.32$, $l_3 = -0.22, u_3 = 0.16$, $l_4 = -0.08, u_4 = 0.12$, $l_5 = 0.34, u_5 = 0.50$, $l_6 = 0.34, u_6 = 1.00$, and $l_7 = 0.60, u_2 = 0.68$.

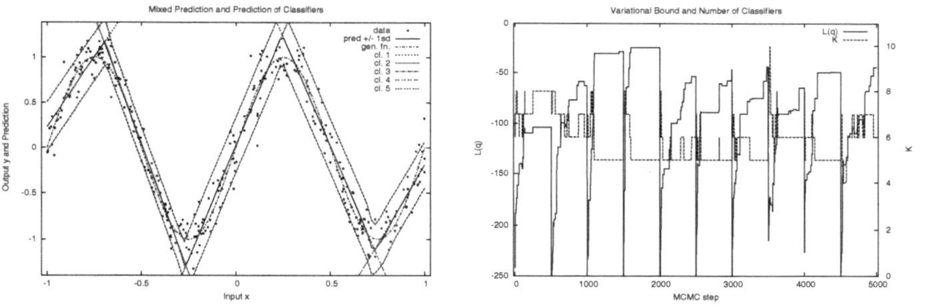

(a) (b)

Fig. 8.15. Plots similar to the ones in Fig. 8.6, using MCMC model structure search applied the noisy sinusoidal function. The best discovered model structure is given by $l_1 = -1.00, u_1 = -0.68$, $l_2 = -0.62, u_2 = -0.30$, $l_3 = -0.24, u_3 = 0.14$, $l_4 = 0.34, u_4 = 0.78$, and $l_5 = 0.74, u_5 = 0.98$.

adequate locations. However, as can be seen in Fig. 8.14(b), the GA initially was operating with 5 classifiers, but was not able to find good interval placements, as the low maximum fitness shows. Once it increased the number of classifiers to 7, at around the 60th iteration, it was able to provide a fitter model structure, but at the cost of an increased number of classifiers. It maintained this model up to the 250th iteration without finding a better one, which indicates that the genetic operators need to be improved and require better tuning to the representation used in order to make the GA perform better model structure search.

That the inappropriate model can be attributed to a weak model structure search rather than a failing optimality criterion becomes apparent when considering the result of the MCMC search with a superior $\mathcal{L}(q) - \ln K! \approx -29.39$, as shown in Fig. 8.15. The discovered model is clearly better, which is also reflected in a higher $p(\mathcal{M}|\mathcal{D})$. Note, however, that this model was not discovered after all restarts of the MCMC algorithm. Rather, model structures with 6 or 7 classifiers were sometimes preferred, as Fig. 8.15(b) shows. This indicates that a further increase of the problem complexity will very likely cause the MCMC search to fail as well.

8.4 Improving Model Structure Search

As previously emphasised, the model structure search procedures introduced in this chapter are naïve in the sense that they are ignorant about a lot of the information that is available in the LCS model. Also, they are only designed for batch learning and as such are unable to handle tasks where incremental learners are required.

Here, a few suggestions are given on how, on one hand, more information can be used to guide the model structure search, and, on the other hand, how the batch learning method can be turned into an incremental learner. The introduced model structure search methods are general such that modifying the LCS model type to a linear LCS, for example, does not invalidate these methods. Guiding the model structure search by information that is extracted from the probabilistic model makes the search procedure depend on the model type. Thus, while it might be more powerful thereafter, it is also only applicable to one particular model structure. The modifications that are suggested here only apply to LCS model types that train their classifiers independently.

Independent of the LCS model type, incremental learning can occur at two different levels: On one hand, one can learn the LCS model parameters incrementally while keeping the model structure fixed. On the other hand, the model structure can be improved incrementally, as done by Michigan-style LCS. Both levels will be discussed here, but as before, they will only be discussed for the LCS model type that is introduced in this book.

8.4.1 Using More Information

Suggestions on how the model structure search can be improved focus exclusively on the GA, as it has the advantage of exploiting building blocks in the LCS model (see Sect. 8.2.3). It can be improved on two levels: i) more information embedded in the LCS model should be used than just the fitness of a model structure, and ii) current theoretical and practical advances in evolutionary computation should be used to improve the GA itself.

With respect to using the information that is available within the model itself, model structure search operates on the classifiers, and in particular on their matching function. Thus, it is of interest to gain more information about a single classifier c_k within a model structure $\mathcal{M}$. Such information could, for example, be gained by evaluating the probability $p(c_k|\mathcal{M}, \mathcal{D})$ of the classifier's model in the light of the available data. By Bayes' rule $p(c_k|\mathcal{M},\mathcal{D}) \propto p(\mathcal{D}|c_k,\mathcal{M})p(c_k|\mathcal{M})$, where the classifier model evidence $p(c_k|\mathcal{M},\mathcal{D})$, is similarly to (7.2) given by

$$p(\mathcal{D}|c_k,\mathcal{M}) = \int p(\mathcal{D}|\boldsymbol{\theta}_k,c_k,\mathcal{M})p(\boldsymbol{\theta}_k|c_k,\mathcal{M})\mathrm{d}\boldsymbol{\theta}_k, \tag{8.14}$$

and matching needs to be taken into account when formulating $p(\mathcal{D}|\boldsymbol{\theta}_k,c_k,\mathcal{M})$. As for good model structures, good classifiers are classifiers for which $p(c_k|\mathcal{M},\mathcal{D})$ is large, or equivalently, for which $p(\mathcal{D}|c_k,\mathcal{M})$ is large, given uniform classifier priors $p(c_k|\mathcal{M})$. Therefore, the mutation operator of the GA can be biased towards changing bad classifiers, and genetic operators can construct new individuals from good classifiers of other individuals, or prune individuals by removing bad classifiers.

From the GA side itself, a more principled approach might be sought from evolutionary computation techniques where variable-length individuals are common. Genetic programming (GP) is one of them, as the program that each individual represents is not fixed in size. However, while the fitness of a program does not necessarily decrease with its complexity, Bayesian model selection penalises overly complex model structures. Thus, GP suffers from the problem of individuals growing out of bound that is naturally avoided in the approach presented here. Nonetheless, some of the theoretical results of GP might still be applicable to improving the GA to search the model structure space.

Another route that can additionally improve the performance of the GA is to use Estimation of Distribution Algorithms (EDAs) [183] that improve the crossover operator by detecting and preserving building blocks. They do so by creating a statistical model of the high-fitness individuals of the current population and draw samples from this model to create new individuals, rather than using standard crossover operators. Good building blocks are expected to appear in several individuals and consequently receive additional support in the model. The Bayesian Optimization Algorithm (BOA), for example, models the alleles of selected individuals by a Bayesian network that is samples thereafter to produce new individuals [182].

Currently, there exists no EDA that can handle variable-length individuals adequately [150]. The problem is that the space of possible classifiers, that is, the space of possible matching function parametrisations, is too large to frequently have similar classifiers within the population. The chance of having the same building blocks is even exponentially smaller [43, 150]. Despite these issues it is still worth trying to construct an EDA that can be used with the population structure at hand, at least to provide a more principled crossover operator. What needs to be considered when constructing the statistical population model is the irrelevance of the location of a classifiers within an individual. The only thing that matters is the classifier itself, and if it frequently co-occurs with the same other classifiers. This should allow modelling and preserving building blocks within a set of classifiers. An additional enhancement of the model is to take the nature of the matching function into account, as done for Michigan-style LCS by Butz and Pelikan [54] and Butz et al. [55].

8.4.2 Incremental Implementations

An even more challenging task is to turn the developed batch implementation into an incremental learner. Incremental learning can be performed on two levels, each of which will be discussed separately: i) the model parameters $\boldsymbol{\theta}$ can be updated incrementally, while holding the model structure $\mathcal{M}$ fixed ii) the model structure can be updated incrementally under the assumption that the correct model parameters are known immediately after each update. The aim of an incremental learner is to perform incremental learning on both levels. To do this successfully, however, we firstly have to be able to handle incremental learning on each of the two levels separately.

Incremental learning on the model parameter level alone is sufficient to handle reinforcement learning tasks. Incrementally learning the model structure, on the other hand, is computationally more efficient as it only requires working with a single model structure at a time (making it a Michigan-style LCS) rather than having to maintain several model structures at once (as is the case for Pittsburgh-style LCS). Thus, performing incremental learning on either level alone is already a useful development.

Incremental Model Parameter Update

Having provided a Bayesian LCS model for a fixed model structure $\mathcal{M}$, one could assume that it automatically provides the possibility of training its parameters incrementally by using the posterior of one update as the prior of the next update. However, due to the use of hyperpriors, this does not always apply.

Assuming independent classifier training, let us initially focus on the classifiers. The classification model that was used does not use a hyperprior and thus can be easily updated incrementally. The update (7.129) of its only parameter $\boldsymbol{\alpha}_k^*$ is a simple sum over all observations, which can be performed for each observation separately.

Classifier models for regression, on the other hand, have several interlinked parameters $\mathbf{W}, \boldsymbol{\tau}$ and $\boldsymbol{\alpha}$ that are to be updated in combination. Let us consider the posterior weight (7.98) and precision (7.97) of the classifier model, which also results from performing matching-weighted ridge regression with ridge complexity $\mathbb{E}_\alpha(\alpha_k)$ (see Sect. 7.3.1). As shown in Sect. 5.3.5, ridge regression can, due to its formal equivalence to RLS, be performed incrementally. Note, however, that the ridge complexity is set by the expectation of the prior on α_k that is modelled by the hyperprior (7.9) and is updated together with the classifier model parameters. A direct change of the ridge complexity after having performed some RLS updates is not feasible. However, there remain two possibilities for an incremental update of these parameters: one could fix the prior parameters by specifying α_k directly rather than modelling it by a hyperprior. Potentially good values for α_k are given in Sect. 7.2.3. Alternatively, one can incrementally update $\sum_n m(\mathbf{x}_n)\mathbf{x}_n\mathbf{x}_n^T$ and recover $\boldsymbol{\Lambda}_k^*$ after each update by using (7.97) directly, which requires a matrix inversion of complexity $\mathcal{O}(D_\mathcal{X}^3)$ rather than the $\mathcal{O}(D_\mathcal{X}^2)$ of the RLS algorithm. Thus, either the bias of the model or the computational complexity of the update is increased. Using uninformative priors, the first approach might be the one to prefer. From inspecting (7.99) and (7.100) it can be seen that both parameters of the noise precision model can be updated incrementally without any modifications.

Even though a least squares approximation could be used to train the mixing model, analogous to Sect. 6.1.2, the results in Chap. 6 have shown that it is possible to design heuristics that outperform this approximation. Additionally, these heuristics might not require any parameters to be updated, besides the parameters of the classifiers themselves. Given that similar parameter-less heuristics exist for the Bayesian model, they can be immediately used in incremental implementations, as no parameters need to be updated. Possible approaches where already outlined in Sect. 6.4.

Incremental Model Structure Search

The GA in Michigan-style LCS has strong parallels to cooperative co-evolutionary algorithms (for example [235]). In these, the fitness of an individual depends on the other individuals in the population. Equally, the fitness of a classifier in a Michigan-style LCS depends on the other classifiers in the set of classifiers as they cooperate to form the solution. Note that while in Pittsburgh-style LCS an individual is a set of classifiers that provides a candidate solution, in Michigan-style each classifier is an individual and the whole population forms the solution.

Having defined a fitness for a set of classifiers by the model structure probability, the aim is to design an algorithm that is able to increase the fitness of this population by modifying separate classifiers. Expressed differently, we want to design a GA for which the fixed point of its operators is the optimal set of classifiers such that $p(\mathcal{M}|\mathcal{D})$ is maximised. While this is not a trivial problem, an obvious approach is to attempt to design a cooperative co-evolutionary algorithm with such operators, or to modify existing LCS, like XCS(F), to aim at satisfying the optimality criterion. However, the lack of theoretical understanding of either method does not make the approach any simpler [171].

Here, an alternative based on Replicator Dynamics (for example, [108]) is suggested: assume that the number of possible matching function parametrisations is given by a finite P (for any finite $\mathcal{X}$ and a sensible representation this is always the case) and that $C_1, \dots, C_P$ enumerate each possible type of matching function. Each C_i stands for a classifier type that is a possible replicator in a population. Let $\mathbf{c} = (c_1, \dots, c_P)^T$ denote the frequency of each of the classifier types. Assuming an infinite population model, c_i gives the proportion of classifiers of C_i in the population. As the c_i's satisfy $0 \leq c_i \leq 1$ and $\sum_i c_i = 1$, $\mathbf{c}$ is an element of the P-dimensional simplex S_P.

The fitness $f_i(\mathbf{c})$ of C_i is a function of all classifiers in the population, described by $\mathbf{c}$. The rate of increase $\dot{c}_i/c_i$ of classifier type C_i is a measure of its evolutionary success and may be expressed as the difference between the fitness of C_i and the average fitness $\bar{f}(\mathbf{c}) = \sum_i c_i f_i(\mathbf{c})$, which results in the *replicator equation*

$$\dot{c}_i = c_i \left(f_i(\mathbf{c}) - \bar{f}(\mathbf{x}) \right) . \tag{8.15}$$

Thus, the frequency of classifier type C_i only remains unchanged if there is no such classifier in the current population, or if its fitness equals the average fitness of the current population. The population is stable only if this applies to all its classifiers.

One wants to define a fitness function for each classifier such that the stable population is the optimal population according to the optimality criterion. Currently $\mathcal{L}(q)$ by (7.96) cannot be fully split into one component per classifier due to the term $\ln |\mathbf{\Lambda}_V^{*\,-1}|$ in $\mathcal{L}_M(q)$ that results from the mixing model. Replacing this mixing model by heuristics should make such a split possible. Even then it is for each classifier a function of all classifiers in the current population, as the mixing coefficients assigned to a single classifier for some input depend on other classifiers that match the same input, which conforms to the above definition of the fitness of a classifier type being a function of the frequency of all classifier types.

The stable state of the population is given if a classifier's fitness is equal to the average fitness of all classifiers. This seems very unlikely to result naturally from splitting $\mathcal{L}(q)$ into the classifier components, and thus either (8.15) needs to be modified, or the fitness function needs to be tuned so that this is the case. If and how this can be done cannot be answered before the fitness function is available. Furthermore, (8.15) does not allow the emergence of classifiers that initially have a frequency of 0. As initialising the population with all possible classifiers is not feasible even for rather small problems, new classifier types need to be added stochastically and periodically. To make this possible, (8.15) needs to be modified to take this into account, resulting in a stochastic equation.

Obviously, a lot more work is required to see if the replicator dynamics approach can be used to design Michigan-style LCS. If it can, the approach opens the door to applying the numerous tools designed to analyse replicator dynamics to the analysis of the classifier dynamics in Michigan-style LCS.

8.5 Summary

In this chapter it was demonstrated how to the optimality criterion that was introduced in the previous chapter can be applied by implementing variational Bayesian inference together with some model structure search procedure. Four simple regression tasks were used to demonstrate that the optimality criterion based on model selection yields adequate results.

A set of function were provided that perform variational Bayesian inference to approximate the model probability $p(\mathcal{M}|\mathcal{D})$ and act as a basis for evaluating the quality of a set of classifiers. More specifically, the function `ModelProbability` takes the model structure $\mathcal{M}$ and the data $\mathcal{D}$ as arguments and returns an approximation to the unnormalised model probability. Thus, in addition to the theoretical treatment of variational inference in the previous chapter, it was shown here how to implement it for the regression case. Due to required complex procedure of finding the mixing weight vectors to combine the localised classifier models to a global model, the described implementation scales unfavourably with the number of classifiers K. This complexity might be reduced by replacing the generalised softmax function by well-tuned heuristics, but further research is required to design such heuristics.

Two methods to find the $\mathcal{M}$ that maximises $p(\mathcal{M}|\mathcal{D})$ have been introduced to emphasise that in theory any global optimisation procedure can be used to find the best set of classifiers. On one hand, a GA was described that operates in a Pittsburgh-style LCS way, and on the other hand, an MCMC was employed that samples $p(\mathcal{M}|\mathcal{D})$ and thus acts like a stochastic hill-climber. Both methods are rather crude, but sufficient to demonstrate the abilities of the optimality criterion.

Using the introduced optimisation algorithms, it was demonstrated on a set of regression tasks that the definition of the best set of classifiers i) is able to differentiate between patterns in the data and noise, ii) prefers simpler model structures over more complex ones, and iii) can handle data where the level of noise differs for different areas of the input space. These features have not been available in any LCS before, without the requirement of manually tuning system parameters that influence not only the model structure search procedure but also the definition of what resembles a good set of classifiers. Being able to handle different levels of noise is a feature that has possibly not been available in any LCS before, regardless of how the system parameters are tuned.

At last, the model structure search has been discussed in more detail, to point out how it might be improved and modified to meet different requirements. Currently, none of the two model structure search procedures facilitate any form of information that is available from the probabilistic LCS model other than an approximation to $p(\mathcal{M}|\mathcal{D})$. Thus, the power of these methods can be improved by using this additional information and by facilitating recent developments that improve on the genetic operators.

Another downside of the presented methods is that they currently only support batch learning. Incremental learning can be implemented on both the model parameter and the model structure level, either of which were discussed separately. While on the parameter level only minor modifications are required, providing an incremental implementation on the model structure level, which effectively results in a Michigan-style LCS, is a major challenge. Its solution will finally bridge the gap between Pittsburgh-style and Michigan-style LCS, which are, as presented here, just different implementations with the same aim of finding the set of classifiers that explains the data best. Up until now, there was no formally well-defined definition of this aim, and providing this definition is the first step towards a solution to that challenge.

9 Towards Reinforcement Learning with LCS

Having until now concentrated on how LCS can handle regression and classification tasks, this chapter returns to the prime motivator for LCS, which are sequential decision tasks. There has been little theoretical LCS work that concentrates on these tasks (for example, [30, 224]) despite some obvious problems that need to be solved [11, 12, 77]. At the same time, other machine learning methods have constantly improved their performance in handling these tasks [126, 28, 204], based on extensive theoretical advances. In order to catch up with these methods, LCS need to refine their theory if they want to be able to feature competitive performance. This chapter provides a strong basis for further theoretical development within the MDP framework, and discusses some currently relevant issues.

Sequential decision tasks are, in general, characterised by having a set of states and actions, where an action performed in a particular state causes a transition to the same or another state. Each transition is mediated by a scalar reward, and the aim is to perform actions in particular states such that the sum of rewards received is maximised in the long run. How to choose an action for a given state is determined by the *policy*. Even though the space of possible policies could be searched directly, a more common and more efficient approach is to learn for each state the sum of future rewards that one can expect to receive from that state, and derive the optimal policy from that knowledge.

The core of Dynamic Programming (DP) is how to learn the mapping between states and their associated expected sum of rewards, but to do so requires a model of the transition probabilities and the rewards that are given. Reinforcement Learning (RL), on the other hand, aims at learning this mapping, known as the *value function*, at the same time as performing the actions, and as such improves the policy simultaneously. It can do so either without any model of the transitions and rewards – known as *model-free* RL – or by modelling the transitions and rewards from observations and then using DP methods based on these models to improve the policy – known as *model-based* RL. Here, we mainly focus on model-free RL as it is the variant that has been used most frequently in LCS.

J. Drugowitsch: Des. & Anal. of Learn. Class. Sys.: A Prob. Approach, SCI 139, pp. 203–235, 2008.
springerlink.com

If the state space is large or even continuous then the value function is not learned for each state separately but rather modelled by some function approximation technique. However, this limits the quality of the discovered policy by how close the approximated value function is to the real value function. Furthermore, the shape of the value function is not known beforehand, and so the function approximation technique has to be able to adjust its resources adaptively. Considering that LCS provide such adaptive regression models, they seem to be a key candidate for approximating the value function of RL methods; and this is in fact exactly what LCS are used for when applied to sequential decision tasks: they act as adaptive value function approximation methods to aid learning the value function of RL methods.

Due to early LCS pre-dating common RL methods, they have not always been characterised as approximating the value function. In fact, the first comparison between RL and LCS was performed by Dorigo and Bersini [74] to show that a Very Simple CS without generalisation and a slightly modified implicit bucket brigade is equivalent to tabular Q-Learning. A more general study shows how evolutionary computation can be used for reinforcement learning [172], but ignores the development of XCS [237], where Wilson explicitly uses Q-Learning as the RL component.

Recently, there has been some confusion [47, 223, 142] about how to correctly implement RL in XCS(F), and this has caused XCS(F) to be modified in various ways. Using the model-based stance, variants of Q-Learning that use LCS function approximation from first principles will be derived and show that XCS(F) already performs correct RL without the need for modifications. Also, it demonstrates how to correctly combine RL methods and LCS function approximation, as an illustration of a general principle, applied to the LCS model type that was introduced in the previous chapters.

Using RL with any form of value function approximation might case the RL method to become unstable and possibly diverge. Only certain forms of function approximation can be used with RL – an issue that will be discussed in detail in a later section, where the compatibility of the introduced LCS model and RL is analysed. Besides stability, learning policies that require a long sequence of actions is another issue that needs special consideration, as function approximation might cause the policy to become sub-optimal. This, and the exploration/exploitation dilemma will be discussed, where the latter concerns the trade-off between exploiting current knowledge in forming the policy and performing further exploration to gain more knowledge.

Appropriately linking LCS into RL firstly requires a formal basis for RL, which is formed by various DP methods. Their introduction is kept brief, and a longer LCS-related version is available as a technical report [79]. Nonetheless, we discuss some stability issues that RL is known to have when the value function is approximated, as these are particularly relevant – though mostly ignored – when combining RL with LCS. Hence, after showing how to derive the use of Q-Learning with LCS from first principles in Sect. 9.3 and discussing the recent confusion around XCS(F), Sect. 9.4 shows how to analyse the stability of RL when used with LCS. Learning of long action sequences is another issue that XCS is known to struggle with [11], and a previously proposed solution [12] only applies to certain problem types. If the introduced optimality criterion provides a potential solution is still an open question, but the outlook is good, as will be discussed before providing an LCS-related overview of the exploration/exploitation dilemma. But firstly, let us define sequential decision tasks more formally in Sect. 9.1, and introduce DP and RL methods that provide solutions to such tasks in Sect. 9.2.

9.1 Problem Definition

The sequential decision tasks that will be considered are the ones describable by a Markov Decision Process (MDP) (see Sect. 2.1). To stay close to the notation that is common in the literature [17, 209], some of the previously used symbols will be assigned a new meaning. The definitions given in this section are similar to the ones used by Bertsekas and Tsitsiklis [17] and Drugowitsch and Barry [79].

9.1.1 Markov Decision Processes

Let $\mathcal{X}$ be the set of states $\mathbf{x} \in \mathcal{X}$ of the problem domain, that is assumed to be of finite size[1] N, and hence is mapped into the natural numbers $\mathbb{N}$. $\mathcal{X}$ was previously defined to be the input space, but as the states are identified by the input that is determined by the environmental state, *state* and *input* are used interchangeably. In every state $\mathbf{x}_i \in \mathcal{X}$, an action a out of a finite set $\mathcal{A}$ is performed and causes a state transition to $\mathbf{x}_j$. The probability of getting to state $\mathbf{x}_j$ after performing action a in state $\mathbf{x}_i$ is given by the transition function $p(\mathbf{x}_j|\mathbf{x}_i, \mathbf{a})$, which is a probability distribution over $\mathcal{X}$, conditional on $\mathcal{X} \times \mathcal{A}$. Each such transition is meditated by a scalar reward $r_{x_i x_j}(a)$, defined by the reward function $r : \mathcal{X} \times \mathcal{X} \times \mathcal{A} \to \mathbb{R}$. The positive discount factor $\gamma \in \mathbb{R}$ with $0 < \gamma \leq 1$ determines the preference of immediate reward over future reward. Therefore, the MDP that describes the problem is defined by the quintuple $\{\mathcal{X}, \mathcal{A}, p, r, \gamma\}$[2]. The involved variables are illustrated in Fig. 2.1(b), which is reproduced in Fig. 9.1 for convenience. Previously, γ denoted the step size for gradient-based incremental methods in Chap. 5. In this chapter, the step size will be denoted by α to conform to the RL literature [209].

The aim is for every state to choose the action that maximises the reward in the long run, where future rewards are possibly valued less that immediate

[1] Assuming a finite state space simplifies the presentation. Extending it to a continuous state space requires considerably more technical work. For examples of an analysis of reinforcement learning in continuous state spaces see Konda and Tsitsiklis [129] and Ormoneit and Sen [177].

[2] The problem definition and with it the solution to the problem changes when the discount rate γ is changed. Thus, it is important to consider the discount rate γ as part of the problem rather than a tunable parameter. This fact is ignored in some LCS research, where the discount rate is modified to make the task seemingly easier to learn, when, in fact, the task itself is changed.

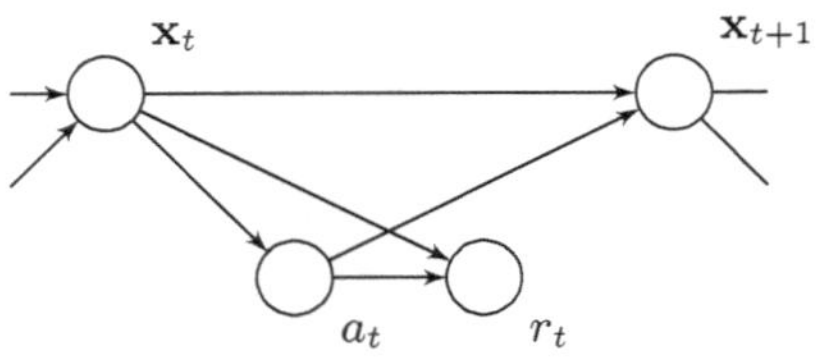

Fig. 9.1. The variables of an MDP involved in a single transition from state $\mathbf{x}_t$ to state $\mathbf{x}_{t+1}$ after the agent performed action a_t and received reward r_t

rewards. A possible solution is represented by a *policy* $\mu : \mathcal{X} \rightarrow \mathcal{A}$, which returns the chosen action $a = \mu(\mathbf{x})$ for any state $\mathbf{x} \in \mathcal{X}$. With a fixed policy μ, the MDP is reduced to a Markov chain with transition probabilities $p^{\mu}(\mathbf{x}_j|\mathbf{x}_i) = p(\mathbf{x}_j|\mathbf{x}_i, a = \mu(\mathbf{x}_i))$, and rewards $r^{\mu}_{x_i x_j} = r_{x_i x_j}(\mu(\mathbf{x}_i))$. In such cases it is common to operate with the expected reward $r^{\mu}_{x_i} = \sum_j p^{\mu}(\mathbf{x}_j|\mathbf{x}_i) r^{\mu}_{x_i,x_j}$. This reward is the one expected to be received in state $\mathbf{x}_i$ when actions are chosen according to policy μ.

9.1.2 The Value Function, the Action-Value Function and Bellman's Equation

The approach taken by dynamic programming (DP) and reinforcement learning (RL) is to define a value function $V : \mathcal{X} \rightarrow \mathbb{R}$ that expresses for each state how much reward we can expect to receive in the long run. While V was previously used to denote the mixing weight vectors, those will not be referred to in this chapter, and hence any ambiguity is avoided. Let $\mu = \{\mu_0, \mu_1, \dots\}$ be a sequence of policies where we use policy μ_t at time t, starting at time $t = 0$. Then, the reward that is accumulated after n steps when starting at state $\mathbf{x}$, called the *n-step return* V^{μ}_n for state $\mathbf{x}$, is given by

$$V^{\mu}_n(\mathbf{x}) = \mathbb{E}\left(\gamma^n R(\mathbf{x}_n) + \sum_{t=0}^{n-1} \gamma^t r^{\mu_t}_{x_t x_{t+1}} | \mathbf{x}_0 = \mathbf{x}\right), \tag{9.1}$$

where $\{\mathbf{x}_0, \mathbf{x}_1, \dots\}$ is the sequence of states, and $R(\mathbf{x}_n)$ denotes the expected return that will be received when starting from state $\mathbf{x}_n$. The *return* differs from the reward in that it implicitly considers future reward.

In *finite horizon cases*, where $n < \infty$, the optimal policy μ is the one that maximises the expected return for each state $\mathbf{x} \in \mathcal{X}$, giving the optimal n-step return $V^*_n(\mathbf{x}) = \max_{\mu} V^{\mu}_n(\mathbf{x})$. Finite horizon cases can be seen as a special case of *infinite horizon cases* with zero-reward absorbing states [17]. For infinite horizon cases, the expected return when starting at state $\mathbf{x}$ is analogously to (9.1) given by

$$V^{\mu}(\mathbf{x}) = \lim_{n \rightarrow \infty} \mathbb{E}\left(\sum_{t=0}^{n-1} \gamma^t r^{\mu_t}_{x_i x_{i+1}} | \mathbf{x}_0 = \mathbf{x}\right). \tag{9.2}$$

The optimal policy is the one that maximises this expected return for each state $\mathbf{x} \in \mathcal{X}$, and results in the optimal value function $V^*(\mathbf{x}) = \max_\mu V^\mu(\mathbf{x})$. Therefore, knowing V^*, we can infer the optimal policy by

$$\mu^*(\mathbf{x}) = \underset{a \in \mathcal{A}}{\operatorname{argmax}} \mathbb{E}\left(r_{xx'}(a) + \gamma V^*(\mathbf{x}')|\mathbf{x}, a\right). \tag{9.3}$$

Thus, the optimal policy is given by choosing the action that maximises the expected sum of immediate reward and the discounted expected optimal return of the next state. This reduces the goal of finding the policy that maximises the reward in the long run to learning the optimal value function, which is the approach taken by DP and RL. In fact, Sutton conjectures that

> "All efficient methods for solving sequential decision problems determine (learn or compute) value functions as an intermediate step."

which he calls the "Value-Function Hypothesis" [206].

In some cases, such as one does not have a model of the transition function, the expectation in (9.3) cannot be evaluated. Then, it is easier to work with the action-value function $Q : \mathcal{X} \times \mathcal{A} \to \mathbb{R}$ that estimates the expected return $Q(\mathbf{x}, a)$ when taking action a in state $\mathbf{x}$, and is for some policy μ defined by

$$\begin{aligned} Q^\mu(\mathbf{x}, a) &= \lim_{n \to \infty} \mathbb{E}\left(r_{x_0 x_1}(a) + \gamma \sum_{t=1}^{n-1} \gamma^t r^\mu_{x_t x_{t+1}} | \mathbf{x}_0 = \mathbf{x}, a \right) \\ &= \mathbb{E}(r_{xx'}(a) + \gamma V^\mu(\mathbf{x}')|\mathbf{x}, a). \end{aligned} \tag{9.4}$$

V^μ is recovered from Q^μ by $V^\mu(\mathbf{x}) = Q^\mu(\mathbf{x}, \mu(\mathbf{x}))$. Given that the optimal action-value function Q^* is known, getting the optimal policy μ^* is simplified from (9.3) to

$$\mu^*(\mathbf{x}) = \underset{a \in \mathcal{A}}{\operatorname{argmax}} Q^*(\mathbf{x}, a), \tag{9.5}$$

that is, by choosing the action a in state $\mathbf{x}$ that maximises the expected return given by $Q^*(\mathbf{x}, a)$.

Note that V^* and Q^* are related by $V^*(\mathbf{x}) = Q^*(\mathbf{x}, \mu^*(\mathbf{x})) = \max_{a \in \mathcal{A}} Q^*(\mathbf{x}, a)$. Combining this relation with (9.4) gives us *Bellman's Equation*

$$V^*(\mathbf{x}) = \max_{a \in \mathcal{A}} \mathbb{E}(r_{xx'}(a) + \gamma V^*(\mathbf{x}')|\mathbf{x}, a), \tag{9.6}$$

which relates the optimal values of different states to each other, and to which finding the solution forms the core of DP. Similarly, *Bellman's equation for a fixed policy* μ is given by

$$V^\mu(\mathbf{x}) = \mathbb{E}(r^\mu_{xx'} + \gamma V^\mu(\mathbf{x}')|\mathbf{x}). \tag{9.7}$$

An example for a problem that can be described by an MDP, together with its optimal value function and one of its optimal policies is shown in Fig. 2.2.

9.1.3 Problem Types

The three basic classes of infinite horizon problems are stochastic shortest path problems, discounted problems, and average reward per step problems, all of which are well described by Bertsekas and Tsitsiklis [17]. Here, only discounted problems and stochastic shortest path problems are considered, where for the latter only *proper* policies that are guaranteed to reach the desired terminal state are assumed. As the analysis of stochastic shortest path problems is very similar to discounted problems, only discounted problems are considered explicitly. These are characterised by $\gamma < 1$ and a bounded reward function to make the values $V^\mu(\mathbf{x})$ well defined.

9.1.4 Matrix Notation

Rather than representing the value function for each state explicitly, it is convenient to exploit the finiteness of $\mathcal{X}$ and collect the values for each state into a vector, which also simplifies the notation. Let $\mathbf{V} = (V(\mathbf{x}_1), \dots, V(\mathbf{x}_N))^T$ be the vector of size N that contains the values of the value function V for each state $\mathbf{x}_n$. Let $\mathbf{V}^*$ and $\mathbf{V}^\mu$ denote the vectors that contain the optimal value function V^* and the value function V^μ for policy μ, respectively. Similarly, let $\mathbf{P}^\mu = (p(\mathbf{x}_j|\mathbf{x}_i))$ denote the transition matrix of the Markov chain for a fixed policy μ, and let $\mathbf{r}^\mu = (r^\mu_{x_1}, \dots, r^\mu_{x_N})^T$ be the vector consisting of the expected rewards when following this policy. With these definitions, we can rewrite Bellman's Equation for a fixed policy (9.7) by

$$\mathbf{V}^\mu = \mathbf{r}^\mu + \gamma \mathbf{P}^\mu \mathbf{V}^\mu. \tag{9.8}$$

This notation is used extensively in further developments.

9.2 Dynamic Programming and Reinforcement Learning

Recall that in order to find the optimal policy μ^*, we aim at learning the optimal value function V^* by (9.6), or the optimal action-value function Q^* for cases where the expectation in (9.6) and (9.3) is hard or impossible to evaluate.

In this section, some common RL methods are introduced, that learn these functions while traversing the state space without building a model of the transition and reward function. These methods are simulation-based approximations to DP methods, and their stability is determined by the stability of the corresponding DP method. These DP methods are introduced firstly, after which RL methods are derived from them.

9.2.1 Dynamic Programming Operators

Bellman's Equation (9.6) is a set of equations that cannot be solved analytically. Fortunately, several methods have been developed that make finding its solution easier, all of which are based on the DP operators T and T_μ.

The operator T is given a value vector $\mathbf{V}$ and returns a new value vector that is based on Bellman's Equation (9.6). The ith element $(\mathrm{T}\mathbf{V})_i$ of the resulting vector $\mathrm{T}\mathbf{V}$ is given by

$$(\mathrm{T}\mathbf{V})_i = \max_{a \in \mathcal{A}} \sum_{x_j \in \mathcal{X}} p(\mathbf{x}_j|\mathbf{x}_i, a) \left(r_{x_i x_j}(a) + \gamma \mathbf{V}_j \right) . \tag{9.9}$$

Similarly, for a fixed policy μ the operator T_μ is based on (9.7), and is given by

$$(\mathrm{T}_\mu \mathbf{V})_i = \sum_{x_j \in \mathcal{X}} p^\mu(\mathbf{x}_j|\mathbf{x}_i) \left(r^\mu_{x_i x_j} + \gamma \mathbf{V}_j \right) , \tag{9.10}$$

which, in matrix notation, is $\mathrm{T}_\mu \mathbf{V} = \mathbf{r}^\mu + \gamma \mathbf{P}^\mu \mathbf{V}$.

The probably most important property of both T and T_μ is that they form a contraction mapping to the maximum norm [17]; that is, given two arbitrary vectors $\mathbf{V}, \mathbf{V}'$, we have

$$\|\mathrm{T}\mathbf{V} - \mathrm{T}\mathbf{V}'\|_\infty \leq \gamma \|\mathbf{V} - \mathbf{V}'\|_\infty, \quad \text{and} \tag{9.11}$$

$$\|\mathrm{T}_\mu \mathbf{V} - \mathrm{T}_\mu \mathbf{V}'\|_\infty \leq \gamma \|\mathbf{V} - \mathbf{V}'\|_\infty, \tag{9.12}$$

where $\|\mathbf{V}\|_\infty = \max_i |\mathbf{V}_i|$ is the maximum norm of $\mathbf{V}$. Thus, every update with T or T_μ reduces the maximum distance between $\mathbf{V}$ and $\mathbf{V}'$ by at least the factor γ. Applying them repeatedly will therefore lead us to some fixed point $\mathrm{T}\mathbf{V} = \mathbf{V}$ or $\mathrm{T}_\mu \mathbf{V} = \mathbf{V}$, that is, according to the *Banach Fixed Point Theorem* [230], unique.

Further properties of the DP operators are that the optimal value vector $\mathbf{V}^*$ and the value vector $\mathbf{V}^\mu$ for policy μ are the unique vectors that satisfy $\mathrm{T}\mathbf{V}^* = \mathbf{V}^*$ and $\mathrm{T}_\mu \mathbf{V}^\mu = \mathbf{V}^\mu$, respectively, which follows from Bellman's Equations (9.6) and (9.7). As these vectors are the fixed points of T and T_μ, applying the operators repeatedly causes convergence to these vectors, that is, $\mathbf{V}^* = \lim_{n \to \infty} \mathrm{T}^n \mathbf{V}$, and $\mathbf{V}^\mu = \lim_{n \to \infty} \mathrm{T}^n_\mu \mathbf{V}$ for an arbitrary $\mathbf{V}$, where T^n and T^n_μ denote n applications of T and T_μ, respectively. A policy μ is optimal if and only if $\mathrm{T}_\mu \mathbf{V}^* = \mathrm{T}\mathbf{V}^*$. Note that, even though $\mathbf{V}^*$ is unique, there can be several optimal policies [17].

9.2.2 Value Iteration and Policy Iteration

The method of value iteration is a straightforward application of the contraction property of T and is based on applying T repeatedly to an initially arbitrary value vector $\mathbf{V}$ until it converges to the optimal value vector $\mathbf{V}^*$. Convergence can only be guaranteed after an infinite number of steps, but the value vector $\mathbf{V}$ is usually already close to $\mathbf{V}^*$ after few iterations.

As an alternative to value iteration, policy iteration will converge after a finite number of *policy evaluation* and *policy improvement* steps. Given a fixed policy μ_t, policy evaluation finds the value vector for this policy by solving $\mathrm{T}_{\mu_t} \mathbf{V}^{\mu_t} = \mathbf{V}^{\mu_t}$. The policy improvement steps generates a new policy μ_{t+1} based on the current $\mathbf{V}^{\mu_t}$, such that $\mathrm{T}_{\mu_{t+1}}\mathbf{V}^{\mu_t} = \mathrm{T}\mathbf{V}^{\mu_t}$. Starting with an initially random policy μ_0, the sequence of policies $\{\mu_0, \mu_1, \dots\}$ generated by iterating policy evaluation and policy improvement is guaranteed to converge to the optimal policy within a finite number of iterations [17].

Various variants to these methods exist, such as asynchronous value iteration, that at each application of T only updates a single state of $\mathbf{V}$. Modified policy iteration performs the policy evaluation step by approximating $\mathbf{V}^{\mu}$ by $\mathrm{T}_{\mu}^{n}\mathbf{V}$ for some small n. Asynchronous policy iteration mixes asynchronous value iteration with policy iteration by at each step either i) updating some states of $\mathbf{V}$ by asynchronous value iteration, or ii) improving the policy of some set of states by policy improvement. Convergence criteria for these variants are given by Bertsekas and Tsitsiklis [17].

9.2.3 Approximate Dynamic Programming

If N is large, we prefer to approximate the value function rather than representing the value for each state explicitly. Let $\tilde{\mathbf{V}}$ denote the vector that holds the value function approximations for each state, as generated by a function approximation technique as an approximation to $\mathbf{V}$. Approximate value iteration is performed by approximating the value iteration update $\mathbf{V}_{t+1} = \mathrm{T}\mathbf{V}_t$ by

$$\tilde{\mathbf{V}}_{t+1} = \Pi \mathrm{T} \tilde{\mathbf{V}}_t, \tag{9.13}$$

where Π is the approximation operator that, for the used function approximation technique, returns the value function estimate approximation $\tilde{\mathbf{V}}_{t+1}$ that is closest to $\mathbf{V}_{t+1} = \mathrm{T}\tilde{\mathbf{V}}_t$ by $\tilde{\mathbf{V}}_{t+1} = \mathrm{argmin}_{\tilde{V}} \|\tilde{\mathbf{V}} - \mathbf{V}_{t+1}\|$. Even though conceptually simple, approximate value iteration was shown to diverge even when used in combination with the simplest function approximation techniques [25]. Thus, special care needs to be take when applying this method, as will be discussed in more detail in Sect. 9.4.

Approximate policy iteration, on the other hand, has less stability problems, as the operator T_{μ} used for the policy evaluation step is linear. While the policy improvement step is performed as for standard policy iteration, the policy evaluation step is based on an approximation of $\mathbf{V}^{\mu}$. As T_{μ} is linear, there are several possibilities of how to perform the approximation, which are outlined by Schoknecht [193]. Here, the only approximation that will be considered is the one most similar to approximation value iteration and is the temporal-difference solution which aims at finding the fixed point $\tilde{\mathbf{V}}^{\mu} = \Pi \mathrm{T}_{\mu} \tilde{\mathbf{V}}^{\mu}$ by the update $\tilde{\mathbf{V}}^{\mu}_{t+1} = \Pi \mathrm{T}_{\mu} \tilde{\mathbf{V}}^{\mu}_{t}$ [194, 195].

9.2.4 Temporal-Difference Learning

Even thought temporal-difference (TD) learning is an incremental method for policy evaluation that was initially developed by Sutton [207] as a modification of the Widrow-Hoff rule [234], we here only concentrate the TD(λ) operator $\mathrm{T}_{\mu}^{(\lambda)}$ as it forms the basis of SARSA(λ), and gives us some necessary information about T_μ. For more information on temporal-difference learning, the interested reader is referred to the work of Bertsekas and Tsitsiklis [17] and Drugowitsch and Barry [79].

The temporal-difference learning operator $T_\mu^{(\lambda)}$ is parametrised by $0 \leq \lambda \leq 1$, and, when applied to $\mathbf{V}$ results in [215]

$$(\mathrm{T}_\mu^{(\lambda)}\mathbf{V})_i = (1-\lambda)\sum_{m=0}^{\infty}\lambda^m \mathbb{E}\left(\sum_{t=0}^{m}\gamma^t r^\mu_{x_t x_{t+1}} + \gamma^{m+1}\mathbf{V}_{m+1}|\mathbf{x}_0 = \mathbf{x}_i\right), \quad (9.14)$$

for $\lambda < 1$. The definition for $\lambda = 1$ is given in [79]. The expectation in the above expression is equivalent to the n-step return V_n^μ (9.1), which shows that the temporal-difference update is based on mixing returns of various lengths, where the mixing coefficients are controlled by λ. To implement the above update incrementally, Sutton uses *eligibility traces* that propagate current temporal differences to previously visited states [207].

Its most interesting property for our purpose is that $\mathrm{T}_\mu^{(\lambda)}$ forms a contraction mapping with respect to the weighted norm $\|\cdot\|_D$, which is defined as given in Sect. 5.2, and the diagonal weight matrix $\mathbf{D}$ is given by the steady-state distribution of the Markov chain $\mathbf{P}^\mu$ that corresponds to policy μ [215, 17]. More formally, we have for any $\mathbf{V}, \mathbf{V}'$,

$$\|\mathrm{T}_\mu^{(\lambda)}\mathbf{V} - \mathrm{T}_\mu^{(\lambda)}\mathbf{V}'\|_D \leq \frac{\gamma(1-\lambda)}{1-\gamma\lambda}\|\mathbf{V}-\mathbf{V}'\|_D \leq \gamma\|\mathbf{V}-\mathbf{V}'\|_D. \quad (9.15)$$

Note that $\mathrm{T}_\mu \equiv \mathrm{T}_\mu^{(0)}$, and therefore T_μ also forms a contraction mapping with respect to $\|\cdot\|_D$. This property can be used to analyse both convergence and stability of the method, as shown in Sect. 9.4.

9.2.5 SARSA(λ)

Coming to the first reinforcement learning algorithm, SARSA stands for State-Action-Reward-State-Action, as SARSA(0) requires only information on the current and next state/action pair and the reward that was received for the transition. Its name was coined by Sutton [208] for an algorithm that was developed by Rummery and Nirahnja [192] in its approximate form, which is very similar to Wilson's ZCS [236], as discussed by Sutton and Barto [209, Chap. 6.10].

It conceptually performs policy iteration and uses TD(λ) to update its action-value function Q. More specifically it performs optimistic policy iteration, where in contrast to standard policy iteration the policy improvement step is based on an incompletely evaluated policy.

Consider being in state $\mathbf{x}_t$ at time t and performing action a_t, leading to the next state $\mathbf{x}_{t+1}$ and reward r_t. The current action-value function estimates are given by $\hat{Q}_t$. These estimates are to be updated for $(\mathbf{x}_t, a_t)$ to reflect the newly observed reward. The basis of policy iteration, as described by T_μ (9.10), is to update the estimate of the value function of one particular state by relating it to all potential next states and the expected reward for these transitions. In SARSA(0), the actually observed transition replaces the potential transitions, such that the target value of the estimate $\hat{Q}(\mathbf{x}_t, a_t)$ becomes $Q(\mathbf{x}_t, a_t) = r_{x_t x_{t+1}}(a_t) + \gamma \hat{Q}_t(\mathbf{x}_{t+1}, a_{t+1})$. Note that the value of the next state is approximated by the current action-value function estimate $\hat{Q}_t$ and the assumption that current policy is chosen when choosing the action in the next state, such that $\hat{V}(\mathbf{x}_{t+1}) \approx \hat{Q}_t(\mathbf{x}_{t+1}, a_{t+1})$.

Using $\hat{Q}_{t+1}(\mathbf{x}_t, a_t) = Q(\mathbf{x}_t, a_t)$ would not lead to good results as it makes the update highly dependent on the quality of the policy that is used to select a_t. Instead, the LMS algorithm (see Sect. 5.3.3) is used to minimise the squared difference between the estimate $\hat{Q}_{t+1}$ and its target Q, such that the action-value function estimate is updated by

$$\hat{Q}_{t+1}(\mathbf{x}_t, a_t) = \hat{Q}_t(\mathbf{x}_t, a_t) + \alpha_t \left(r_{x_t x_{t+1}}(a_t) + \gamma \hat{Q}_t(\mathbf{x}_{t+1}, a_{t+1}) - \hat{Q}_t(\mathbf{x}_t, a_t) \right), \tag{9.16}$$

where α_t denotes the step-size of the LMS algorithm at time t. For all state/action pairs $\mathbf{x} \neq \mathbf{x}_t$, $a \neq a_t$, the action-value function estimates remain unchanged, that is $\hat{Q}_{t+1}(\mathbf{x}, a) = \hat{Q}_t(\mathbf{x}, a)$.

The actions can be chosen according to the current action-value function estimate, such that $a_t = \text{argmax}_a \hat{Q}_t(\mathbf{x}_t, a_t)$. This causes SARSA(0) to always perform the action that is assumed to be the reward-maximising one according to the current estimate. Always following such a policy is not advisable, as it could cause the method to get stuck in a local optimum by not sufficiently exploring the whole state space. Thus, a good balance between exploiting the current knowledge and exploring the state space by performing seemingly sub-optimal actions is required. This explore/exploit dilemma is fundamental to RL methods and will hence be discussed in more detail in a later section. For now let us just note that the update of $\hat{Q}$ is based on the state trajectory of the current policy, even when sub-optimal actions are chosen, such that SARSA is called an *on-policy* method.

SARSA(λ) for $\lambda > 0$ relies on the operator $\mathrm{T}_\mu^{(\lambda)}$ rather than T_μ. A detailed discussion of the consequences of this change is beyond the scope of this book, but more details are given by Sutton [207] and Sutton and Barto [209].

9.2.6 Q-Learning

The much-celebrated Q-Learning was developed by Watkins [228] as a result of combining TD-learning and DP methods. It is similar to SARSA(0), but rather than using $Q(\mathbf{x}_t, a_t) = r_{x_t x_{t+1}}(a_t) + \gamma \hat{Q}_t(\mathbf{x}_{t+1}, a_t)$ as the target value for $\hat{Q}(\mathbf{x}_t, a_t)$, it uses $Q(\mathbf{x}_t, a_t) = r_{x_t x_{t+1}}(a_t) + \gamma \max_a \hat{Q}_t(\mathbf{x}_{t+1}, a)$, and thus approximates value iteration rather than policy iteration. SARSA(0) and Q-Learning are equivalent if both always follow the greedy action $a_t = \text{argmax}_a \hat{Q}_t(\mathbf{x}_t, a)$, but this would ignore the explore/exploit dilemma. Q-Learning is called an *off-policy* method as the value function estimates $\hat{V}(\mathbf{x}_{t+1}) \approx \max_a \hat{Q}_t(\mathbf{x}_{t+1}, a)$ are independent of the actions that are actually performed.

For a sequence of states $\{\mathbf{x}_1, \mathbf{x}_2, \dots\}$ and actions $\{a_1, a_2, \dots\}$, the Q-values are updated by

$$Q_{t+1}(\mathbf{x}_t, a_t) = Q_t(\mathbf{x}_t, a_t) + \alpha_t \left(r_{x_t x_{t+1}}(a_t) + \gamma \max_{a \in \mathcal{A}} Q_t(\mathbf{x}_{t+1}, a) - Q_t(\mathbf{x}_t, a_t) \right), \tag{9.17}$$

where α_t denotes the step size at time t. As before, the explore/exploit dilemma applies when selecting actions based on the current $\hat{Q}$.

A variant of Q-Learning, called Q(λ), is an extension that uses eligibility traces like TD(λ) as long as it performs on-policy actions [229]. As soon as an off-policy action is chosen, all traces are reset to zero, as the off-policy action breaks the temporal sequence of predictions. Hence, the performance increase due to traces depends significantly on the policy that is used, but is usually marginal. In a study by Drugowitsch and Barry [77] it was shown that, when used in XCS, it performs even worse than standard Q-Learning.

9.2.7 Approximate Reinforcement Learning

Analogous to approximate DP, RL can handle large state spaces by approximating the action-value function. Given some estimator $\hat{Q}$ that approximates the action-value function, this estimator is, as before, to be updated after receiving reward r_t for a transition from $\mathbf{x}_t$ to $\mathbf{x}_{t+1}$ when performing action a_t. The estimator's target value is $Q(\mathbf{x}_t, a_t) = r_{x_t x_{t+1}}(a_t) + \gamma \hat{V}(\mathbf{x}_{t+1})$, where $\hat{V}(\mathbf{x}_{t+1})$ is the currently best estimate of the value of state $\mathbf{x}_{t+1}$. Thus, at time t, the aim is to find the estimator $\hat{Q}$ that minimises some distance between itself and all previous target values, which, when assuming the squared distance, results in minimising

$$\sum_{m=1}^{t} \left(\hat{Q}(\mathbf{x}_m, a_n) - \left(r_{x_m x_{m+1}}(a_m) + \gamma \hat{V}_t(\mathbf{x}_{m+1}) \right) \right)^2 . \tag{9.18}$$

As previously shown, Q-Learning uses $\hat{V}_t(\mathbf{x}) = \max_a \hat{Q}_t(\mathbf{x}, a)$, and SARSA(0) relies on $\hat{V}_t(\mathbf{x}) = \hat{Q}_t(\mathbf{x}, a)$, where in case of the latter, a is the action performed in state $\mathbf{x}$.

Tabular Q-Learning and SARSA(0) are easily extracted from the above problem formulation by assuming that each state/action pair is estimated separately by $\hat{Q}(\mathbf{x}, a) = \theta_{x,a}$. Under this assumption, applying the LMS algorithm to minimising (9.18) directly results in (9.16) or (9.17), depending on how $\hat{V}_t$ is estimated.

The next section shows from first principles how the same approach can be applied to performing RL with LCS, that is, when $\hat{Q}$ is an estimator that is given by an LCS.

9.3 Reinforcement Learning with LCS

Performing RL with LCS means to use LCS to approximate the action-value function estimate. RL methods upgrade this estimate incrementally, and we can only use LCS with RL if the LCS implementation can handle incremental model parameter updates. Additionally, while approximating the action-value function is a simple univariate regression task, the function estimate to approximate is non-stationary due to its sequential update. Thus, in addition to incremental learning, the LCS implementation needs to be able to handle non-stationary target functions.

This section demonstrates how to derive Q-Learning with the LCS model as introduced in Chap. 4, to act as a template for how any LCS model type can be used for RL. Some of the introduced principles and derivations are specific to the LCS model with independently trained classifiers, but the underlying ideas also transfer to other LCS model types. The derivations themselves are performed from first principles to make explicit the usually implicit design decisions. Concentrating purely on incremental model parameter learning, the model structure $\mathcal{M}$ is assumed to be constant. In particular, the derivation focuses on the classifier parameter updates, as these are the most relevant with respect to RL.

Even though the Bayesian update equations from Chap. 7 protect against overfitting, this section falls back to maximum likelihood learning that was the basis for the incremental methods described in Chaps. 5 and 6. The resulting update equations conform exactly to XCS(F), which reveals its design principles and should clarify some of the recent confusion about how to implement gradient descent in XCS(F). An additional bonus is a more accurate noise precision update method for XCS(F) based on the methods developed in Chap. 5.

Firstly, the LCS approximation operator is introduced, that conforms to the LCS model type of this work. This is followed by discussing how the principle of independent classifier training relates to how DP and RL update the value and action-value function estimates, which is essential for the use of this LCS model type to perform RL. As Q-Learning is based on asynchronous value iteration, it will be firstly shown how LCS can perform asynchronous value iteration, followed by the derivation of two Q-Learning variants – one based on LMS, and the other on RLS. Finally, these derivations are related to the recent controversy about how XCS(F) correctly performs Q-Learning with gradient descent.

9.3.1 Approximating the Value Function

Given a value vector $\mathbf{V}$, LCS approximates it by a set of K localised models $\{\hat{\mathbf{V}}_k\}$ that are combined to form a global model $\hat{\mathbf{V}}$. The localised models are provided by the classifiers, and the mixing model is used to combine these to the global model.

Each classifier k matches a subset of the state space that is determined by its matching function m_k which returns for each state $\mathbf{x}$ the probability $m_k(\mathbf{x})$ of matching it. Let us for now assume that we approximate the value function V rather than the action-value function Q. Then, classifier k provides the probabilistic model $p(V|\mathbf{x}, \boldsymbol{\theta}_k)$ that gives the probability of the expected return of state $\mathbf{x}$ having the value V. Assuming linear classifiers (5.3), this model is given by

$$p(V|\mathbf{x}, \boldsymbol{\theta}_k) = \mathcal{N}(V|\mathbf{w}_k^T\mathbf{x}, \tau_k^{-1}), \tag{9.19}$$

where we assume $\mathbf{x}$ to be the vector of size $D_{\mathcal{X}}$ that represents the features of the corresponding input, $\mathbf{w}_k$ denotes the weight vector of size $D_{\mathcal{X}}$, and τ_k is the scalar non-negative noise precision. As shown in (5.10), following the principle of maximum likelihood results in the estimator of the mean of $p(V|\mathbf{x}, \boldsymbol{\theta}_k)$,

$$\tilde{\mathbf{V}}_k = \boldsymbol{\Pi}_k \mathbf{V}, \tag{9.20}$$

where $\boldsymbol{\Pi}_k = \mathbf{X}(\mathbf{X}^T\mathbf{M}_k\mathbf{X})^{-1}\mathbf{X}^T\mathbf{M}_k$ is the projection matrix that provides the matching-weighted maximum likelihood estimate approximation to $\mathbf{V}$, and $\mathbf{X}$ and $\mathbf{M}_k$ denote the state matrix by (3.4) and the diagonal matching matrix $\mathbf{M}_k = \mathrm{diag}(m_k(\mathbf{x}_1), \dots, m_k(\mathbf{x}_2))$, respectively. Thus, $\boldsymbol{\Pi}_k$ can be interpreted as the approximation operator for classifier k that maps the value function vector $\mathbf{V}$ to its approximation $\hat{\mathbf{V}}_k$.

Given the classifier approximations $\{\hat{\mathbf{V}}_1, \dots, \hat{\mathbf{V}}_K\}$, the mixing model combines them to a global approximation. For a particular state $\mathbf{x}$, the global approximation is given by $\tilde{V}(\mathbf{x}) = \sum_k g_k(\mathbf{x})\hat{V}_k(\mathbf{x})$, where the functions $\{g_k\}$ are determined by the chosen mixing model. Possible mixing models and their training are discussed in Chap. 6, and we will only assume that the used mixing model honours matching by $g_k(\mathbf{x}) = 0$ if $m_k(\mathbf{x}) = 0$, and creates a weighted average of the local approximations by $g_k(\mathbf{x}) \geq 0$ for all $\mathbf{x}$, k, and $\sum_k g_k(\mathbf{x}) = 1$ for all $\mathbf{x}$. Thus, the global approximation $\hat{\mathbf{V}}$ of $\mathbf{V}$ is given by

$$\hat{\mathbf{V}} = \Pi\mathbf{V}, \qquad \text{with } \Pi\mathbf{V} = \sum_k \mathbf{G}_k\boldsymbol{\Pi}_k\mathbf{V}, \tag{9.21}$$

where the $\mathbf{G}_k$'s are diagonal $N \times N$ matrices that specify the mixing model and are given by by $\mathbf{G}_k = \mathrm{diag}(g_k(\mathbf{x}_1), \dots, g_k(\mathbf{x}_N))$. The approximation operator Π in (9.21) defines how LCS approximate the value function, given a fixed model structure.

9.3.2 Bellman's Equation in the LCS Context

Any DP or RL method is based on relating the expected return estimate for the current state to the expected return estimate of any potential next state. This can be seen when inspecting Bellman's Equation (9.6), where the value of $\mathbf{V}^*(\mathbf{x})$ is related to the values $\mathbf{V}^*(\mathbf{x}')$ for all $\mathbf{x}'$ that are reachable from $\mathbf{x}$. Similarly, Q-Learning (9.17) updates the action-value $Q(\mathbf{x}_t, a_t)$ by relating it to the action-value $\max_{a \in \mathcal{A}} Q(\mathbf{x}_{t+1}, a)$ of the next state that predicts the highest expected return.

According to the LCS model as given in Chap. 4, each classifier models the value function over its matched area in the state space independently of the other classifiers. Let us consider a single transition from state $\mathbf{x}$ to state $\mathbf{x}'$ by performing action a. Given that classifier k matches both states, it could update its local model of the value function $\hat{V}_k(\mathbf{x})$ for $\mathbf{x}$ by relating it to its own local model of the value function $\hat{V}_k(\mathbf{x}')$ for $\mathbf{x}'$. However, what happens if $\mathbf{x}'$ is not matched by classifier k? In such a case we cannot rely on its approximation

$\hat{V}_k(\mathbf{x}')$ as the classifier does not aim at modelling the value for this state. The most reliable model in such a case is in fact given by the global model $\tilde{V}(\mathbf{x}')$.

Generally, the global model will be used for all updates, regardless of whether the classifier matches the next state or not. This is justified by the observation that the global model is on average more accurate that the local models, as was established in Chap. 6. Based on this principle, Bellman's Equation $\mathbf{V}^* = \mathrm{T}\mathbf{V}^*$ can be reformulated for LCS with independent classifiers to

$$\hat{\mathbf{V}}_k^* = \mathbf{\Pi}_k \mathrm{T} \hat{\mathbf{V}}^* = \mathbf{\Pi}_k \mathrm{T} \sum_k \mathbf{G}_k \hat{\mathbf{V}}_k^*, \qquad k = 1, \dots, K, \tag{9.22}$$

where $\mathbf{\Pi}_k$ expresses the approximation operator for classifier k, that does not necessarily need to describe a linear approximation. By adding $\sum_k \mathbf{G}_k$ to both sides of the first equality of (9.22) and using (9.21), we get the alternative expression $\hat{\mathbf{V}}^* = \Pi \mathrm{T} \hat{\mathbf{V}}^*$, which shows that (9.22) is in fact Bellman's Equation with LCS approximation. Nonetheless, the relation is preferably expressed by (9.22), as it shows what the classifiers model rather than what the global model models. For a fixed model structure $\mathcal{M}$, any method that performs DP or RL with the here described LCS model type should aim at finding the solution to (9.22).

9.3.3 Asynchronous Value Iteration with LCS

Let us consider approximate value iteration before its asynchronous variant is derived: as given in Sect. 9.2.3, approximate value iteration is performed by the iteration $\mathbf{V}_{t+1} = \Pi \mathrm{T} \mathbf{V}_t$. Therefore, using (9.21), value iteration with LCS is given by the iteration

$$\tilde{\mathbf{V}}_{k,t+1} = \mathbf{\Pi}_k \mathbf{V}_{t+1}, \qquad \text{with } \mathbf{V}_{t+1} = \mathrm{T} \sum_k \mathbf{G}_{k,t} \tilde{\mathbf{V}}_{k,t}, \tag{9.23}$$

which has to be performed by each classifier separately. The iteration was split into two components to show that firstly one finds the updated value vector $\mathbf{V}_{t+1}$ by applying the T operator to the global model, which is then approximated by each classifier separately. The subscript $\cdot_t$ is added to the mixing model to express that it might depend on the current approximation and might therefore change with each iteration. Note that the fixed point of (9.23) is the desired Bellman Equation in the LCS context (9.22).

The elements of the updated value vector $\mathbf{V}_{t+1}$ are based on (9.23) and (9.9), which results in

$$V_{t+1}(\mathbf{x}_i) = \max_{a \in \mathcal{A}} \sum_{\mathbf{x}_j \in \mathcal{X}} p(\mathbf{x}_j | \mathbf{x}_i, a) \left(r_{x_i x_j}(a) + \gamma \sum_k g_{k,t}(\mathbf{x}_j) \tilde{V}_{k,t}(\mathbf{x}_j) \right), \tag{9.24}$$

where $V_{t+1}(\mathbf{x}_i)$ denotes the ith element of $\mathbf{V}_{t+1}$, and $\tilde{V}_{k,t}(\mathbf{x}_j)$ denotes the jth element of $\tilde{\mathbf{V}}_{k,t}$. Subsequently, each classifier is trained by batch learning, based on $\mathbf{V}_{t+1}$ and its matching function, as described in Sect. 5.2. This completes one iteration of LCS approximate value iteration.

The only modification introduced by the asynchronous variant is that rather than updating the value function for all states at once, a single state is picked per iteration, and the value function is updated for this state, as already described in Sect. 9.2.2. Let $\{\mathbf{x}_{i_1}, \mathbf{x}_{i_2}, \dots\}$ be the sequence of states that determine with state is updated at which iteration. Thus in the tth iteration we compute $V_t(\mathbf{x}_{i_t})$ by (9.24), which results in the sequence $\{V_1(\mathbf{x}_{i_1}), V_2(\mathbf{x}_{i_2}), \dots\}$ that can be used to incrementally train the classifiers by a method of choice from Sect. 5.3. For the asynchronous variant we cannot use batch learning anymore, as not all elements of $\mathbf{V}_{t+1}$ are available at once.

9.3.4 Q-Learning by Least Mean Squares

So far it was assumed that the transition and reward function of the given problem are known. To perform model-free RL, the LCS model needs to be adjusted to handle action-value function estimates rather than value function estimates by redefining the input state to be the space of all state/action pairs. Thus, given state $\mathbf{x}$ and action a, the matching probability of classifier k is given by $m_k(\mathbf{x}, a)$, and the approximation of its action-value by $\hat{Q}_k(\mathbf{x}, a)$. Mixing is also based on state and action, where the mixing coefficient for classifier k is given by $g_k(\mathbf{x}, a)$. This results in the global approximation of the action-value for state $\mathbf{x}$ and action a to be given by

$$\hat{Q}(\mathbf{x}, a) = \sum_k g_k(\mathbf{x}, a)\hat{Q}_k(\mathbf{x}, a). \tag{9.25}$$

As describes in Sect. 9.2.7, approximate Q-Learning in based on minimising (9.18). For independently trained classifiers, each classifier minimises this cost independently, but with the value function estimate $\hat{V}$ of the next state based on the global estimate. Thus the target for $\hat{Q}_k$ for the transition from $\mathbf{x}_t$ to $\mathbf{x}_{t+1}$ under action a_t is

$$Q_{t+1}(\mathbf{x}_t, a_t) = r_{x_t x_{t+1}}(a_t) + \gamma \max_{a \in \mathcal{A}} \hat{Q}_t(\mathbf{x}_{t+1}, a), \tag{9.26}$$

given that classifier k matches $(\mathbf{x}_t, a_t)$. Using the linear classifier model $\hat{Q}_k(\mathbf{x}, a) = \mathbf{w}_k^T \mathbf{x}$, each classifier k aims at finding $\mathbf{w}_k$ that, by (9.18) after t steps, minimises

$$\sum_{m=0}^{T} m_k(\mathbf{x}_m, a_m) \left(\mathbf{w}_k^T \mathbf{x} - Q_{m+1}(\mathbf{x}_m, a_m)\right)^2, \tag{9.27}$$

where $m_k(\mathbf{x}_m, a_m)$ was introduced to only consider matched state/action pairs. This standard linear least squares problem can be handled incrementally with any of the methods discussed in Chap. 5. It cannot be trained by batch learning as the target function relies on previous updates and is thus non-stationary.

Using the normalised least mean squared (NLMS) algorithm as described in Sect. 5.3.4, the weight vector estimate update for classifier k is given by

$$\hat{\mathbf{w}}_{k,t+1} = \hat{\mathbf{w}}_{k,t} + \alpha m_k(\mathbf{x}_t, a_t) \frac{\mathbf{x}_t}{\|\mathbf{x}_t\|^2} \left(Q_{t+1}(\mathbf{x}_t, a_t) - \hat{\mathbf{w}}_k^T \mathbf{x}_t \right), \tag{9.28}$$

where α denotes the step size, and $Q_{t+1}(\mathbf{x}_t, a_t)$ is given by (9.26). As discussed in more detail in Sect. 9.3.6, this is the weight vector update of XCSF.

The noise variance of the model can be estimate by the LMS algorithm, as described in Sect. 5.3.7. This results in the update equation

$$\hat{\tau}_{k,t+1}^{-1} = \hat{\tau}_{k,t}^{-1} + \alpha m_k(\mathbf{x}_t, a_t) \left(\left(\hat{\mathbf{w}}_{k,t+1}^T \mathbf{x}_t - Q_{t+1}(\mathbf{x}_t, a_t) \right)^2 - \hat{\tau}_{k,t}^{-1} \right), \tag{9.29}$$

where α is again the scalar step size, and $Q_{t+1}(\mathbf{x}_t, a_t)$ is given by (9.26).

9.3.5 Q-Learning by Recursive Least Squares

As shown in Chap. 5, incremental methods based on gradient descent might suffer from slow convergence rates. Thus, despite their higher computational and space complexity, methods based on directly tracking the least squares solution are to be preferred. Consequently, rather than using NLSM, this section shown how to apply recursive least squares (RLS) and direct noise precision tracking to Q-Learning with LCS.

The non-stationarity of the action-value function estimate needs to be take into account by using a recency-weighed RLS variant that puts more weight on recent observation. This was not an issue for the NLMS algorithm, as it performs recency-weighting implicitly.

Minimising the recency-weighted variant of the sum of squared errors (9.27), the update equations are according to Sect. 5.3.5 given by

$$\hat{\mathbf{w}}_{k,t+1} = \lambda^{m_k(\mathbf{x}_t, a_t)} \hat{\mathbf{w}}_{k,t} + m_k(\mathbf{x}_t, a_t) \mathbf{\Lambda}_{k,t+1}^{-1} \mathbf{x}_t \left(Q_{t+1}(\mathbf{x}_t, a_t) - \hat{\mathbf{w}}_{k,t}^T \mathbf{x}_t \right) \tag{9.30}$$

$$\mathbf{\Lambda}_{k,t+1}^{-1} = \lambda^{-m_k(\mathbf{x}_t, a_t)} \mathbf{\Lambda}_{k,t}^{-1}, \tag{9.31}$$

$$- m_k(\mathbf{x}_t, a_t) \lambda^{-m_k(\mathbf{x}_t, a_t)} \frac{\mathbf{\Lambda}_{k,t}^{-1} \mathbf{x}_t \mathbf{x}_t^T \mathbf{\Lambda}_{k,t}^{-1}}{\lambda^{m_k(\mathbf{x}_t, a_t)} + m_k(\mathbf{x}_t, a_t) \mathbf{x}_t^T \mathbf{\Lambda}_{k,t}^{-1} \mathbf{x}_t},$$

where $Q_{t+1}(\mathbf{x}_t, a_t)$ is given by (9.26), and $\hat{\mathbf{w}}_{k,0}$ and $\mathbf{\Lambda}_{k,0}^{-1}$ are initialised by $\hat{\mathbf{w}}_{k,0} = \mathbf{0}$ and $\mathbf{\Lambda}_{k,0} = \delta \mathbf{I}$, where δ is a large scalar. λ determines the recency weighting, which is strongest for $\lambda = 0$, where only the last observation is considered, and deactivated when $\lambda = 1$.

Using the RLS algorithm to track the least squares approximation of the action-values for each classifier allows us to directly track the classifier's model noise variance, as described in Sect. 5.3.7. More precisely, we track the sum of squared errors, denoted by $s_{k,t}$ for classifier k at time t, and can the compute the noise precision by (5.63). By (5.69), the sum of squared errors is updated by

$$s_{k,t+1} = \lambda^{m(\mathbf{x}_t,a_t)} s_{k,t} \qquad (9.32)$$
$$+m_k(\mathbf{x}_t,a_t)(\hat{\mathbf{w}}_{k,t}^T\mathbf{x}_t - Q_{t+1}(\mathbf{x}_t,a_t))(\hat{\mathbf{w}}_{k,t+1}^T\mathbf{x}_t - Q_{t+1}(\mathbf{x}_t,a_t)),$$

starting with $s_{k,0} = 0$.

Even though XCS has already been used with RLS by Lanzi et al. [142, 143], they have never applied it to sequential decision tasks. We have already investigated the incremental noise precision update as given in this chapter for simple regression tasks [155], but its empirical evaluation when applied to sequential decision tasks is still pending.

9.3.6 XCS with Gradient Descent

Some recent work by Butz et al. [47, 45] has caused some confusion over how XCS performs Q-Learning, and how this can be enhanced by the use of gradient descent [223, 224, 142, 140, 139]. This section clarifies that XCS(F) in its current form already performs gradient descent and does not need to be modified, as it updates the classifiers' weight vectors by (9.28), which is a gradient-based update algorithm. As XCSF is (besides the MAM update) equivalent to XCS if $D_{\mathcal{X}} = 1$, the following discussion only considers XCSF.

To show the equivalence between XCSF and (9.28), let us have a closer look at the weight vector update of XCSF: upon arriving at state $\mathbf{x}_t$, XCSF forms a *match set* that contains all classifiers for that $m_k(\mathbf{x}_t, a) > 0$, independent of the action a. The match set is then partitioned into one subset per possible action, resulting in $|\mathcal{A}|$ subsets. The subset associated with action a contains all classifiers for that $m_k(\mathbf{x}, a) > 0$, and for each of these subsets the action-value estimate $\hat{Q}_t(\mathbf{x}_t, a) = \sum_k g_k(\mathbf{x}_t, a)\hat{Q}_{k,t}(\mathbf{x}_t, a)$ is calculated, resulting in the *prediction vector* $(\hat{Q}_t(\mathbf{x}_t, a_1), \dots, \hat{Q}_t(\mathbf{x}_t, a_{|\mathcal{A}|}))$ that predicts the expected return for the current state $\mathbf{x}_t$ and each possible action that can be performed. Based on this prediction vector, an action a_t is chosen and performed, leading to the next state $\mathbf{x}_{t+1}$ and reward $r_{x_t x_{t+1}}(a_t)$. The subset of the match set that promoted the chosen action becomes the *action set* that contains all classifiers such that $m_k(\mathbf{x}_t, a_t) > 0$. At the same time, a new prediction vector $(\hat{Q}_t(\mathbf{x}_{t+1}, a_1), \dots, \hat{Q}_t(\mathbf{x}_{t+1}, a_{|\mathcal{A}|}))$ for state $\mathbf{x}_{t+1}$ is formed, and its largest element is chosen, giving $\max_{a\in\mathcal{A}} \tilde{Q}_t(\mathbf{x}_{t+1}, a)$. Then, all classifiers in the action set are updated by the *modified delta rule* (which is equivalent to the NLMS algorithm) with the target value $r_{x_t x_{t+1}}(a_t) + \gamma \max_{a\in\mathcal{A}} \hat{Q}_t(\mathbf{x}_{t+1}, a)$. The update in (9.28) uses exactly this target value, as given by (9.26), and updates the weight vector of each classifier for which $m_k(\mathbf{x}_t, a_t) > 0$, which are the classifiers in the action set. This shows that (9.28) describes the weight vector update as it is performed in XCSF, and therefore XCS(F) performs gradient descent without any additional modification.

9.4 Stability of RL with LCS

An additional challenge when using LCS for sequential decision tasks is that some combinations of DP and function approximation can lead to instabilities and divergence, and the same applies to RL. In fact, as RL is based on DP, convergence of RL with value function approximation is commonly analysed by showing that the underlying DP method is stable when used with this function approximation methods, and that the difference between the RL and the DP methods asymptotically converges to zero (for example, [215, 17, 16, 129]).

In this section we investigate whether the LCS approximation architecture of our LCS model type is stable when used with DP. While value iteration is certainly the most critical method, as Q-Learning is based on it, the use of LCS with policy iteration is also discussed. No conclusive answers are provided, but initial results are presented that can lead to such answers.

Firstly, let us have a closer look at the compatibility of various function approximation architecture with value iteration and policy iteration, followed by a short discussion on the issue of stability on learning model parameters and model structure of the LCS model. This is followed by a more detailed analysis of the compatibility of the LCS model structure with value and policy iteration, to act as the basis of further investigations of RL with LCS. Note that in the following, a method that is known not to diverge is automatically guaranteed to converge. Thus, showing stability of RL with LCS implies that this combination converges.

9.4.1 Stability of Approximate Dynamic Programming

Approximate value iteration (9.13) is based on the operator conjunction $\Pi \mathrm{T}$, where T is a nonlinear operator. As shown by Boyan and Moore [25], this procedure might diverge when used with even the most common approximation architectures, such as linear or quadratic regression, local weighted regression, or neural networks. Gordon [96] has shown that stability is guaranteed if the approximation Π is – just like T – a non-expansion to the maximum norm, that is, if for any two $\mathbf{V}, \mathbf{V}'$ we can guarantee $\|\Pi \mathbf{V} - \Pi \mathbf{V}'\|_\infty \leq \|\mathbf{V} - \mathbf{V}'\|_\infty$. This is due to the fact that combining a contraction and a non-expansion to the same norm results in a contraction. The class of *averagers* satisfy this requirement, and contain "[...] local weighted averaging, k-nearest neighbour, Bézier patches, linear interpolation, bilinear interpolation on a square (or cubical, etc.) mesh, as well as simpler methods like grids and other state aggregations." [96].

Due to the linearity of T_μ, approximate policy iteration has less stability problems that approximate value iteration. Just as T, T_μ is a contraction with respect to the maximum norm, and is thus guaranteed to be stable if used in combination with an approximation Π that is a non-expansion to the same norm. Also, note that $\mathrm{T}_\mu^{(\lambda)}$ forms a contraction mapping with respect to $\|\cdot\|_D$, and $\mathrm{T}_\mu \equiv \mathrm{T}_\mu^{(0)}$. Thus, another stable option is for Π to be a non-expansion with respect to $\|\cdot\|_D$ rather than $\|\cdot\|_\infty$. This property was used to show that approximate policy iteration is guaranteed to converge, as long as the states are sampled according to the steady-state distribution of the Markov chain $\mathbf{P}^{\mu}$ [215]. As long as the states are visited by performing actions according to μ, such a sampling regime is guaranteed. On the other hand, counterexamples where sampling of the states does not follow this distribution were shown to potentially lead to divergence [8, 25, 96, 214].

The same stability issues also apply to the related RL methods: Q-Learning was shown to diverge in some cases when used with linear approximation architectures [27], analogous to approximate value iteration. Thus, special care needs to be taken when Q-Learning is used in LCS.

To summarise, if Π is a non-expansion with respect to $\| \cdot \|_{\infty}$, its use for approximate value iteration and policy iteration is guaranteed to be stable. If it is a non-expansion with respect to $\| \cdot \|_D$, then its is stable when used for approximate policy iteration, but its stability with approximate value iteration is not guaranteed. Even if the function approximation method is a non-expansion to neither of these norms, this does not necessarily mean that it will diverge when used with approximate DP. However, one needs to resort to other approaches than contraction and non-expansion to analyse its stability.

9.4.2 Stability on the Structure and the Parameter Learning Level

Approximating the action-value function with LCS requires on one hand to find a good set of classifiers, and on the other hand to correctly learn the parameters of that set of classifiers. In other words, we want to find a good model structure $\mathcal{M}$ and the correct model parameter $\boldsymbol{\theta}$ for that structure, as discussed in Chap. 3.

Incremental learning can be performed on the structure level as well as the parameter level (see Sect. 8.4.2). Similarly, stability of using LCS with DP can be considered at both of these levels.

Stability on the Structure Learning Level

Divergence of DP with function approximation is expressed by the values of the value function estimate rapidly growing out of bounds (for example, [25]). Let us assume that for some fixed LCS model structure, the parameter learning process diverges when used with DP, and that there exist model structures for which this is not the case.

Divergence of the parameters usually happens locally, that is, not for all classifiers at once. Therefore, it can be detected by monitoring the model error of single classifiers, which, for linear classifier models as given in Chap. 5, would be the model noise variance. Subsequently, divergent classifiers can be detected and replaced until the model structure allows the parameter learning to be compatible with the used DP method.

XCSF uses linear classifier models and Q-Learning, but such combinations are known to be unstable [25]. However, XCSF has never been reported to show divergent behaviour. Thus, it is conjectured that it provides stability on the model structure level by replacing divergent classifiers with potentially better ones.

Would the classifier set optimality criterion that was introduced in Chap. 7 also provide us with a safeguard against divergence at the model structure level; that is, would divergent classifiers be detected? In contrast to XCS(F), the criterion that was presented does not assume a classifier to be a bad local model as soon as its model error is above a certain threshold. Rather, the localisation of a classifier is inappropriate if its model is unable to capture the apparent pattern that is hidden in the noisy data. Therefore, it is not immediately clear if the criterion would detect the divergent model as a pattern that the classifier cannot model, or if it would assume it to be noise.

In any case, providing stability on the model structure level is to repair the problem of divergence *after* it occurred, and relies on the assumption that changing the model structure does indeed provide us with the required stability. This is not a satisfactory solution, which is why the focus should be on preventing the problem from occurring at all, as discussed in the next section.

Stability on the Parameter Learning Level

Given a fixed model structure $\mathcal{M}$, the aim is to provide parameter learning that is guaranteed to converge when used with DP methods. Recall that both value iteration and policy iteration are guaranteed to converge if the approximation architecture is a non-expansion with respect to the maximum norm $\|\cdot\|_\infty$. It being a non-expansion with respect to the weighted norm $\|\cdot\|_D$, on the other hand, is sufficient for the convergence of the policy evaluation step of policy iteration, but not value iteration. In order to guarantee stability of either method when using LCS, the LCS approximation architecture needs to provide such a non-expansion.

Observe that having a single classifier that matches all states is a valid model structure. In order for this model structure to provide a non-expansion, the classifier model itself must form a non-expansion. Therefore, to ensure that the LCS model provides the non-expansion property for any model structure, every classifier model needs to form a non-expansion, and any mixture of a set of localised classifiers that forms the global LCS model needs to form a non-expansion as well. Formally, if $\|\cdot\|$ denotes the norm in question, we need

$$\|\Pi\mathbf{V} - \Pi\mathbf{V}'\| \leq \|\mathbf{V} - \mathbf{V}'\| \tag{9.33}$$

to hold for any two $\mathbf{V}, \mathbf{V}'$, where Π is the approximation operator of a given LCS model structure. If the model structure is formed by a single classifier that matches all states,

$$\|\Pi_k\mathbf{V} - \Pi_k\mathbf{V}'\| \leq \|\mathbf{V} - \mathbf{V}'\| \tag{9.34}$$

needs to hold for any two $\mathbf{V}, \mathbf{V}'$, where Π_k is the approximation operator of a single classifier. These requirements are independent of the LCS model type.

Returning to the LCS model structure with independently trained classifiers, the next two sections concentrate on its non-expansion property, firstly with respect to $\|\cdot\|_\infty$, and then with respect to $\|\cdot\|_D$.

9.4.3 Non-expansion with Respect to $\| \cdot \|_\infty$

In the following, $\pi(\mathbf{x})$ denotes the sampling probability for state $\mathbf{x}$ according to the steady state distribution of the Markov chain $\mathbf{P}^\mu$, and $\pi_k(\mathbf{x}) = m_k(\mathbf{x})\pi(\mathbf{x})$ denotes this distribution augmented by the matching of classifier k. Also, $\mathbf{M}_k = \text{diag}(m_k(\mathbf{x}_1), \dots, m_k(\mathbf{x}_N))$ is the matching matrix, as in Chap. 5, $\mathbf{D} = \text{diag}(\pi(\mathbf{x}_1), \dots, \pi(\mathbf{x}_N))$ is the sampling distribution matrix, and $\mathbf{D}_k = \mathbf{M}_k\mathbf{D}$ is the sampling distribution matrix with respect to classifier k. Each classifier k aims at finding the weight vector $\mathbf{w}_k$ that minimises $\|\mathbf{X}\mathbf{w}_k - \mathbf{V}\|_{D_k}$, which is given by $\hat{\mathbf{w}}_k = (\mathbf{X}^T\mathbf{D}_k\mathbf{X})^{-1}\mathbf{X}^T\mathbf{D}_k\mathbf{V}$. Thus, a classifier's approximation operator is the projection matrix $\boldsymbol{\Pi}_k = \mathbf{X}(\mathbf{X}^T\mathbf{D}_k\mathbf{X})^{-1}\mathbf{X}^T\mathbf{D}_k$. such that $\hat{\mathbf{V}} = \mathbf{X}\hat{\mathbf{w}} = \boldsymbol{\Pi}_k\mathbf{V}$.

It cannot be guaranteed that general linear models form a non-expansion with respect to $\| \cdot \|_\infty$. Gordon, for example, has shown that this is not the case for straight line models [96]. Averagers, on the other hand, are a form of linear model, but provide a non-expansion with respect to $\| \cdot \|_\infty$ and thus can be used for both value iteration and policy iteration.

With respect to LCS, each single classifier, as well as the whole set of classifiers need to conform to the non-expansion property. This rules out the general use of linear model classifiers. Instead, only averaging classifiers (see Ex. 5.2) will be discussed, as their model provides a non-expansion with respect to $\| \cdot \|$:

Lemma 9.1. *The model of averaging classifiers forms a non-expansion with respect to the maximum norm.*

Proof. As for averaging classifiers $\mathbf{X} = (1, \dots, 1)^T$, their approximation operator is the same for all states and is given by

$$(\boldsymbol{\Pi}_k\mathbf{V})_j = \text{Tr}(\mathbf{D}_k)^{-1} \sum_{x \in \mathcal{X}} \pi_k(\mathbf{x})V(\mathbf{x}). \tag{9.35}$$

Let $\mathbf{V}, \mathbf{V}'$ be two vectors such that $\mathbf{V} \leq \mathbf{V}'$, which implies that the vector $\mathbf{a} = \mathbf{V}' - \mathbf{V}$ is non-negative in all its elements. Thus, we have for any i,

$$(\boldsymbol{\Pi}_k\mathbf{V})_i = \text{Tr}(\mathbf{D}_k)^{-1} \sum_{x \in \mathcal{X}} \pi_k(\mathbf{x})V'(\mathbf{x}) - \text{Tr}(\mathbf{D}_k)^{-1} \sum_{x \in \mathcal{X}} \pi_k(\mathbf{x})a \leq (\boldsymbol{\Pi}_k\mathbf{V})_i, \tag{9.36}$$

due to the non-negativity of the elements of $\mathbf{D}_k\mathbf{a}$. Also, for any scalar b and vector $\mathbf{e} = (1, \dots, 1)^T$,

$$(\boldsymbol{\Pi}_k(\mathbf{V} + b\mathbf{e}))_i = (\boldsymbol{\Pi}_k\mathbf{V})_i + b \tag{9.37}$$

holds.

Let $\mathbf{V}, \mathbf{V}'$ now be two arbitrary vectors, not bound to $\mathbf{V} \leq \mathbf{V}'$, and let $c = \|\mathbf{V} - \mathbf{V}'\|_\infty = \max_i |V_i - V_i'|$ be their maximum absolute distance. Thus, for any i,

$$V_i - c \leq V_i' \leq V_i + c \tag{9.38}$$

holds. Applying $\boldsymbol{\Pi}_k$ and using the above properties gives

$$(\boldsymbol{\Pi}_k\mathbf{V})_i - c \leq (\boldsymbol{\Pi}_k\mathbf{V}')_i \leq (\boldsymbol{\Pi}_k\mathbf{V})_i + c, \tag{9.39}$$

and thus

$$|(\mathbf{\Pi}_k\mathbf{V})_i - (\mathbf{\Pi}_k\mathbf{V})_i| \leq c. \quad (9.40)$$

As this holds for any i, we have

$$\|\mathbf{\Pi}_k\mathbf{V} - \mathbf{\Pi}_k\mathbf{V}'\|_\infty \leq \|\mathbf{V} - \mathbf{V}'\|_\infty. \quad (9.41)$$

which completes the proof.

Thus, averaging classifiers themselves are compatible with value iteration and policy iteration. Does this, however, still hold for a set of classifiers that provides its prediction by (9.21)? Let us first consider the special case when the mixing functions are constant:

Lemma 9.2. *Let the global model $\hat{\mathbf{V}}$ be given by $\hat{\mathbf{V}} = \sum_k \mathbf{G}_k\mathbf{\Pi}_k\mathbf{V}$, where $\mathbf{\Pi}_k$ is the approximation operator of averaging classifiers, and the $\mathbf{G}_k$'s are constant diagonal mixing matrices with non-negative elements such that $\sum_k \mathbf{G}_k = \mathbf{I}$ holds. Then, this model forms a non-expansion with respect to the maximum norm.*

Proof. The method to proof the above statement is similar to the one used to prove Lemma 9.1: firstly, we show that for any two vectors $\mathbf{V}, \mathbf{V}'$ such that $\mathbf{V} \leq \mathbf{V}'$, an arbitrary scalar b, and $\mathbf{e} = (1, \dots, 1)^T$ we have $\mathbf{\Pi V} \leq \mathbf{\Pi V}'$ and $\mathbf{\Pi}(\mathbf{V} + b\mathbf{e}) = (\mathbf{\Pi V}) + b\mathbf{e}$, where $\mathbf{\Pi} = \sum_k \mathbf{G}_k\mathbf{\Pi}_k$. Then, non-expansion is shown by applying these properties to the maximum absolute difference between two arbitrary vectors.

$\mathbf{\Pi V} \leq \mathbf{\Pi V}'$ follows from observing that for the vector $\mathbf{a} = \mathbf{V}' - \mathbf{V}$ with non-negative elements, all elements of $\sum_k \mathbf{G}_k\mathbf{\Pi}_k\mathbf{a}$ are non-negative due to the non-negativity of the elements of $\mathbf{G}_k$, and $\mathbf{\Pi}_k\mathbf{a} \geq \mathbf{0}$, and thus

$$\mathbf{\Pi V} = \sum_k \mathbf{G}_k\mathbf{\Pi}_k\mathbf{V}' - \sum_k \mathbf{G}_k\mathbf{\Pi}_k\mathbf{a} \leq \mathbf{\Pi V}'. \quad (9.42)$$

Also, as $\mathbf{\Pi}_k(\mathbf{V} + b\mathbf{e}) = (\mathbf{\Pi}_k\mathbf{V}) + b\mathbf{e}$ and $\sum_k \mathbf{G}_k(\mathbf{V}_k + b\mathbf{e}) = b\mathbf{e} + \sum_k \mathbf{G}_k\mathbf{V}_k$ for any K vectors $\{\mathbf{V}_k\}$, we have

$$\mathbf{\Pi}(\mathbf{V} + b\mathbf{e}) = \mathbf{\Pi V} + b\mathbf{e}. \quad (9.43)$$

Let $\mathbf{V}, \mathbf{V}'$ now be to arbitrary vectors, not bound to $\mathbf{V} \leq \mathbf{V}'$, and let $c = \|\mathbf{V} - \mathbf{V}'\|_\infty = \max_i |V_i - V_i'|$ be their maximum absolute distance. Thus, for any i,

$$V_i - c \leq V_i' \leq V_i + c. \quad (9.44)$$

Given the properties of $\mathbf{\Pi}$ it follows that

$$(\mathbf{\Pi V})_i - c \leq (\mathbf{\Pi V}')_i \leq (\mathbf{\Pi V})_i + c, \quad (9.45)$$

and therefore

$$|(\mathbf{\Pi V})_i - (\mathbf{\Pi V}')_i| \leq c, \quad (9.46)$$

from which follows that

$$\|\mathbf{\Pi V} - \mathbf{\Pi V}'\|_\infty \leq \|\mathbf{V} - \mathbf{V}'\|_\infty, \quad (9.47)$$

which completes the proof.

This shows that it is save to use LCS with independently trained averaging classifiers for both value iteration and policy iteration, given that the mixing weights are fixed. Fixing these weights, however, does not allow them to react to the quality of a classifier's approximation. As discussed in Chap. 6, it is preferable to adjust the mixing weights inversely proportional to the classifier's prediction error.

To show that the mixing weights are relevant when investigating the non-expansion property of the LCS model, consider the following: given two states $\mathcal{X} = \{\mathbf{x}_1, \mathbf{x}_2\}$ that are sampled with equal frequency, $\pi(\mathbf{x}_1) = \pi(\mathbf{x}_2) = 1/2$, and two classifiers of which both match $\mathbf{x}_2$, but only the first one matches $\mathbf{x}_1$, we have $m_1(\mathbf{x}_1) = m_1(\mathbf{x}_2) = m_2(\mathbf{x}_2) = 1$ and $m_2(\mathbf{x}_1) = 0$. Let the two target vectors be $\mathbf{V} = (0, 1)^T$ and $\mathbf{V}' = (2, 4)$. As the classifiers are averaging classifiers, they will give the estimates $\hat{\mathbf{V}}_1 = 1/2$, $\hat{\mathbf{V}}_2 = 1$, $\hat{\mathbf{V}}_1^{'} = 3$, $\hat{\mathbf{V}}_2^{'} = 4$. For $\mathbf{x}_1$ the global prediction is given by classifier 1. For $\mathbf{x}_2$, on the other hand, the predictions of the classifiers are mixed and thus, the global prediction will be in the range $[1/2, 1]$ for $V(\mathbf{x}_2)$ and within $[3, 4]$ for $V'(\mathbf{x}_2)$. Note that $\|\mathbf{V} - \mathbf{V}'\|_\infty = |V(\mathbf{x}_2) - V'(\mathbf{x}_2)| = 3$. Choosing arbitrary mixing weights, classifier 2 can be assigned full weights for $V'(\mathbf{x}_2)$, such that $\hat{V}'(\mathbf{x}_2) = 4$. As a results, $3 \leq \|\hat{\mathbf{V}} - \hat{\mathbf{V}}^{'}\|_\infty \leq 3.5$, depending on how $\hat{V}_1(\mathbf{x}_2)$ and $\hat{V}_2(\mathbf{x}_2)$ are combined to $\hat{V}(\mathbf{x}_2)$. Thus, for a particular set of mixing weights that assign non-zero weights to $\hat{V}_1(\mathbf{x}_2)$, the non-expansion property is violated, which shows that mixing weights are relevant when considering this property.

In the above example, the non-expansion property was violated by using different mixing schemes for $\mathbf{V}$ and $\mathbf{V}'$. In the case of $\mathbf{V}'$, the more accurate Classifier 2 has been assigned full weights. For $\mathbf{V}$, on the other hand, some weight was assigned to less accurate Classifier 1. Assigning full weight to Classifier 2 in both cases would have preserved the non-expansion property. This puts forward the question if using a consistent mixing scheme, like mixing by inverse prediction error, guarantees a non-expansion with respect to the maximum norm and thus convergence of the algorithm? More generally, which are the required properties of the mixing scheme such that non-expansion of $\mathbf{\Pi}$ can be guaranteed?

The proof of Lemma 9.2 relies on the linearity of $\mathbf{\Pi}$, based on the constant mixing model, such that $\mathbf{\Pi V} - \mathbf{\Pi V}' = \mathbf{\Pi}(\mathbf{V} - \mathbf{V}')$. Making the mixing model depend on the classifier predictions violates this linearity and requires the use of a different method for the analysis of its properties. Besides some conjectures [80, 81], the question of which mixing weights guarantee a non-expansion with respect to $\|\cdot\|_\infty$ is still open and requires further investigation.

9.4.4 Non-expansion with Respect to $\|\cdot\|_D$

Recall that the diagonal of $\mathbf{D}$ is the sampling distribution π over $\mathcal{X}$ with respect to a particular policy μ, and is given by the steady-state probabilities of the Markov chain $\mathbf{P}^\mu$. Following this Markov chain by performing actions according to μ guarantees that the states are sampled according to π. In the following, it is assumed that this is the case.

Given the linear approximation $\mathbf{\Pi}_D = \mathbf{X}(\mathbf{X}^T\mathbf{D}\mathbf{X})^{-1}\mathbf{X}^T\mathbf{D}$ that returns the estimate $\hat{\mathbf{V}} = \mathbf{\Pi}_D\mathbf{V}$ that minimises the sampling-weighted distance $\|\mathbf{X}\mathbf{w}-\mathbf{V}\|_D$, this approximation operator is a non-expansion with respect to $\|\cdot\|_D$:

Lemma 9.3. *The linear approximation operator* $\mathbf{\Pi}_D = \mathbf{X}(\mathbf{X}^T\mathbf{D}\mathbf{x})^{-1}\mathbf{X}^T\mathbf{D}$ *defines a non-expansion mapping with respect to the weighted norm* $\|\cdot\|_D$.

Proof. Note that $\mathbf{D} = \sqrt{\mathbf{D}}\sqrt{\mathbf{D}}$, and thus we have $\sqrt{\mathbf{D}}\mathbf{\Pi}_D = \mathbf{\Pi}'_D\sqrt{\mathbf{D}}$, where $\mathbf{\Pi}'_D = \sqrt{\mathbf{D}}\mathbf{X}(\mathbf{X}^T\mathbf{D}\mathbf{X})^{-1}\mathbf{X}^T\sqrt{\mathbf{D}}$ is also a projection matrix. Therefore, for any two vectors $\mathbf{V}$, $\mathbf{V}'$, using the induced matrix norm $\|\mathbf{A}\| = \max\{\|\mathbf{A}\mathbf{x}\| : \text{with } \|\mathbf{x}\| \leq 1\}$, and the property $\|\mathbf{\Pi}'_D\| \leq 1$ of projection matrices,

$$\begin{aligned}
\|\mathbf{\Pi}_D\mathbf{V} - \mathbf{\Pi}_D\mathbf{V}'\|_D &= \|\sqrt{D}\mathbf{\Pi}_D(\mathbf{V} - \mathbf{V}')\| \\
&= \|\mathbf{\Pi}'_D\sqrt{D}(\mathbf{V} - \mathbf{V}')\| \\
&\leq \|\mathbf{\Pi}'_D\|\|\sqrt{D}(\mathbf{V} - \mathbf{V}')\| \\
&\leq \|\mathbf{V} - \mathbf{V}'\|_D,
\end{aligned} \tag{9.48}$$

which shows that $\mathbf{\Pi}_D$ is a non-expansion with respect to $\|\cdot\|_D$.

This shows that linear models are compatible with approximate policy iteration [215]. However, the LCS model discussed here is non-linear due to the independent training of the classifiers. Also, these classifiers are not trained according to the sampling distribution π if they do not match all states. From the point-of-view of classifier k, the states are sampled according to $\mathrm{Tr}(\mathbf{D}_k)^{-1}\pi_k$, where π_k needs to be normalised by $\mathrm{Tr}(\mathbf{D}_k)^{-1}$ as $\sum_x \pi_k(x) \leq 1$ and therefore π_k is not guaranteed to be a proper distribution. This implies, that the approximation operator $\mathbf{\Pi}_k$ is a non-expansion mapping with respect to $\|\cdot\|_{D_k}$ rather than $\|\cdot\|_D$, and $\|\mathbf{\Pi}_k\mathbf{z}\|_{D_k} \leq \|\mathbf{z}\|_{D_k}$ for any vector $\mathbf{z}$. However, as $\sqrt{\mathbf{D}_k} = \sqrt{\mathbf{M}_k}\sqrt{\mathbf{D}}$, we have

$$\|\mathbf{z}\|_{D_k} = \|\sqrt{\mathbf{D}_k}\mathbf{z}\| = \|\sqrt{\mathbf{M}_k}\sqrt{\mathbf{D}}\mathbf{z}\| \leq \|\sqrt{\mathbf{M}_k}\|\|\sqrt{\mathbf{D}}\mathbf{z}\| \leq \|\mathbf{z}\|_D. \tag{9.49}$$

The second inequality is based on the matrix norm of a diagonal matrix being given by its largest diagonal element, and thus $\|\sqrt{\mathbf{M}_k}\| = \max_x \sqrt{m_k(\mathbf{x})} \leq 1$. This implies that, for any two $\mathbf{V}$, $\mathbf{V}'$,

$$\|\mathbf{\Pi}_k\mathbf{V} - \mathbf{\Pi}_k\mathbf{V}'\|_D \geq \|\mathbf{\Pi}_k\mathbf{V} - \mathbf{\Pi}_k\mathbf{V}\|_{D_k} \leq \|\mathbf{V} - \mathbf{V}'\|_{D_k} \leq \|\mathbf{V} - \mathbf{V}'\|_D. \tag{9.50}$$

Due to the first inequality having the wrong direction, we cannot state that $\mathbf{\Pi}_k$ is a non-expansion with respect to $\|\cdot\|_D$. In fact, it becomes rather unlikely[3]. Nonetheless, to be sure about either outcome, further investigation is required.

Not having a clear result for single classifiers, expanding the investigation to sets of classifiers is superfluous. In any case, it is certain that given stable classifier models, the non-expansion property of a whole set of classifiers is, as for $\|\cdot\|_\infty$, determined by the properties of the mixing model.

[3] We have previously stated that $\mathbf{\Pi}_k$ is a non-expansion with respect to $\|\cdot\|_D$ [79]. While showing this, however, a flawed matrix equality was used, which invalidates the result.

9.4.5 Consequences for XCS and XCSF

Both XCS and XCSF use Q-Learning as their reinforcement learning component. To show that this combination is stable, the first step to take is to show the stability of value iteration with the model structure underlying XCS and XCSF.

XCS uses averaging classifiers, which were shown to be stable with value iteration. Stability at the global model level is very likely, but depends on the mixing model, and definite results are still pending.

XCSF in its initial implementation [240, 241], on the other hand, uses classifiers that model straight lines, and such models are known to be unstable with value iteration. Thus, XCSF does not even guarantee stability at the classifier level, and therefore neither at the global model level. As previously mentioned, it is conjectured that XCSF provides its stability at the structure learning level instead, which is not considered as being a satisfactory solution. Instead, one should aim at replacing the classifier models such that stability at the parameter level can be guaranteed. Averaging RL seems to be a good starting point, but how exactly this can be done still needs to be investigated.

9.5 Further Issues

Besides the stability concerns when using LCS to perform RL, there are still some further issues to consider, two of which will be discussed in this section: the learning of long paths, and how to best handle the explore/exploit dilemma.

9.5.1 Long Path Learning

The problem of long path learning is to find the optimal policy in sequential decision tasks when the solution requires learning of action sequences of substantial length. As identified by Barry [11, 12], XCS struggles with such tasks due to the generalisation method that it uses.

While a solution was proposed to handle this problem [12], it was only designed to work for a particular problem class, as will be shown after discussing how XCS fails at long path learning. The classifier set optimality criterion from Chap. 7 might provide better results, but in general, long path learning remains an open problem.

Long path learning is not only an issue for LCS, but for approximate DP and RL in general. It arises from the used approximation glossing over small differences in the value or action-value function, which causes the policy that is derived from this function to become sub-optimal. This effect is amplified by weak discounting (that is, for γ close to 1) which causes the expected return to differ less between adjacent states.

XCS and Long Path Learning

Consider the problem that is shown in Fig. 9.2. The aim is to find the policy that reaches the terminal state x_6 from the initial state x_{1a} in the shortest number of

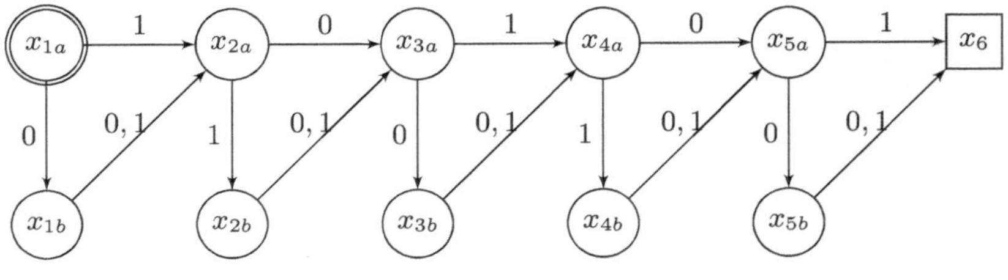

Fig. 9.2. A 5-step corridor finite state world. The circles represent the states of the problem, and the arrows the possible state transitions. The numbers next to the arrows are the actions that cause the transitions, showing that the only available actions are 0 and 1. The state x_{1a} is the initial state in which the task starts, and the square state x_6 is the terminal state in which the task ends.

steps. In RL terms, this aim is described by giving a reward of 1 upon reaching the terminal state, and a reward of 0 for all other transitions[4]. The optimal policy is to alternately choose actions 0 and 1, starting with action 1 in state x_{1a}.

The optimal value function V^* over the number of steps to the terminal state is for a 15-step corridor finite state world shown in Fig. 9.3(a). As can be seen, the difference of the values of V^* between two adjacent states decreases with the distance from the terminal state.

Recall that, as described in Sect. 7.1.1, XCS seeks for classifiers that feature the mean absolute error ϵ_0, where ϵ_0 is the same for all classifiers. Thus, with increasing ϵ_0, XCS will start generalising over states that are further away from the terminal state, as due to their similar value they can be modelled by a single classifier while keeping its approximation error below ϵ_0. On the other hand, ϵ_0 cannot be set too small, as otherwise the non-stationarity of the function to model would make all classifiers seem inaccurate. Generalising over states x_{ia} for different i's, however, causes the policy in these areas to be sub-optimal, as choosing the same action in two subsequent steps in the corridor finite state world causes at least one sidestep to one of the x_{ib} states[5].

To summarise, XCS struggles in learning the optimal policy for tasks where the difference in value function between two successive states is very small and might be modelled by the same classifier, and where choosing the same action for both states leads to a sub-optimal policy. The problem was identified by Barry [11, 12], and demonstrated by means of different-length corridor finite state worlds in. Due to the same classifier accuracy criterion, XCSF can be

[4] More precisely, the reward 1 that is received upon reaching the terminal state was modelled by adding a transition that, independent of the chosen action, leads from the terminal state to an absorbing state and is rewarded by 1. Each transition from the absorbing state leads to itself, with a reward of 0.

[5] It should be noted that while the classifiers in standard implementations of XCS(F) can match several states, they always match and thus promote a single action.

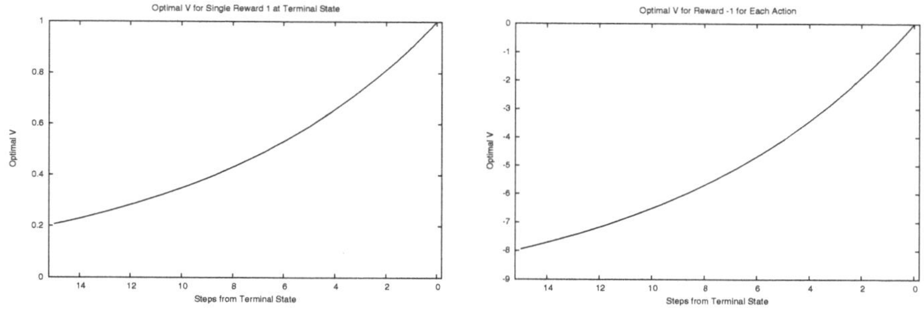

(a) (b)

Fig. 9.3. Plots showing the optimal value function for the 15-step corridor finite state world for $\gamma = 0.9$. The value function in (a) results from describing the task by giving a single reward 1 upon reaching the terminal state, and a reward of 0 for all other transitions. In (b) the values are based on a task description that gives a reward of -1 for all transitions. Note that in both cases the optimal policy is the same, but in (a) all values are positive, and in (b) they are negative.

expected to suffer from the same problem, even though that remains to be shown empirically.

Using the Relative Error

Barry proposed two preliminary approaches to handle the problem in long path learning in XCS, both based on making the error calculation of a classifier relative to its prediction of the value function [12]. The first approach is to estimate the distance of the matched states to the terminal state and scale the error accordingly, but this approach suffers from the inaccuracy of predicting this distance.

A second, more promising alternative proposed in his study is to scale the measured prediction error by the inverse absolute magnitude of the prediction. The underlying assumption is that the difference in optimal values between two successive states is proportional to the absolute magnitude of these values, as can be see in Fig. 9.3(a). Consequently, the relative error is larger for states that are further away from the terminal state, and overly general classifiers are identified as such. This modification allows XCS to find the optimal policy in the 15-step corridor finite state world, which it fails to do without the modification.

Where it Fails

The problem of finding the shortest path to the terminal state can also be defined differently: rather than giving a single reward of 1 upon reaching the terminal state, one can alternatively punish each transition with a reward of -1. As the reward is to be maximised, the number of transitions is minimised, and therefore the optimal policy is the same as before. Fig. 9.3(b) shows the optimal value function for the modified problem definition.

Observe that, in contrast to Fig. 9.3(a), all values of V^* are negative or zero, and their absolute magnitude grows with the distance from the terminal state. The difference in magnitude between two successive state, on the other hand, still decreases with the distance from the terminal state. This clearly violates the assumption that this difference is proportional to the absolute magnitude of the values, as the modified problem definition causes exactly the opposite pattern. Hence, the relative error approach will certainly fail, as it was not designed to handle such cases.

To create a task where the relative error measure fails, the problem had to be redefined such that the value function takes exclusively negative values. While it might be possible to do the opposite and redefine each problem such that it conforms to the assumption that the relative error measure is based on, an alternative that does not require modification of the problem definition is preferable.

A Possible Alternative?

It was shown in Sect. 8.3.4 that the optimality criterion that was introduced in Chap. 7 is able to handle problem where the noise differs in different areas of the input space. Given that it is possible to use this criterion in an incremental implementation, will such an implementation be able to perform long path learning?

As previously discussed (see Sect. 5.1.2 and 7.2.2), a linear classifier model attributes all observed deviation from its linear model to measurement noise (implicitly including the stochasticity of the data-generating process). In reinforcement learning, and additional component of stochasticity is introduced by updating the value function estimates which makes them non-stationary. Thus, in order for the LCS model to provide a good representation of the value function estimate, it needs to be able to handle both the measurement noise and the update noise – a differentiation that is absent Barry's work [11, 12].

Let us assume that the optimality criterion causes the size of the area of the input space that is matched by a classifier to be proportional to the level of noise in the data, such that the model is refined in areas where the observations are known to accurately represent the data-generating process. Considering only measurement noise, when applied to value function approximation this would lead to having more specific classifiers in states where the difference in magnitude of the value function for successive states is low, as in such areas this noise is deemed to be low. Therefore, the optimality criterion should provide an adequate value function approximation of the optimal value function, even in cases where long action sequences need to be represented.

Also considering update noise, its magnitude is related to the magnitude of the optimal value function, as demonstrated in Fig. 9.4. Therefore, the noise appears to be largest where the magnitude of the optimal value function is large. Due to this noise, the model in such areas will most likely be coarse. With respect to the corridor finite state world, for which the optimal value function is shown

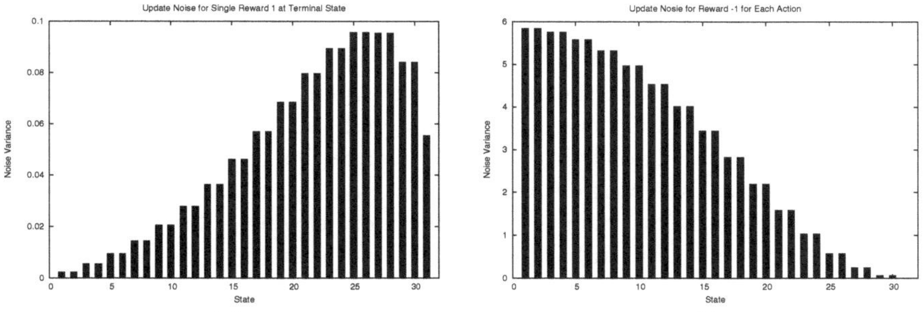

(a) (b)

Fig. 9.4. Update noise variance for value iteration performed on 15-step corridor finite state world. Plot (a) shows the variance when a reward of 1 is given upon reaching the terminal state, and 0 for all other transitions. Plot (b) shows the same when rewarding each transition with -1. The states are enumerated in the order $x_{1a}, x_{1b}, x_{2a}, \dots, x_{15b}, x_{16}$. The noise variance is determined by initialising the value vector to 0 for each state, and storing the value vector after each iteration of value iteration, until convergence. The noise variance is the variance of the values of each state over all iterations. It clearly shows that this variance is higher for states which have a larger absolute optimal value. The optimal values are shown in Fig. 9.3.

in Fig. 9.3(b), this would have the effect of providing an overly coarse model for states that are distant from the terminal state, and thus might cause the policy to be sub-optimal, just as in XCS. However, this depends heavily on the dynamic interaction between the RL method and the incremental LCS implementation. Thus, definite statements needs to be postponed until such an implementation is available.

Overall, the introduced optimality criterion seems to be a promising approach to handle long path learning in LCS, when considering only measurement noise. Given the additional update noise, however, the criterion might suffer from the same problems as the approach based on the relative error. The significance of its influence cannot be evaluated before an incremental implementation is available. Alternatively, it might be possible to seek for RL approaches that allow for the differentiation between measurement and update noise, which makes it possible for the model itself to only concentrate on the measurement noise. If such an approach is feasible still needs to be investigated.

9.5.2 Exploration and Exploitation

Maintaining the balance between exploiting current knowledge to guide action selection and exploring the state space to gain new knowledge is an essential problem for reinforcement learning. Too much exploration implies the frequent selection of sub-optimal actions and causes the accumulated reward to decrease. Too much emphasis on exploitation of current knowledge, on the other hand, might cause the agent to settle on a sub-optimal policy due to insufficient knowledge of the reward distribution [228, 209]. Keeping a good balance is important as it has a significant impact on the performance of RL methods.

There are several approaches to handling exploration and exploitation: one can choose a sub-optimal action every now and then, independent of the certainty of the available knowledge, or one can take this certainty into account to choose actions that increase it. A variant of the latter is to use Bayesian statistics to model this uncertainty, which seems the most elegant solution but is unfortunately also the least tractable. All of these variants and their applicability to RL with LCS are discussed below.

Undirected Exploration

A certain degree of exploration can be achieved by selecting a sub-optimal action every now and then. This form of exploration is called undirected as it does not take into account the certainty about the current value or action-value function estimate. The probably most popular instances of this exploration type are the ε-greedy policy and softmax action selection.

The greedy policy is the one the chooses the action that is to the current knowledge optimal at each step, as is thus given by $\mu(\mathbf{x}) = \max_a \hat{Q}(\mathbf{x}, a)$. In contrast, the ε-greedy policy selects a random sub-optimal action with probability ε, and the greedy action otherwise. Its stochastic policy is given by

$$\mu(a|\mathbf{x}) = \begin{cases} 1 - \varepsilon & \text{if } a = \text{argmax}_{a \in \mathcal{A}} \hat{Q}(\mathbf{x}, a), \\ \varepsilon / (|\mathcal{A}| - 1) & \text{otherwise.} \end{cases}, \tag{9.51}$$

where $\mu(a|\mathbf{x})$ denotes the probability of choosing action a in state $\mathbf{x}$.

ε-greedy does not consider the magnitude of the action-value function when choosing the action and thus does not differentiate between actions that are only slightly sub-optimal and ones that are significantly. This is accounted for by softmax action selection, where actions are chosen in proportion to the magnitude of the estimate of their associated expected return. One possible implementation is to sample actions from the Boltzmann distribution, given by

$$\mu(a|\mathbf{x}) = \frac{\exp(\hat{Q}(\mathbf{x}, a)/T)}{\sum_{a' \in \mathcal{A}} \exp(\hat{Q}(\mathbf{x}, a')/T)}, \tag{9.52}$$

where T is the temperature that allows regulating the strength with which the magnitude of the expected return is taken into account. A low temperature $T \to 0$ causes greedy action selection. Raising the temperature $T \to \infty$, on the other hand, makes the stochastic policy choose all actions with equal probability.

XCS(F) also uses indirect exploration, but with neither of the above policies. Instead, it alternates between exploration and exploitation trials, where a single trial is a sequence of transitions until a goal state is reached or the maximum number of steps is exceeded. Exploration trials feature a uniform random action selection, without considering the action-value estimates. In exploitation trials, on the other hand, all actions are chosen greedily. This causes significant exploration of the search space, which facilitates learning of the optimal policy. A drawback of this approach is that on average, the received reward is lower than if a more reward-oriented policy, like ε-greedy or softmax action selection, is used. In any case, undirected policies should only be used, if directed exploration is too costly to implement.

Directed Exploration

Directed exploration is significantly more efficient than undirected exploration by taking into account the uncertainty of the action-value function estimate. This allows it to perform sub-optimal actions in order to reduce this uncertainty, until it is certain that no further exploration is required to find the optimal policy. The result is less, but more intelligent exploration.

This strategy is implemented by several methods, both model-based and model-free (for example, [126, 28, 204]). In fact, some of them have shown to be very efficient in the Probably Approximately Correct (PAC) sense (for example, [204]). These, however, require a model of the reward and transition function, and thus they have a larger space complexity than model-free RL methods [205]. Nonetheless, methods that perform intelligent exploration currently outperform all other RL methods [149]. Recently, also a model-free method with intelligent exploration became available [205], but according to Littman [149], it performs "*really* slow" when compared to model-based alternatives. None of the methods will be discussed in detail, but they all share the same concept of performing actions such that the certainty of the action-value function estimate is increased.

A recent LCS approach aimed at providing such intelligent exploration [166], but without considering the RL techniques that are available. These techniques could be used in LCS even without having models of the transition and reward function by proceeding in a manner similar to [69], and building a model at the same time as using it to learn the optimal policy. Anticipatory LCS [202, 48, 40, 91, 90] are already used to model at least the transition function, and can easily be modified to include the reward function. An LCS that does both has already been developed to perform model-based RL [92, 89], but as it uses heuristics rather than evolutionary computation techniques for model structure search, some LCS workers did not consider it as being an LCS. In any case, having such a model opens the path towards using new exploration control methods to improve their efficiency and performance.

Bayesian Reinforcement Learning

The intelligent exploration methods discussed above either consider the certainty of the estimate only implicitly, or maintain it by some form of confidence interval. A more elegant approach is to facilitate Bayesian statistics and maintain complete distributions over each of the estimates.

For model-free RL, this means to model the action-value function estimates by probability distributions for each state/action pair. Unfortunately, this approach is not analytically tractable, as the distributions are strongly correlated due to the state transitions. This leads to complex posterior distributions that cannot be expressed analytically. A workaround is to use various assumptions and approximations that make the method less accurate but analytically and computationally tractable. This workaround was used to develop Bayesian Q-Learning [68] that, amongst other things, assumes the independence of all action-value function estimates, and uses an action selection scheme that maximises the information gain. Its performance increase when compared to methods based on confidence intervals is noticeable but moderate.

Bayesian model-based RL is more popular as it provides cleaner implementations. It is based on modelling the transition and reward function estimates by probability distributions that are updated with new information. This results in a problem that can be cast as a POMDP, and can be solved with the same methods [84]. Unfortunately, this implies that it comes with the same complexity, which makes it unsuitable for application to large problems. Nonetheless, some implementations have been devices (for example, [185]), and research in Bayesian RL is still very active. It is to hope that its complexity can be reduced by the use of approximation, but without losing too much accuracy and maintaining full distributions that are the advantage of the Bayesian approach.

So far, the only form of Bayesian RL that has been used with LCS is Bayesian Q-Learning by using Bayesian classifier models within a standard XCS(F), with the result of more effective and stable action selection, when compared to XCS(F) [1]. This approach could be extended to use the full Bayesian model that was introduced here, once an incremental implementation is available. The use of model-based Bayesian RL requires anticipatory LCS, but is immediate contribution is questionable due to the high complexity of the RL method itself.

9.6 Summary

Despite sequential decision tasks being the prime motivator for LCS, they are still the ones which LCS handle least successfully. This chapter provides a primer on how to use dynamic programming and reinforcement learning to handle such tasks, and on how LCS can be combined with either approach from first principles. Also, some important issues regarding such combinations, as stability, long path learning, and the exploration/exploitation dilemma were discussed.

An essential part of the LCS type discussed in this book is that classifiers are trained independently. This is not completely true when using LCS with reinforcement learning, as the target values that the classifiers are trained on are based on the global prediction, which is formed by all matching classifiers in combination. In that sense, classifiers interact when forming their action-value function estimates. Still, besides combining classifier predictions to form the target values, independent classifier training still forms the basis of this model type, even when used in combination with RL. Thus, the update equations developed in Chap. 5 can be used, and result in weight vector updates that resemble those of XCS(F). On the side, this also demonstrates that XCS(F) performs gradient descent without the need to be modified.

Regarding stability, it has been discussed which properties the approximation operator provided by LCS has to satisfy in order to guarantee convergence with approximate value iteration and policy iteration. These properties are all based on a non-expansion to some norm, where the norm determines which method can be applied. An initial analysis has been provided, but no conclusive answers have been given, pending further research.

Related to stability is also the issue of learning long action sequences, which was shown to cause problems in XCS due to its accuracy definition. While a preliminary modification to XCS solves this issue for particular problem types [12], it is not universally applicable. The introduced optimality criterion seems more promising with this respect, but definite results have to wait until an incremental LCS implementation is available that satisfies this criterion.

Overall, using LCS to approximate the value or action-value function in RL is appealing as LCS dynamically adjust to the form of this function and thus might provide a better approximation than standard function approximation techniques. It should be noted, however, that the field of RL is moving quickly, and that Q-Learning is by far not the best method that is currently available. Hence, in order for LCS to be a competitive approach to sequential decision tasks, they also need to keep track with new developments in RL, some of which were discussed when detailing the exploration/exploitation dilemma that is an essential component of RL.

In summary, it is obvious that there is still plenty of work to be done until LCS can provide the same formal development as RL currently does. Nonetheless, the initial formal basis is provided in this chapter, upon which other research can build further analysis and improvements to how LCS handles sequential decision tasks effectively, competitively, and with high reliability.

10 Concluding Remarks

Reflecting back on the aim, let us recall that it was to "develop a formal framework for LCS that lets us design, analyse, and interpret LCS" (see Section 1.3). Defining LCS in terms of the model structure that they use to model the data clearly provides a new interpretation to what LCS are. Their design is to be understood in terms of the algorithms that result from the application of machine learning methods to train this model in the light of the available data. Their analysis arises "for free" from the application of those methods and the knowledge of their theoretical properties.

Regarding the theoretical basis of LCS, most of the existing theory builds on a facet-wise approach that investigates the properties of sub-components of existing LCS algorithms by means of representing these components by simplified models (see Section 2.4). The underlying assumption is that one can gain knowledge about the operation of an algorithm by understanding its components. While one could question if such an approach is also able to adequately capture the interaction between these components, its main limitation seems to be the focus on the analysis of existing algorithms, which are always just a means to an end.

Here, the focus is on the end itself, which is the solution to the problems that LCS want to solve, and the design of algorithms around it, guided by how LCS were characterised by previous theoretical investigations. The main novelty of this work is the methodology of taking a model-centred view to specifying the structure of LCS and their training. All the rest follows from this approach.

The model-centred view is characterised by first formalising a probabilistic model that represents a set of classifiers, and then using standard machine learning methods to find the model that explains the given data best. This results in a probabilistic model that represents a set of classifiers and makes explicit the usually implicit assumptions that are made about the data. It also provides a definition for the optimal set of classifiers that is general in the sense that it is independent of the representation, suitable for continuous input and output spaces and hardly dependent on any system parameters, given that the priors are sufficiently uninformative. In addition, it bridges the gap between LCS and

J. Drugowitsch: Des. & Anal. of Learn. Class. Sys.: A Prob. Approach, SCI 139, pp. 237–239, 2008.
springerlink.com

machine learning by using the latter to train LCS, and facilitates the good understanding of machine learning to improve the understanding of LCS. Overall, approaching LCS from a different perspective has given us a clearer view of the problems that need to be solved and which tools can be used to solve them.

This approach still leaves high degrees of freedom in how the LCS model itself is formulated. The one provided in this work is inspired by XCS(F) and results in a similar algorithm to update its parameters. One can think of a wide range of other model types that can be considered as LCS but are quite different from the one that was used here, one example being the linear LCS model that might result in algorithms that are similar to ZCS. One thing, however, that is shared by all of these models is what makes them an LCS: a global model that is formed by a combination of replaceable localised models, namely the classifiers.

The model structure search itself might not have received the same attention as common to LCS research. This was on one hand deliberate to emphasise that, as defined here, finding the optimal classifier set is nothing else than an optimisation problem that can be solved with any global optimiser. On the other hand, however, it was only dealt with on the side due to the complexity of the problem itself: most influential LCS research is contributed to the analysis and improvement of the search for good sets of classifiers. Applying a genetic algorithm to the optimisation problem results in a Pittsburgh-style LCS, as in Chap. 8. Designing a Michigan-style LCS is a quite different problem that cannot simply be handled by the application of an existing machine learning algorithm. So far, such LCS never had a clearly defined optimal set of classifier as the basis of their design. Such a definition is now available, and it remains a challenge to further research how Michigan-style LCS can be designed on the basis of this definition.

It needs to be emphasised that the model-centred approach taken in this work is holistic in the sense that rather than handling each LCS component separately, it allows us to deal with function approximation, reinforcement learning and classifier replacement from the same starting point, which is the model.

Is taking this approach really so much better than the ad-hoc approach; that is, does it result in better methods? This question can only be answered by evaluating the performance of a resulting LCS, and needs to be postpones until such an LCS becomes available. Nonetheless, even the model-based perspective by itself provides a new view on LCS. Also, considering that most popular machine learning methods started ad-hoc and were later improved by reformulating them from a model-centred perspective, applying the same methodology to reformulating LCS is very likely to be profitable in the long run.

Another question is whether theoretical advances in a field really help improve its methods. It seems like that founding the theoretical understanding of a method is a sign of its maturity. The method does not necessarily need to be initially developed from the formal perspective, as Support Vector Machines (SVMs) were [219]. Still, providing a theoretical foundation that explains what a method is doing adds significantly to its understanding, if not also to its performance. An example where the understanding was improved is the interpretation of weight decay in neural networks as Gaussian priors on their weights (see Ex. 3.4). The significant performance increase of reinforcement learning through intelligent exploration can almost exclusively be attributed to advances in their theoretical understanding [126, 28, 204]. Correspondingly, while further improvement of the already competitive performance of LCS in supervised learning tasks cannot be guaranteed through advances from the theoretical side, such advances unquestionably increase their understanding and provide a different perspective.

Of course, the presented methodology is by no means supposed to be the ultimate and only approach to design LCS. It is not the aim to stifle the innovation in this field. Rather, its uptake is promoted for well-defined tasks such as regression and classification tasks, due to the obvious advantages that this approach promises. Also, given that Sutton's value-function hypothesis [206] is correct, and value function learning is the only efficient way to handle sequential decision tasks, then these tasks are most likely best approached by taking the model-centred view as well. On the other hand, given that the task does not fall into these categories (for example, [197]), then an ad-hoc approach without strong formal foundations might still be the preferred choice for designing LCS. However, even following the outlined route leaves significant space for design variations in how to formulate the model, and in particular which method to develop or apply to search the space of possible model structures.

Overall, with the presented perspective, the answer to "What is a Learning Classifier System?" is: a family of models that are defined by a global model being formed by a set of localised models known as classifiers, an approach for comparing competing models with respect to their suitability in representing the data, and a method to search the space of sets of classifiers to provide a good model for the problem at hand. Thus, the model was added to the method.

A Notation

The notation used in this work is very similar to the machine learning standard (for example, [19]). The subscript k always refers to the kth classifier, and the subscript n refers to the nth observation. The only exception is Chapter 5 that discusses a single classifier, which makes the use of k superfluous. Composite objects, like sets, vectors and matrices, are usually written in bold. Vectors are usually column vectors and are denoted by a lowercase symbol; matrices are denoted by an uppercase symbol. $\cdot^T$ is the transpose of a vector/matrix. $\hat{\cdot}$ is an estimate. $\cdot^*$ in Chapter 7 denotes the parameters of the variational posterior, and the posterior itself, and in Chapter 9 indicates optimality.

The tables in the next pages give the used symbol in the first column, a brief explanation of its meaning in the second column, and — where appropriate — the section number that is best to consult with respect to this symbol in the third column.

J. Drugowitsch: Des. & Anal. of Learn. Class. Sys.: A Prob. Approach, SCI 139, pp. 241–246, 2008.
springerlink.com

Sets, Functions and Distributions

$\emptyset$	empty set	
$\mathbb{R}$	set of real numbers	
$\mathbb{N}$	set of natural numbers	
$\mathbb{E}_X(X, Y)$	expectation of X, Y with respect to X	
$\mathrm{var}(X)$	variance of X	
$\mathrm{cov}(X, Y)$	covariance between X and Y	
$\mathrm{Tr}(\mathbf{A})$	trace of matrix $\mathbf{A}$	
$\langle \mathbf{x}, \mathbf{y} \rangle$	inner product of $\mathbf{x}$ and $\mathbf{y}$	5.2
$\langle \mathbf{x}, \mathbf{y} \rangle_A$	inner product of $\mathbf{x}$ and $\mathbf{y}$, weighted by matrix $\mathbf{A}$	5.2
$\Vert\mathbf{x}\Vert_A$	norm of $\mathbf{x}$ associated with inner product space $\langle \cdot, \cdot \rangle_A$	5.2
$\Vert\mathbf{x}\Vert$	Euclidean norm of $\mathbf{x}$, $\Vert\mathbf{x}\Vert \equiv \Vert\mathbf{x}\Vert_I$	5.2
$\Vert\mathbf{x}\Vert_\infty$	maximum norm of $\mathbf{x}$	9.2.1
$\otimes, \oslash$	multiplication and division operator for element-wise matrix and vector multiplication/division	8.1
L	loss function, $\mathrm{L} : \mathcal{X} \times \mathcal{X} \rightarrow \mathbb{R}^+$	3.1.1
l	log-likelihood function	4.1.2
$\mathcal{N}(\mathbf{x} \vert \boldsymbol{\mu}, \boldsymbol{\Sigma})$	normal distribution with mean vector $\boldsymbol{\mu}$ and covariance matrix $\boldsymbol{\Sigma}$	4.2.1
$\mathrm{Gam}(x \vert a, b)$	gamma distribution with shape a, scale b	7.2.3
$\mathrm{St}(\mathbf{x} \vert \boldsymbol{\mu}, \boldsymbol{\Lambda}, a)$	Student's t distribution with mean vector $\boldsymbol{\mu}$, precision matrix $\boldsymbol{\Lambda}$, and a degrees of freedom	7.4
$\mathrm{Dir}(\mathbf{x} \vert \boldsymbol{\alpha})$	Dirichlet distribution with parameter vector $\boldsymbol{\alpha}$	7.5
p	probability mass/density	
q	variational probability mass/density	7.3.1
q^*	variational posterior	7.3
Γ	gamma function	7.2.3
ψ	digamma function	7.3.2
$\mathrm{KL}(q \Vert p)$	Kullback-Leibler divergence between q and p	7.3.1
$\mathcal{L}(q)$	variational bound of q	7.3.1
$\mathbf{U}$	set of hidden variables	7.2.6

Data and Model

$\mathcal{X}$	input space	3.1
$\mathcal{Y}$	output space	3.1
$D_{\mathcal{X}}$	dimensionality of $\mathcal{X}$	3.1.2
$D_{\mathcal{Y}}$	dimensionality of $\mathcal{Y}$	3.1.2
N	number of observations	3.1
n	index referring to the nth observation	3.1
$\mathbf{X}$	set/matrix of inputs	3.1, 3.1.2
$\mathbf{Y}$	set/matrix of outputs	3.1, 3.1.2
$\mathbf{x}$	input, $\mathbf{x} \in \mathcal{X}$,	3.1
$\mathbf{y}$	output, $\mathbf{y} \in \mathcal{Y}$	3.1
$\boldsymbol{\upsilon}$	random variable for output $\mathbf{y}$	5.1.1
$\mathcal{D}$	data/training set, $\mathcal{D} = \{\mathbf{X}, \mathbf{Y}\}$	3.1
f	target function, mean of data-generating process, $f : \mathcal{X} \rightarrow \mathcal{Y}$	3.1.1
ϵ	zero-mean random variable, modelling stochasticity of data-generating process and measurement noise	3.1.1
$\mathcal{M}$	model structure, $\mathcal{M} = \{\mathbf{M}, K\}$	3.1.1, 3.2.5
$\boldsymbol{\theta}$	model parameters	3.2.1
$\hat{f}_{\mathcal{M}}$	hypothesis for data-generating process of model with structure $\mathcal{M}$, $\hat{f}_{\mathcal{M}} : \mathcal{X} \rightarrow \mathcal{Y}$	3.1.1
K	number of classifiers	3.2.2
k	index referring to classifier k	3.2.3

Classifier Model

$\mathcal{X}_k$	input space of classifier k, $\mathcal{X}_k \subseteq \mathcal{X}$	3.2.3
m_{nk}	binary matching random variable of classifier k for observation n	4.3.1
m_k	matching function of classifier k, $m_k : \mathcal{X} \rightarrow [0, 1]$	3.2.3
$\mathbf{M}$	set of matching functions, $\mathbf{M} = \{m_k\}$	3.2.5
$\mathbf{M}_k$	matching matrix of classifier k	5.2.1
$\mathbf{M}$	matching matrix for all classifiers	8.1
$\boldsymbol{\theta}_k$	parameters of model of kth classifier	9.1.1
$\mathbf{w}_k$	weight vector of classifier k, $\mathbf{w}_k \in \mathbb{R}^{D_\mathcal{X}}$	4.2.1
$\boldsymbol{\omega}_k$	random vector for weight vector of classifier k	5.1.1
$\mathbf{W}_k$	weight matrix of classifier k, $\mathbf{W} \in \mathbb{R}^{D_\mathcal{Y} \times D_\mathcal{X}}$	7.2
τ_k	noise precision of classifier k, $\tau_k \in \mathbb{R}$	4.2.1
α_k	weight shrinkage prior	7.2
a_τ, b_τ	shape, scale parameters of prior on noise precision	7.2
a_{τ_k}, b_{τ_k}	shape, scale parameters of posterior on noise precision of classifier k	7.3.1
a_α, b_α	shape, scale parameters of hyperprior on weight shrinkage priors	7.2
$a_{\alpha_k}, b_{\alpha_k}$	shape, scale parameters of hyperposterior on weight shrinkage prior of classifier k	7.3.1
$\mathbf{W}$	set of weight matrices, $\mathbf{W} = \{\mathbf{W}_k\}$	7.2
$\boldsymbol{\tau}$	set of noise precisions, $\boldsymbol{\tau} = \{\tau_k\}$	7.2
$\boldsymbol{\alpha}$	set of weight shrinkage priors, $\boldsymbol{\alpha} = \{\alpha_k\}$	7.2
ϵ_k	zero-mean Gaussian noise for classifier k	5.1.1
c_k	match count of classifier k	5.2.2
$\boldsymbol{\Lambda}_k^{-1}$	input covariance matrix (for RLS, input correlation matrix) of classifier k	5.3.5
γ	step size for gradient-based algorithms	5.3
λ_{min} / λ_{max}	smallest / largest eigenvalue of input correlation matrix $c_k^{-1}\mathbf{X}^T\mathbf{M}_k\mathbf{X}$	5.3
T	time constant	5.3
λ	ridge complexity	5.3.5
λ	decay factor for recency-weighting	5.3.5
ζ	Kalman gain	5.3.6

Gating Network / Mixing Model

z_{nk}	binary latent variable, associating observation n to classifier k	4.1
r_{nk}	responsibility of classifier k for observation n, $r_{nk} = \mathbb{E}(z_{nk})$	4.1.3, 7.3.1
$\mathbf{v}_k$	gating/mixing vector, associated with classifier k, $\mathbf{v}_k \in \mathbb{R}^{D_V}$	4.1.2
β_k	mixing weight shrinkage prior, associated with classifier k	7.2
a_β, b_β	shape, scale parameters for hyperprior on mixing weight shrinkage priors	7.2
a_{β_k}, b_{β_k}	shape, scale parameters for hyperposterior on mixing weight shrinkage priors, associated with classifier k	7.3.1
$\mathbf{Z}$	set of latent variables, $\mathbf{Z} = \{z_{nk}\}$	4.1
$\mathbf{V}$	set/vector of gating/mixing vectors	4.1.2
$\boldsymbol{\beta}$	set of mixing weight shrinkage priors, $\boldsymbol{\beta} = \{\beta_k\}$	7.2
D_V	dimensionality of gating/mixing space	6.1
g_k	gating/mixing function (softmax function in Section 4.1.2, any mixing function in Chapter 6, otherwise generalised softmax function), $g_k : \mathcal{X} \rightarrow [0, 1]$	4.1.2, 4.3.1
ϕ	transfer function, $\phi : \mathcal{X} \rightarrow \mathbb{R}^{D_V}$	6.1
$\boldsymbol{\Phi}$	mixing feature matrix, $\boldsymbol{\Phi} \in \mathbb{R}^{N \times D_V}$	8.1
$\mathbf{H}$	Hessian matrix, $\mathbf{H} \in \mathbb{R}^{KD_V \times KD_V}$	6.1.1
E	error function of mixing model, $E : \mathbb{R}^{KD_V} \rightarrow \mathbb{R}$	6.1.1
γ_k	function returning quality metric for model of classifier k for state $\mathbf{x}$, $\gamma_k : \mathcal{X} \rightarrow \mathbb{R}^+$	6.2

Dynamic Programming and Reinforcement Learning

$\mathcal{X}$	set of states	9.1.1
$\mathbf{x}$	state, $\mathbf{x} \in \mathcal{X}$	9.1.1
N	number of states	9.1.1
$\mathcal{A}$	set of actions	9.1.1
a	action, $a \in \mathcal{A}$	9.1.1
$r_{xx'}(a)$	reward function, $r : \mathcal{X} \times \mathcal{X} \times \mathcal{A} \to \mathbb{R}$	9.1.1
$r^{\mu}_{xx'}$	reward function for policy μ	9.1.1
r^{μ}_{x}	reward function for expected rewards and policy μ	9.1.1
$\mathbf{r}^{\mu}$	reward vector of expected rewards for policy μ, $\mathbf{r}^{\mu} \in \mathbb{R}^N$	9.1.1
p^{μ}	transition function for policy μ	9.1.1
$\mathbf{P}^{\mu}$	transition matrix for policy μ, $\mathbf{P}^{\mu} \in [0,1]^{N \times N}$	9.1.4
γ	discount rate, $0 < \gamma \leq 1$	9.1.1
μ	policy, $\mu : \mathcal{X} \to \mathcal{A}$	9.1.1
V	value function, $V : \mathcal{X} \to \mathbb{R}$, V^* optimal, V^{μ} for policy μ, $\tilde{V}$ approximated	9.1.2
$\mathbf{V}$	value vector, $\mathbf{V} \in \mathbb{R}^N$, $\mathbf{V}^*$ optimal, $\mathbf{V}^{\mu}$ for policy μ, $\tilde{\mathbf{V}}$ approximated	9.1.4
$\tilde{\mathbf{V}}_k$	value vector approximated by classifier k	9.3.1
Q	action-value function, $Q : \mathcal{X} \times \mathcal{A} \to \mathbb{R}$, Q^* optional, Q^{μ} for policy μ, $\tilde{Q}$ approximated	9.1.2
$\tilde{Q}_k$	action-value function approximated by classifier k	9.3.4
T	dynamic programming operator	9.2.1
T_{μ}	dynamic programming operator for policy μ	9.2.1
$\mathrm{T}^{(\lambda)}_{\mu}$	temporal-difference learning operator for policy μ	9.2.4
Π	approximation operator	9.2.3
$\mathbf{\Pi}_k$	approximation operator of classifier k	9.3.1
π	steady-state distribution of Markov chain $\mathbf{P}^{\mu}$	9.4.3
π_k	matching-augmented stead-state distribution for classifier k	9.4.3
$\mathbf{D}$	diagonal state sampling matrix	9.4.3
$\mathbf{D}_k$	matching-augmented diagonal state sampling matrix for classifier k	9.4.3
α	step-size for gradient-based incremental algorithms	9.2.6

B XCS and XCSF

As frequently referred to throughout this work, a short account of the functionality of XCS [237, 238] and XCSF [240, 241] is given here from the model-based perspective. The interested reader is referred to Butz and Wilson [57] for a full description of its algorithmic implementation. The description given here focuses on XCSF and only considers XCS explicitly in cases where it differs from XCSF.

Even though XCSF is trained incrementally and is designed to handle sequential decision tasks, it is described here as if it would perform batch learning and univariate regression to relate it more easily to the methods that are described in this work. More information on how XCSF handles sequential decision tasks is given in Section 9.3.

We assume a univariate regression setup as described in Sect. 3.1.2 with N given observations. The description concentrates firstly on the classifier and mixing models, and how to find the model parameters for a fixed model structure $\mathcal{M}$, and then focuses on how the model structure search in XCSF searches for better model structures.

B.1 Classifier Model and Mixing Model

Let us assume a model structure $\mathcal{M} = \{K, \mathbf{M}\}$ with K classifiers and their matching functions $\mathbf{M} = \{m_k : \mathcal{X} \to [0,1]\}$. The classifier models are univariate regression models that are trained independently by maximum likelihood and thus aim at finding weight vectors $\mathbf{w}_k$ that minimise

$$\sum_{n=1}^{N} m_k(\mathbf{x}_n) \left(\mathbf{w}_k^T \mathbf{x}_n - y_n \right)^2, \quad k = 1, \dots, K, \tag{B.1}$$

as described in more detail in Chap. 5. In addition to the weight vector, each classifier maintains its match count c_k, called *experience*, and estimates its mean absolute prediction error ϵ_k, simply called *error*, by

$$\epsilon_k = c_k{-1} \sum_{n=1}^{N} m(\mathbf{x}_n) \left| y_n - \mathbf{w}_k^T \mathbf{x}_n \right| . \tag{B.2}$$

J. Drugowitsch: Des. & Anal. of Learn. Class. Sys.: A Prob. Approach, SCI 139, pp. 247–249, 2008.
springerlink.com

A classifier's *accuracy* is some inverse function $\kappa(\epsilon_k)$ of the classifier error. It was initially given by an exponential [237], but was later [238, 57] redefined to

$$\kappa(\epsilon) = \begin{cases} 1 & \text{if } \epsilon < \epsilon_0, \\ \alpha \left(\frac{\epsilon}{\epsilon_0} \right)^{-\nu} & \text{otherwise,} \end{cases} \tag{B.3}$$

where the constant scalar ϵ_0 is the *minimum error*, the constant α is the scaling factor, and the constant ν is a mixing power factor [57]. The accuracy is constantly 1 up to the error ϵ_0 and then drops off steeply, with the shape of the drop determined by α and ν. The *relative accuracy* is a classifier's accuracy for a single input normalised by the sum of the accuracies of all classifiers matching that input. The *fitness* is the relative accuracy of a classifier averaged over all inputs that it matches, that is

$$F_k = c_k^{-1} \sum_{n=1}^{N} \frac{m_k(\mathbf{x}_n)\kappa(\epsilon_k)}{\sum_{j=1}^{K} m_j(\mathbf{x}_n)\kappa(\epsilon_j)} \tag{B.4}$$

Each classifier additionally maintains an estimate of the *action set size* as_k, which is the average number of classifiers that match the classifier's matched inputs, and is given by

$$\text{as}_k = c_k^{-1} \sum_{n=1}^{N} m_k(\mathbf{x}_n) \sum_{j=1}^{K} m_j(\mathbf{x}_n). \tag{B.5}$$

The error, fitness, and action set size are incrementally updated by the LMS algorithm (see Sect. 5.3.3), using the MAM update (see Sect. 5.4.1). The weight vector is in XCSF updated by the NLMS algorithm (see Sect. 5.3.4), and in XCS updated by the LMS algorithm and the MAM update with $\mathbf{x}_n = 1$ for all n.

The mixing model is the fitness-weighted average of all matching classifiers (see also Sect. 6.2.5), and is formally specified by the mixing function

$$g_k(\mathbf{x}) = \frac{m_k(\mathbf{x}_n)F_k}{\sum_{j=1}^{K} m_j(\mathbf{x}_n)F_j}. \tag{B.6}$$

For both classifier and mixing model training, XCSF aims at minimising the empirical risk rather than the expected risk, regardless of the risk of overfitting that come with this approach. Overfitting is handled at the model structure search level, as will be described in the following section.

B.2 Model Structure Search

The model structure search incrementally improves the model structure by promoting classifiers whose error is close to but not above ϵ_0 (that is, classifiers that are most general but still accurate), and a set of classifiers that is non-overlapping in the input space.

The search is performed by a Michigan-style niche GA that interprets a single classifier as an individual in a population, formed by the current set of classifiers. The set of classifiers that matches the current input is called the *match set*, and its subset that promotes the performed action is called the *action set*[1]. In regression tasks, these two sets are equivalent, as the actions are irrelevant.

Reproduction of classifiers is invoked at regular intervals, based on the time since the last invocation, averaged over the classifiers in the current action set. Upon reproduction, two classifiers from the current action set are selected with probabilities proportional to their fitnesses[2], are then copied, and – after performing crossover and mutation on their condition which represents their matching function – are injected back into the current population. If the number of classifiers in the population reaches a certain preset limit on the population size, deletion occurs. Classifier deletion is not limited to the current action set but, in general[3], classifiers are selected with a probability proportional to their estimated action set size as_k. If unmatched inputs are observed, XCSF induces classifiers into the population that match that input, called *covering*, and additionally deletes other classifiers if the population size grows out of bounds.

As reproduction is performed in the action sets, classifiers which are more general and thus participate in more action sets are more likely to reproduce. Deletion, on the other hand, does not depend on the classifiers' generality but mainly on their action set size estimates. In combination, this causes a preference for more general classifiers that are still considered as being accurate, a GA pressure called the *set pressure* in [53]. Note that due to the *fitness pressure*, classifiers with $\epsilon > \epsilon_0$ will have a very low fitness and are therefore very unlikely to be selected for reproduction. The *deletion pressure* refers to deletion being proportional to the action set size estimates, and causes an even distribution of resources over all inputs. The *mutation pressure* depends on the mutation operator and in general pushes the classifiers towards more generality up to a certain threshold.

In combination, these pressures cause XCSF to evolve classifiers that feature an error ϵ as close to ϵ_0 as possible. Thus, generality of the classifiers is controlled by the parameter ϵ_0. Therefore, overfitting is avoided by the explicit tendency of classifiers to feature some (small) deliberate error. XCSF additionally prefers non-overlapping sets of classifiers, as overlapping classifiers compete for selection within the same action set until either of them dominates. For a further discussion of the set of classifiers that XCSF tends to evolve, see Sect. 7.1.1.

---

1 Initially, XCS as described in [237] performed GA reproduction in the match set, but was later modified to act on the action set [238]. The description given here conforms to the latter version.

2 Selection for reproduction does not need to be with probabilities proportional to classifier fitness. As an alternative, tournament selection has been used [56].

3 Various variations to the described deletion scheme have been proposed and investigated [237, 131, 136].

References

1. Aliprandi, D., Mancastroppa, A., Matteucci, M.: A Bayesian Approach to Learning Classifier Systems in Uncertain Environments. In: Keijzer, et al. [128], pp. 1537–1544
2. Anderson, B.D.O., Moore, J.B.: Optimal Filtering. Information and System Sciences Series. Prentice-Hall, Inc., Englewood Cliffs (1979)
3. Armano, G.: NXCS Experts for Financial Time Series Forecasting. In: Bull [32], pp. 68–91
4. Azran, A.: Data Dependent Risk Bounds and Algorithms for Hierarchical Mixture of Experts Classifiers. Master's thesis, Israel Institute of Technology, Haifa, Israel (June 2004)
5. Azran, A., Meir, R.: Data Dependent Risk Bounds for Hierarchical Mixture of Experts Classifiers. In: Shawe-Taylor, J., Singer, Y. (eds.) COLT 2004. LNCS (LNAI), vol. 3120, pp. 427–441. Springer, Heidelberg (2004)
6. Bacardit, J., Guiu, J.M.G.: Bloat control and generalization pressure using the minimum description length principle for a Pittsburgh approach Learning Classifier System. In: Kovacs, et al. [137], pp. 59–79
7. Bacardit, J., Stout, M., Hirst, J.D., Sastry, K., Llorá, X., Krasnogor, N.: Automated Alphabet Reduction Method with Evolutionary Algorithms for Protein Structure Prediction. In: Thierens, et al. [212], pp. 346–353
8. Baird, L.C.: Residual Algorithms: Reinforcement Learning with Function Approximation. In: International Conference on Machine Learning, pp. 30–37 (1995)
9. Banzhaf, W., Daida, J.M., Eiben, A.E., Garzon, M.H., Honavar, V., Jakiela, M.J., Smith, R.E. (eds.): Proceedings of the Genetic and Evolutionary Computation Conference (GECCO 1999). Morgan Kaufmann, San Francisco (1999)
10. Barry, A.: XCS Performance and Population Structure within Multiple-Step Environments. PhD thesis, Queens University Belfast (2000)
11. Barry, A.M.: The Stability of Long Action Chains in XCS. In: Bull, et al. [37], pp. 183–199
12. Barry, A.M.: Limits in Long Path Learning with XCS. In: Cantú-Paz, et al. [59], pp. 1832–1843
13. Bartlett, P.L., Boucheron, S., Lugosi, G.: Model selection and error estimation. Machine Learning 48, 85–113 (2002)
14. Bartlett, P.L., Mendelson, S.: Rademacher and gaussian complexities: Risk bounds and structural results. Journal of Machine Learning Research 3, 462–482 (2002)

15. Bernardo, J.M., Smith, A.F.M.: Bayesian Theory. Wiley, Chichester (1994)
16. Bertsekas, D.P., Borkas, V.S., Nedić, A.: Improved Temporal Difference Methods with Linear Function Approximation. In: Si, J., Barto, A.G., Powell, W.B., Wunsch, D. (eds.) Handbook of Learning and Approximate Dynamic Programming, ch. 9, pp. 235–260. Wiley Publishers, Chichester (2004)
17. Bertsekas, D.P., Tsitsiklis, J.N.: Neuro-Dynamic Programming. Athena Scientific, Belmont (1996)
18. Beyer, H.-G., O'Reilly, U.-M., Arnold, D.V., Banzhaf, W., Blum, C., Bonabeau, E.W., Cant Paz, E., Dasgupta, D., Deb, K., Foster, J.A., de Jong, E.D., Lipson, H., Llora, X., Mancoridis, S., Pelikan, M., Raidl, G.R., Soule, T., Tyrrell, A., Watson, J.-P., Zitzler, E. (eds.): Proceedings of the Genetic and Evolutionary Computation Conference, GECCO-2005, vol. 2. ACM Press, New York (2005)
19. Bishop, C.M.: Pattern Recognition and Machine Learning. Information Science and Statistics. Springer, Heidelberg (2006)
20. Bishop, C.M., Svensén, M.: Bayesian Hierarchical Mixtures of Experts. In: Proceedings of the 19th Annual Conference on Uncertainty in Artificial Intelligence (UAI-2003), pp. 57–64. Morgan Kaufmann, San Francisco (2003)
21. Booker, L.B.: Triggered rule discovery in classifier systems. In: David Schaffer, J. (ed.) Proceedings of the 3rd International Conference on Genetic Algorithms (ICGA 1989), George Mason University, pp. 265–274. Morgan Kaufmann, San Francisco (1989)
22. Booker, L.B.: Do We Really Need to Estimate Rule Utilities in Classifier Systems?. In: Lanzi, et al. [145], pp. 125–142
23. Booker, L.B.: Approximating value function in classifier systems. In: Bull, Kovacs [36]
24. Booker, L.B.: Personal Communication (May 2006)
25. Boyan, J.A., Moore, A.W.: Generalization in reinforcement learning: Safely approximating the value function. In: Tesauro, G., Touretzky, D.S., Leen, T.K. (eds.) Advances in Neural Information Processing Systems 7, pp. 369–376. MIT Press, Cambridge (1995)
26. Boyd, S., Vandenberghe, L.: Convex Optimization. Cambridge University Press, Cambridge (2004)
27. Bradtke, S.J.: Reinforcement Learning Applied to Linear Quadratic Regulation. In: Advances in Neural Information Processing Systems, vol. 5, Morgan Kaufmann, San Francisco (1993)
28. Brafman, R.I., Tennenholtz, M.: R-max: a General Polynomial Time Algorithm for Near-optimal Reinforcement Learning. In: Proceedings of the 17th International Joint Conference on Artificial Intelligence, pp. 953–958 (2001)
29. Brown, G., Kovacs, T., Marshall, J.: UCSPv: Principled Voting in UCS Rule Populations. In: Thierens, et al. [212], pp. 1774–1782
30. Bull, L.: Simple Markov Models of the Genetic Algorithm in Classifier Systems: Multi-step Tasks. In: Lanzi, et al. [148]
31. Bull, L.: On accuracy-based fitness. Journal of Soft Computing 6(3–4), 154–161 (2002)
32. Bull, L. (ed.): Applications of Learning Classifier Systems. Studies in Fuzziness and Soft Computing, vol. 150. Springer, Heidelberg (2004)
33. Bull, L.: Two Simple Learning Classifier Systems. In: Bull, Kovacs [36], pp. 63–90. YCS part also in TR UWELCSG03–005
34. Bull, L., Hurst, J.: ZCS redux. Evolutionary Computation 10(2), 185–205 (2002)

35. Bull, L., Hurst, J.: A Neural Learning Classifier System with Self-Adaptive Constructivism. In: Proceedings of the 2003 IEEE Congress on Evolutionary Computation, vol. 2, pp. 991–997. IEEE Press, Los Alamitos (2003), Also TR UWELCSG03-003
36. Bull, L., Kovacs, T. (eds.): Foundations of Learning Classifier Systems. Studies in Fuzziness and Soft Computing, vol. 183. Springer, Berlin (2005)
37. Bull, L., Lanzi, P.L., Stolzmann, W. (eds.): Journal of Soft Computing, vol. 6. Elsevir Science Publishers, Amsterdam (2002)
38. Bull, L., O'Hara, T.: A Neural Rule Representation for Learning Classifier Systems. In: Lanzi, et al. [146]
39. Bull, L., Sha'Aban, J., Tomlinson, A., Addison, J.D., Heydecker, B.G.: Towards Distributed Adaptive Control for Road Traffic Junction Signals using Learning Classifier Systems. In: Bull [32], pp. 279–299
40. Butz, M.V.: An Algorithmic Description of ACS2. In: Lanzi, et al. [146], pp. 211–229
41. Butz, M.V.: Kernel-based, Ellipsoidal Conditions in the Real-Valued XCS Classifier System. In: Beyer, et al. [18], pp. 1835–1842
42. Butz, M.V.: Rule-Based Evolutionary Online Learning Systems: A Principled Approach to LCS Analysis and Design. Studies in Fuzziness and Soft Computing, vol. 191. Springer, Heidelberg (2006)
43. Butz, M.V.: Personal Communication (July 2007)
44. Butz, M.V., Goldberg, D.E.: Bounding the population size in XCS to ensure reproductive opportunities. In: Cantú-Paz, et al. [29], pp. 1844–1856
45. Butz, M.V., Goldberg, D.E., Lanzi, P.L.: Gradient Descent Methods in Learning Classifier Systems: Improving XCS Performance in Multistep Problems. Technical Report 2003028, Illinois Genetic Algorithms Laboratory (December 2003)
46. Butz, M.V., Goldberg, D.E., Lanzi, P.L.: Bounding Learning Time in XCS. In: Deb, K., et al. (eds.) GECCO 2004. LNCS, vol. 3102, Springer, Heidelberg (2004)
47. Butz, M.V., Goldberg, D.E., Lanzi, P.L.: Gradient Descent Methods in Learning Classifier Systems: Improving XCS Performance in Multistep Problems. IEEE Transactions on Evolutionary Computation 9(5), 452–473 (2005), Also IlliGAl TR No. 2003028
48. Butz, M.V., Goldberg, D.E., Stolzmann, W.: Introducing a Genetic Generalization Pressure to the Anticipatory Classifier System Part I: Theoretical Approach. In: Proceedings of the 2000 Genetic and Evolutionary Computation Conference (GECCO 2000), pp. 34–41 (2000)
49. Butz, M.V., Goldberg, D.E., Tharakunnel, K.: Analysis and Improvement of Fitness Exploitation in XCS: Bounding Models, Tournament Selection and Bilateral Accuracy. Evolutionary Computation 11, 239–277 (2003)
50. Butz, M.V., Kovacs, T., Lanzi, P.L., Wilson, S.: Toward a Theory of Generalization and Learning in XCS. IEEE Transaction on Evolutionary Computation 8, 28–46 (2004)
51. Butz, M.V., Kovacs, T., Lanzi, P.L., Wilson, S.W.: How XCS Evolves Accurate Classifiers. In: Spector, et al. [20], pp. 927–934
52. Butz, M.V., Lanzi, P.L., Wilson, S.W.: Hyper-ellipsoidal conditions in XCS: Rotation, linear approximation, and solution structure. In: Keijzer, et al. [128], pp. 1457–1464
53. Butz, M.V., Pelikan, M.: Analyzing the Evolutionary Pressures in XCS. In: Spector, et al. [201], pp. 935–942

54. Butz, M.V., Pelikan, M.: Studying XCS/BOA learning in Boolean functions: structure encoding and random Boolean functions. In: Keijzer, et al. [128], pp. 1449–1456
55. Butz, M.V., Pelikan, M., Llorá, X., Goldberg, D.E.: Automated global structure extraction for effective local building block processing in XCS. Evolutionary Computation 14(3) (September 2006)
56. Butz, M.V., Sastry, K., Goldberg, D.E.: Tournament selection: Stable fitness pressure in XCS. In: Cantú-Paz, et al. [59], pp. 1857–1869
57. Butz, M.V., Wilson, S.W.: An Algorithmic Descriprion of XCS. In: Bull, et al. [37], pp. 144–153
58. Butz, M.V., Goldberg, D.E., Lanzi, P.L.: Computational Complexity of the XCS Classifier System. In: Bull, Kovacs [36]
59. Cantú-Paz, E., Foster, J.A., Deb, K., Davis, L., Roy, R., O'Reilly, U.-M., Beyer, H.-G., Standish, R.K., Kendall, G., Wilson, S.W., Harman, M., Wegener, J., Dasgupta, D., Potter, M.A., Schultz, A.C., Dowsland, K.A., Jonoska, N., Miller, J.F. (eds.): GECCO 2003. LNCS, vol. 2723. Springer, Heidelberg (2003)
60. Casillas, J., Carse, B., Bull, L.: Fuzzy-XCS: A Michigan Genetic Fuzzy System. IEEE Transactions on Furrz Systems 15(4) (August 2007)
61. Chalk, K., Smith, G.D.: Multi-Agent Classifier Systems and the Iterated Prisoner's Dilemma. In: Smith, G.D., Steele, N.C., Albrecht, R.F. (eds.) Artificial Neural Networks and Genetic Algorithms, pp. 615–618. Springer, Heidelberg (1997)
62. Chipman, H., George, E.I., McCulloch, R.E.: Bayesian Treed Models. Machine Learning 48(1–3), 299–320 (2002)
63. Chipman, H.A., George, E.I., McCulloch, R.E.: Bayesian CART Model Search. Journal of the American Statistical Association 93(443), 935–948 (1998)
64. Corne, D., Michalewicz, Z., Dorigo, M., Eiben, G., Fogel, D., Fonseca, C., Greenwood, G., Chen, T.K., Raidl, G., Zalzala, A., Lucas, S., Paechter, B., Willies, J., Guervos, J.J.M., Eberbach, E., McKay, B., Channon, A., Tiwari, A., Volkert, L.G., Ashlock, D., Schoenauer, M. (eds.): Proceedings of the 2005 IEEE Congress on Evolutionary Computation, vol. 3. IEEE Press, Los Alamitos (2005)
65. Corne, D., Michalewicz, Z., Dorigo, M., Eiben, G., Fogel, D., Fonseca, C., Greenwood, G., Chen, T.K., Raidl, G., Zalzala, A., Lucas, S., Paechter, B., Willies, J., Guervos, J.J.M., Eberbach, E., McKay, B., Channon, A., Tiwari, A., Volkert, L.G., Ashlock, D., Schoenauer, M. (eds.): Proceedings of the 2005 IEEE Congress on Evolutionary Computation, vol. 1. IEEE Press, Los Alamitos (2005)
66. Coursey, D., Nyquist, H.: On Least Absolute Error Estimation of Linear Regression Models with Dependent Stable Residuals. The Review of Economics and Statistics 65(4), 687–692 (1983)
67. Dam, H.H., Abbass, H.A., Lokan, C.: BCS: Bayesian Learning Classifier System. Technical Report TR-ALAR-200604005, The Artificial Life and Adaptic Robotics Laboratory, School of Information Technology and Electrical Engineering, University of New South Wales (2006)
68. Dearden, R., Friedman, N., Russel, S.: Bayesian Q-Learning. In: Proceedings of the 15th National Conference on Artificial Intelligens, Menlo Park, CA, USA (1998)
69. Degris, T., Sigaud, O.P.-H., Wuillemin: Learning the Structure of Factored Markov Decision Processes in Reinforcement Learning Problems. In: Proceedings of the 23rd International Conference on Machine Learning (ICML 2006), CMU, Pennsylvania, USA, pp. 257–264 (2006)
70. DeGroot, M.H.: Lindley's Paradox: Comment. Journal of the American Statistical Association 77(378), 337–339 (1982)

71. Dempster, A.P., Laird, N.M., Rubin, D.B.: Maximum likelihood from incomplete data via the EM algorithm. Journal of the Royal Statistical Society B 39, 1–38 (1977)
72. Denison, D.G.T., Holmes, C.C., Mallick, B.K., Smith, A.F.M.: Bayesian Methods for Nonlinear Classification and Regression. Wiley Series in Probability and Statistics. John Wiley & Sons, Ltd., Chichester (2002)
73. Donoho, D.L., Johnstone, I.M.: Ideal spatial adaptation by wavelet shrinkage. Biometrika 81, 425–455 (1994)
74. Dorigo, M., Bersini, H.: A Comparison of Q-Learning and Classifier Systems. In: Cliff, D., Husbands, P., Meyer, J.-A., Wilson, S.W. (eds.) From Animals to Animats 3. Proceedings of the Third International Conference on Simulation of Adaptive Behavior (SAB 1994), A Bradford Book, pp. 248–255. MIT Press, Cambridge (1994)
75. Dorigo, M., Schnepf, U.: Genetic-based Machine Learning and Behaviour Based Robotics: A New Synthesis. IEEE Transactions on Systems, Man and Cybernetics 23(1) (1993)
76. Douglas, S.C.: A Family of Normalized LMS Algorithms. IEEE Signal Processing Letters SPL-1(3), 49–51 (1994)
77. Drugowitsch, J., Barry, A.M.: XCS with Eligibility Traces. In: Beyer, et al. [18], pp. 1851–1858
78. Drugowitsch, J., Barry, A.M.: A Formal Framework and Extensions for Function Approximation in Learning Classifier Systems. Technical Report 2006–01, University of Bath, U.K. (January 2006)
79. Drugowitsch, J., Barry, A.M.: A Formal Framework for Reinforcement Learning with Function Approximation in Learning Classifier Systems. Technical Report 2006–02, University of Bath, U.K. (January 2006)
80. Drugowitsch, J., Barry, A.M.: Towards Convergence of Learning Classifier Systems Value Iteration. Technical Report 2006–03, University of Bath, U.K. (April 2006)
81. Drugowitsch, J., Barry, A.M.: Towards Convergence of Learning Classifier Systems Value Iteration. In: Proceedings of the 9th International Workshop on Learning Classifier Systems, pp. 16–20 (2006)
82. Drugowitsch, J., Barry, A.M.: Generalised Mixtures of Experts, Independent Expert Training, and Learning Classifier Systems. Technical Report 2007–02, University of Bath (April 2007)
83. Drugowitsch, J., Barry, A.M.: Mixing independent classifiers. In: Thierens, et al. [212], pp. 1596–1603, Also TR CSBU-2006-13
84. Duff, M.: Optimal learning: Computational procedures for Bayes-adaptive Markov decision processes. PhD thesis, University of Massachusetts Amherst (2002)
85. Odeh, M., Kharbat, F., Bull, L.: Revisiting genetic selection in the XCS learning classifier system. In: Corne, et al. [64], pp. 2061–2068
86. Fisher, R.A.: The use of multiple measurements in taxonomic problems. Annual Eugenics 7(2), 179–188 (1963)
87. Fogarty, T.C., Bull, L., Carse, B.: Evolving Multi-Agent Systems. In: Periaux, J., Winter, G. (eds.) Genetic Algorithms in Engineering and Computer Science, pp. 3–22. John Wiley & Sons, Chichester (1995)
88. Forrest, S., Miller, J.H.: Emergent behavior in classifier systems. In: Forrest, S. (ed.) Emergent Computation. Proceedings of the Ninth Annual International Conference of the Center for Nonlinear Studies on Self-organizing, Collective, and Cooperative Phenomena in Natural and Artificial Computing Networks. A special

issue of Physica D, vol. 42, pp. 213–217. Elsevier Science Publishers, Amsterdam (1990)
89. Gérard, P., Meyer, J.-A., Sigaud, O.: Combining Latent Learning with Dynamic Programming in MACS. European Journal of Operational Research 160, 614–637 (2005)
90. Gérard, P., Sigaud, O.: Adding a Generalization Mechanism to YACS. In: Spector, et al. [201], pp. 951–957
91. Gérard, P., Sigaud, O.: YACS: Combining Anticipation and Dynamic Programming in Classifier Systems. In: Lanzi, et al. [148], pp. 52–69
92. Gérard, P., Sigaud, O.: Designing Efficient Exploration with MACS: Modules and Function Approximation. In: Cantú-Paz, et al. [59], pp. 1882–1893
93. Gibbs, M.N.: Bayesian Gaussian Processes for Regression and Classification. PhD thesis, University of Cambridge (1997)
94. Girosi, F., Jones, M., Poggio, T.: Regularization Theory and Neural Networks Architectures. Neural Computation 7, 219–269 (1995)
95. Goldberg, D.E.: Genetic Algorithms in Search, Optimisation, and Machine Learning. Addison-Wesley, Reading (1989)
96. Gordon, G.J.: Stable Function Approximation in Dynamic Programming. In: Prieditis, A., Russell, S. (eds.) Proceedings of the Twelfth International Conference on Machine Learning, pp. 261–268. Morgan Kaufmann, San Francisco (1995)
97. Graybill, F.A.: An Introduction to Linear Statistical Models, vol. 1. McGraw-Hill Education, New York (1961)
98. Greenyer, A.: The use of a learning classifier system JXCS. In: van der Putten, P., van Someren, M. (eds.) CoIL Challenge 2000: The Insurance Company Case, Leiden Institute of Advanced Computer Science (June 2000), Technical report 2000-09
99. Grefenstette, J.J. (ed.): Proceedings of the 2nd International Conference on Genetic Algorithms (ICGA 1987), Cambridge, MA, July 1987. Lawrence Erlbaum Associates, Mahwah (1987)
100. Grefenstette, J.J.: Evolutionary Algorithms in Robotics. In: Jamshedi, M., Nguyen, C. (eds.) Robotics and Manufacturing: Recent Trends in Research, Education and Applications, v5. Proc. Fifth Intl. Symposium on Robotics and Manufacturing, ISRAM 1994, pp. 65–72. ASME Press, New York (1994), `http://www.ib3.gmu.edu/gref/`
101. Grünwald, P.D.: A tutorial introduction to the minimum description length. In: Grünwald, P., Myung, J., Pitt, M.A. (eds.) Advances in Minimum Description Length Theory and Applications, ch. 1 & 2. Neural Information Processing Series, pp. 3–79. MIT Press, Cambridge (2005)
102. Hastie, T., Tibshirani, R., Friedman, J.: The Elements of Statistical Learning: Data Mining, Inference, and Prediction. Springer Series in Statistics. Springer, Heidelberg (2001)
103. Hastings, W.K.: Monte Carlo sampling using Markov chains and their applications. Biometrika 57, 97–109 (1970)
104. Haykin, S.: Neural Networks: A Comprehensive Foundation, 2nd edn. Prentice Hall International, Upper Saddle River (1999)
105. Haykin, S.: Adaptive Filter Theory, 4th edn. Information and System Sciences Series. Prentice Hall, Upper Saddle River (2002)
106. Hertz, J.A., Palmer, R.G.: Introduction to the Theory of Neural Computation. Westview Press (1991)

107. Hoeting, J.A., Madigan, D., Raftery, A.E., Volinsky, C.T.: Bayesian Model Averaging: A Tutorial. Statistical Science 14(4), 382–417 (1999)
108. Hofbauer, J., Sigmund, K.: Evolutionary Games and Replicator Dynamics. Cambridge University Press, Cambridge (1998)
109. Holland, J.H.: Hierachical descriptions of universal spaces and adaptive systems. Technical Report ORA Projects 01252 and 08226, University of Michigan (1968)
110. Holland, J.H.: Processing and processors for schemata. In: Jacks, E.L. (ed.) Associative Information Processing, pp. 127–146. American Elsevier, New York (1971)
111. Holland, J.H.: Adaptation in Natural and Artificial Systems. University of Michigan Press, Ann Arbor (1975), Republished by the MIT press (1992)
112. Holland, J.H.: Properties of the bucket brigade. In: Grefenstette, J.J. (ed.) Proceedings of the 1st International Conference on Genetic Algorithms and their Applications (ICGA 1985), pp. 1–7. Lawrence Erlbaum Associates, Pittsburgh (1985)
113. Holland, J.H.: A Mathematical Framework for Studying Learning in Classifier Systems. Physica D 22, 307–317 (1986)
114. Holland, J.H.: Escaping Brittleness: The Possibilities of General-Purpose Learning Algorithms Applied to Parallel Rule-Based Systems. In: Mitchell, Michalski, Carbonell (eds.) Machine Learning, an Artificial Intelligence Approach, ch. 20, vol. II, pp. 593–623. Morgan Kaufmann, San Francisco (1986)
115. Holland, J.H., Booker, L.B., Colombetti, M., Dorigo, M., Goldberg, D.E., Forrest, S., Riolo, R.L., Smith, R.E., Lanzi, P.L., Stolzmann, W., Wilson, S.W.: What is a Learning Classifier System?. In: Lanzi, et al. [145], pp. 3–32
116. Holland, J.H., Reitman, J.S.: Cognitive systems based on adaptive algorithms. In: Waterman, D.A., Hayes-Roth, F. (eds.) Pattern-directed Inference Systems, Academic Press, New York (1978); Reprinted in: Evolutionary Computation. The Fossil Record Fogel, D.B. (ed.), IEEE Press, Los Alamitos (1998), ISBN: 0-7803-3481-7
117. Hyndman, R.J.: Computing and graphing highest density regions. The American Statistician 50(2), 120–126 (1996)
118. Jaakkola, T.S.: Tutorial on variational approximation methods. In: Opper, M., Saad, D. (eds.) Advanced Mean Field Methods, pp. 129–160. MIT Press, Cambridge (2001)
119. Jaakkola, T.S., Jordan, M.I.: Bayesian parameter estimation via variational methods. Statistics and Computing 10(1), 25–37 (2000)
120. Jacobs, R.A., Jordan, M.I., Nowlan, S., Hinton, G.E.: Adaptive mixtures of local experts. Neural Computation 3, 1–12 (1991)
121. Jordan, M.I., Jacobs, R.A.: Hierarchical mixtures of experts and the EM algorithm. Neural Computation 6, 181–214 (1994)
122. Kaelbling, L.P., Littman, M.L., Cassandra, A.R.: Planning and Acting in Partially Observable Stochastic Domains. Artificial Intelligence 101, 99–134 (1998)
123. Kalman, R.E.: A New Approach to Linear Filtering and Prediction Problems. Transactions of the ASME–Journal of Basic Engineering 82(Series D), 35–45 (1960)
124. Kalman, R.E., Bucy, R.S.: New results in linear filtering and prediction theory. Transactions ASME, Part D (J. Basic Engineering) 83, 95–108 (1961)
125. Kearns, M.J., Mansour, Y., Ng, A.Y., Ron, D.: An experimental and theoretical comparison of model selection methods. Machine Learning 27, 7–50 (1997)
126. Kearns, M.J., Singh, S.: Near-optimal Reinforcement Learning in Polynomial Time. In: Proceedings of the 15th International Conference on Machine Learning, pp. 260–268. Morgan Kaufmann, San Francisco (1998)

127. Kearns, M.J., Vazirani, U.V.: An Introduction to Computational Learning Theory. MIT Press, Cambridge (1994)
128. Keijzer, M., Cattolico, M., Arnold, D., Babovic, V., Blum, C., Bosman, P., Butz, M.V., Coello Coello, C., Dasgupta, D., Ficici, S.G., Foster, J., Hernandez-Aguirre, A., Hornby, G., Lipson, H., McMinn, P., Moore, J., Raidl, G., Rothlauf, F., Ryan, C., Thierens, D. (eds.): GECCO 2006: Proceedings of the 8th annual conference on Genetic and evolutionary computation, Seattle, Washington, USA. ACM Press, New York (2006)
129. Konda, V.R., Tsitsiklis, J.N.: On actor-critic algorithms. SIAM Journal on Control and Optimization 42(4), 1143–1166 (2003)
130. Kovacs, T.: Evolving Optimal Populations with XCS Classifier Systems. Master's thesis, School of Computer Science, University of Birmingham, Birmingham, U.K. (1996), Also technical report CSR-96-17 and CSRP-96-17, `ftp://ftp.cs.bham.ac.uk/pub/tech-reports/1996/CSRP-96-17.ps.gz`
131. Kovacs, T.: Deletion schemes for classifier systems. In : Banzhaf, et al. [9], pp. 329–336. Also TR CSRP-99-08, School of Computer Science, University of Beirmingham
132. Kovacs, T.: Strength or accuracy? A comparison of two approaches to fitness calculation in learning classifier systems. In: Wu, A.S. (ed.) Proceedings of the 1999 Genetic and Evolutionary Computation Conference Workshop Program, pp. 258–265 (1999)
133. Kovacs, T.: A Comparison and Strength and Accuracy-based Fitness in Learning Classifier Systems. PhD thesis, University of Birmingham (2002)
134. Kovacs, T.: Two views of classifier systems. In: Lanzi, et al. [146], pp. 74–87
135. Kovacs, T.: What should a classifier systems learn and how should we measure it? In: Bull, et al. [37], pp. 171–182
136. Kovacs, T., Bull, L.: Towards a better understanding of rule initialisation and deletion. In: Thierens, et al. [212], pp. 2777–2780
137. Kovacs, T., Llorá, X., Takadama, K., Lanzi, P.L., Stolzmann, W., Wilson, S.W.: Learning Classifier Systems: International Workshops, IWLCS 2003–2005. LNCS (LNAI), vol. 4399. Springer, Heidelberg (2007)
138. Lanzi, P.L.: Learning Classifier Systems from a Reinforcement Learning Perspective. In: Bull, et al. [37], pp. 162–170
139. Lanzi, P.L., Butz, M.V., Goldberg, D.E.: Empirical Analysis of Generalization and Learning in XCS with Gradient Descent. In: Thierens, et al. [212], pp. 1814–1821
140. Lanzi, P.L., Loiacono, D.: Standard and averaging reinforcement learning in XCS. In: Keijzer, et al. [128], pp. 1489–1496
141. Lanzi, P.L., Loiacono, D., Wilson, S.W., Goldberg, D.E.: Extending XCSF Beyond Linear Approximation. In: Beyer, et al. [18], pp. 1827–1834
142. Lanzi, P.L., Loiacono, D., Wilson, S.W., Goldberg, D.E.: Generalization in the XCSF Classifier Systems: Analysis, Improvement, and Extenstion. Technical Report 2005012, Illinois Genetic Algorithms Laboratory (March 2005)
143. Lanzi, P.L., Loiacono, D., Wilson, S.W., Goldberg, D.E.: Generalization in the XCSF Classifier System: Analysis, Improvement, and Extension. Evolutionary Computation 15(2), 133–168 (2007)
144. Lanzi, P.L., Perrucci, A.: Extending the Representation of Classifier Conditions Part II: From Messy Coding to S-Expressions. In: Banzhaf, W., Daida, J.M., Eiben, A.E., Garzon, M.H., Honavar, V., Jakiela, M.J., Smith, R.E. (eds.) Proceedings of the Genetic and Evolutionary Computation Conference (GECCO 1999), pp. 253–345. Morgan Kaufmann, San Francisco (1999)

145. Lanzi, P.L., Stolzmann, W., Wilson, S.W. (eds.): IWLCS 1999. LNCS (LNAI), vol. 1813. Springer, Heidelberg (2000)
146. Lanzi, P.L., Stolzmann, W., Wilson, S.W. (eds.): IWLCS 2001. LNCS (LNAI), vol. 2321. Springer, Heidelberg (2002)
147. Lanzi, P.L., Wilson, S.W.: Using convex hulls to represent classifier conditions. In: Keijzer, et al. [128], pp. 1481–1488
148. Lanzi, P.L., Stolzmann, W., Wilson, S.W. (eds.): IWLCS 2000. LNCS (LNAI), vol. 1996. Springer, Heidelberg (2001)
149. Littman, M.: Personal Communication (September 2006)
150. Llorá, X.: Personal Communication (July 2007)
151. Llorá, X., Garrell, J.M.: Knowledge-Independent Data Mining with Fine-Grained Parallel Evolutionary Algorithms. In: Spector, et al. [201], pp. 461–468
152. Llorá, X., Reddy, R., Matesic, B., Bhargava, R.: Towards Better than Human Capability in Diagnosing Prostate Cancer Using Infrared Spectroscopic Imaging. In: Thierens, et al. [212], pp. 2098–2105
153. Llorá, X., Sastry, K., Goldberg, D.E.: The Compact Classifier System: Motivation, Analysis and First Results. In: Corne, et al. [65], pp. 596–603, Also IlliGAl TR No. 2005019
154. Llorá, X., Sastry, K., Goldberg, D.E., de la Ossa, L.: The χ-ary Extended Compact Classifier System: Linkage Learning in Pittsburgh LCS. In: Proceedings of the International Workshop on Learning Classifier Systems (IWLCS-2006) (to appear), Also IlliGAl TR No. 2006015
155. Loiacono, D., Drugowitsch, J., Barry, A.M., Lanzi, P.L.: Improving Classifier Error Estimate in XCSF. In: Proceedings of the 9th International Workshop on Learning Classifier Systems (2006)
156. Loiacono, D., Lanzi, P.L.: Neural Networks for Classifier Prediction in XCSF. In: Cagnoni, S., Collet, P., Nicosia, G., Vanneschi, L. (eds.) Proceeding of the Workshop on Evolutionary Computation $(EC)^2$AI, pp. 36–40 (August 2006)
157. Loiacono, D., Marelli, A., Lanzi, P.L.: Support Vector Regression for Classifier Prediction. In: Thierens, et al. [212], pp. 1806–1813
158. Luke, S., Panait, L.: A comparison of bloat control methods for genetic programming. Evolutionary Computation 14(3), 309–344 (2006)
159. MacKay, D.J.C.: Bayesian interpolation. Neural Computation 4(3), 415–447 (1992)
160. MacQueen, J.: Some methods for classification and analysis of multivariate observations. In: Proceedings of the Fifth Berkeley Symposium on Mathematical Statistics and Probability, vol. 1, pp. 281–297. University of Claifornia Press (1967)
161. Mansilla, E.B., Guiu, J.M.G.: Accuracy-based learning classifier systems: Models, analysis and applications to classification tasks. Evolutionary Computation 11(3), 209–238 (2003)
162. Marshall, J.A.R., Brown, G., Kovacs, T.: Bayesian estimation of rule accuracy in UCS. In: GECCO 2007: Proceedings of the 2007 GECCO conference companion on Genetic and evolutionary computation, pp. 2831–2834. ACM Press, New York (2007)
163. Marshall, J.A.R., Kovacs, T.: A representational ecology for learning classifier systems. In: Keijzer, et al. [128], pp. 1529–1536
164. Maybeck, P.S.: Stochastic Models, Estimation, and Control. Mathematics in Science and Engineering, vol. 141. Academic Press, Inc., New York (1979)
165. McCullach, P., Nelder, J.A.: Generalized Linear Models. Monographs on Statistics and Applied Probability. Chapman and Hall, Boca Raton (1983)

166. McMahon, A., Scott, D., Browne, W.: An autonomous explore/exploit strategy. In: GECCO 2005: Proceedings of the 2005 workshops on Genetic and evolutionary computation, pp. 103–108. ACM Press, New York (2005)
167. Mitchell, M.: An Introduction to Genetic Algorithms. MIT Press, Cambridge (1998)
168. Mitchell, T.: Machine Learning. McGraw Hill, New York (1997)
169. Mitlöhner, J.: Classifier systems and economic modelling. In: APL 1996. Proceedings of the APL 1996 Conference on Designing the Future, vol. 26(4), pp. 77–86 (1996)
170. Mook, D.J., Junkins, J.L.: Minimum Model Error Estimation for Poorly Modeled Dynamic Systems. Journal of Guidance, Control and Dynamics 11(3), 256–261 (1988)
171. Moraglio, A.: Personal Communication (November 2006)
172. Moriarty, D.E., Schultz, A.C., Grefenstette, J.J.: Evolutionary Algorithms for Reinforcement Learning. Journal of Artificial Intelligence Research 11, 199–229 (1999), `http://www.ib3.gmu.edu/gref/papers/moriarty-jair99.html`
173. Nabney, I.T.: Netlab: Algorithms for Pattern Recognition. Springer, Heidelberg (2002)
174. Neal, R., Hinton, G.E.: A View of the EM Algorithm that Justifies Incremental, Sparse, and other Variants. In: Jordan, M.I. (ed.) Learning in Graphical Models, pp. 355–368. MIT Press, Cambridge (1999)
175. O'Hara, T., Bull, L.: A Memetic Accuracy-based Neural Learning Classifier System. In: Corne, et al. [64], pp. 2040–2045
176. O'Hara, T., Bull, L.: Backpropagation in Accuracy-based Neural Learning Classifier Systems. In: Kovacs, et al. [137], pp. 26–40
177. Ormoneit, D., Sen, S.: Kernel-Based Reinforcement Learning. Machine Learning 49(2-3), 161–178 (2002)
178. Orriols-Puig, A., Bernadó-Mansilla, E.: Class Imbalance Problem in the UCS Classifier System: Fitness Adaptation. In: Corne, et al. [65], pp. 604–611
179. Orriols-Puig, A., Bernadó-Mansilla, E.: Bounding XCS's Parameters for Unbalanced Datasets. In: Keijzer, et al. [128], pp. 1561–1568
180. Orriols-Puig, A., Goldberg, D.E., Sastry, K., Mansilla, E.B.: Modeling XCS in Class Imbalances: Population Size and Parameter Settings. In: Thierens, et al. [212], pp. 1838–1846
181. Orriols-Puig, A., Satary, K., Lanzi, P.L., Goldberg, D.E., Mansilla, E.B.: Modeling Selection Pressure in XCS for Proportionate and Tournament Selection. In: Thierens, et al. [212], pp. 1846–1854
182. Pelikan, M.: Hierarchical Bayesian Optimization Algorithm: Toward a New Generation of Evolutionary Algorithms. Studies in Fuzziness and Soft Computing. Springer, Heidelberg (2005)
183. Pelikan, M., Sastry, K., Cantu-Paz, E. (eds.): Scalable Optimization via Probabilistic Modeling: From Algorithms to Applications. Studies in Computational Intelligence. Springer, Heidelberg (2006)
184. Piater, J.H., Cohen, P.R., Zhang, X., Atighetchi, M.: A Randomized ANOVA Procedure for Comparing Performance Curves. In: ICML 1998: Proceedings of the Fifteenth International Conference on Machine Learning, pp. 430–438. Morgan Kaufmann Publishers Inc., San Francisco (1998)
185. Poupart, P., Vlassis, N., Hoey, J., Regan, K.: An analytic solution to discrete Bayesian reinforcement learning. In: Proceeding of the 23rd international conference on machine learning. ACM International Conference Proceeding Series, vol. 148, pp. 697–704 (2006)

186. Riolo, R.L.: Bucket Brigade Performance: I. Long Sequences of Classifiers. In: Grefenstette [99], pp. 184–195
187. Riolo, R.L.: Bucket Brigade Performance: II. Default Hierarchies. In: Grefenstette [99], pp. 196–201
188. Rissanen, J.: Modeling by the shortest data description. Automatica 14, 465–471 (1978)
189. Rissanen, J.: A universal prior for integers and estimation by minimum description length. Annals of Statistics 11, 416–431 (1983)
190. Rissanen, J.: Stochastic Complexity in Statistical Inquiry. World Scientific, Singapore (1989)
191. Rissanen, J.: Fisher information and stochastic complexity. IEEE Transactions on Information Theory 42(1), 40–47 (1996)
192. Rummery, G., Niranja, M.: On-line Q-Learning using Connectionist Systems. Technical Report 166, Engineering Department, University of Cambridge (1994)
193. Schoknecht, R.: Optimality of Reinforcement Learning Algorithms with Linear Function Approximation. In: Proceedings of the 15th Neural Information Processing Systems conference, pp. 1555–1562 (2002)
194. Schoknecht, R., Merke, A.: Convergent Combinations of Reinforcement Learning with Linear Function Approximation. In: Proceedings of the 15th Neural Information Processing Systems conference, pp. 1579–1586 (2002)
195. Schoknecht, R., Merke, A.: TD(0) Converges Provably Faster than the Residual Gradient Algorithm. In: ICML 2003: Proceedings of the twentieth international conference on Machine Learning, pp. 680–687 (2003)
196. Smith, R.E.: Memory Exploitation in Learning Classifier Systems. Evolutionary Computation 2(3), 199–220 (1994)
197. Smith, R.E., Dike, B.A., Ravichandran, B., El-Fallah, A., Mehra, R.K.: The Fighter Aircraft LCS: A Case of Different LCS Goals and Techniques. In: Lanzi, et al. [145], pp. 283–300
198. Smith, S.F.: A Learning System Based on Genetic Adaptive Algorithms. PhD thesis, University of Pittsburgh (1980)
199. Smith, S.F.: Flexible Learning of Problem Solving Heuristics through Adaptive Search. In: Proceedings Eight International Joint Conference on Artificial Intelligence, pp. 422–425 (1983)
200. Smith, S.F.: Adaptive learning systems. In: Forsyth, R. (ed.) Expert Systems: Principles and Case Studies, pp. 169–189. Chapman and Hall, Boca Raton (1984)
201. Spector, L., Goodman, E.D., Wu, A., Langdon, W.B., Voigt, H.-M., Gen, M., Sen, S., Dorigo, M., Pezeshk, S., Garzon, M.H., Burke, E. (eds.): GECCO-2001: Proceedings of the Genetic and Evolutionary Computation Conference. Morgan Kaufmann, San Francisco (2001)
202. Stolzmann, W.: Anticipatory Classifier Systems. In: Koza, J.R., Banzhaf, W., Chellapilla, K., Deb, K., Dorigo, M., Fogel, D.B., Garzon, M.H., Goldberg, D.E., Iba, H., Riolo, R. (eds.) Genetic Programming, pp. 658–664. Morgan Kaufmann Publishers, Inc., San Francisco (1998)
203. Stone, C., Bull, L.: For real! XCS with continuous-valued inputs. Evolutionary Computation 11(3), 299–336 (2003), Also UWE TR UWELCSG02-007
204. Strehl, A.L.: Model-Based Reinforcement Learning in Factored MDPs. In: IEEE Symposium on Approximate Dynamic Programming, pp. 103–110 (2007)
205. Strehl, A.L., Li, L., Wiewiora, E., Langford, J., Littman, M.L.: PAC Model-Free Reinforcement Learning. In: Proceedings of the 23rd International Conference on Machine Learning (ICML 2006), Pittsburgh, PA, USA, pp. 881–888 (2006)

206. Sutton, R.S.: Value-function hypothesis. From Reinforcement Learning and Artificial Intelligence,
`http://rlai.cs.ualberta.ca/RLAI/valuefunctionhypothesis.html`
207. Sutton, R.S.: Learning to predict by the method of temporal differences. Machine Learning 3, 9–44 (1988)
208. Sutton, R.S.: Generalization in Reinforcement Learning: Successful Examples Using Sparse Coarse Coding. In: Touretzky, D.S., Mozer, M.C., Hasselmo, M.E. (eds.) Advances in Neural Information Processing Systems, vol. 8, pp. 1038–1044. MIT Press, Cambridge (1996)
209. Sutton, R.S., Barto, A.G.: Reinforcement Learning: An Introduction. MIT Press, Cambridge (1998)
210. Syswerda, G.: Uniform Crossover in Genetic Algorithms. In: Proceedings of the 3rd International Conference on Genetic Algorithms, pp. 2–9. Morgan Kaufmann Publishers Inc., San Francisco (1989)
211. Tamee, K., Bull, L., Pinngern, O.: Towards Clustering with XCS. In: Thierens, et al. [212], pp. 1854–1860
212. Thierens, D., Beyer, H.-G., Birattari, M., Bongard, J., Branke, J., Clark, J.A., Cliff, D., Congdon, C.B., Deb, K., Doerr, B., Kovacs, T., Kumar, S., Miller, J.F., Moore, J., Neumann, F., Pelikan, M., Poli, R., Sastry, K., Stanley, K.O., Stützle, T., Watson, R.A., Wegener, I. (eds.): GECCO-2007: Proceedings of the 9th Annual Conference on Genetic and Evolutionary Computation Congerece 2007, vol. 2. ACM Press, New York (2007)
213. Tikhonov, A.N., Arsenin, V.Y.: Solutions of Ill-posed Problems. Winston (1977)
214. Tsitsiklis, J., Van Roy, B.: Feature-Based Methods for Large Scale Dynamic Programming. Machine Learning 22, 59–94 (1996)
215. Tsitsiklis, J., Van Roy, B.: An Analysis of Temporal-Difference Learning with Function Approximation. IEEE Transactions on Automatic Control 42(5), 674–690 (1997)
216. Ueda, N., Ghahramani, Z.: Bayesian model search for mixture models based on optimizing variational bounds. Neural Networks 15, 1223–1241 (2002)
217. van Laarhoven, P.J., Aarts, E.H.: Simulated Annealing: Theory and Applications. Springer, Heidelberg (1987)
218. Vapnik, V.N.: An Overview of Statistical Learning Theory. IEEE Transactions on Neural Networks 10(5), 988–999 (1999)
219. Vapnik, V.N.: The Nature of Statistical Learning Theory. Springer, Heidelberg (1999)
220. Venturini, G.: Apprentissage Adaptatif et Apprentissage Supervisé par Algorithme Génétique. PhD thesis, Université de Paris-Sud (1994)
221. Vriend, N.: Self-Organization of Markets: An Example of a Computational Approach. Computational Economics 8(3), 205–231 (1995)
222. Wada, A., Takadama, K., Shimohara, K.: Counter Example for Q-Bucket-Brigade under Prediction Problema. In: Kovacs, et al. [137], pp. 130–145
223. Wada, A., Takadama, K., Shimohara, K., Katai, O.: Is Gradient Descent Method Effective for XCS? Analysis of Reinforcement Process in XCSG. In: Stolzmann, W., et al. (eds.) Proceedings of the Seventh International Workshop on Learning Classifier Systems, 2004, Seattle, WA. LNCS (LNAI), Springer, Heidelberg (2004)
224. Wada, A., Takadama, K., Shimohara, K., Katai, O.: Learning Classifier System with Convergence and Generalisation. In: Bull, Kovacs [36]
225. Wainwright, M., Jaakkola, T., Willsky, A.: A new class of upper bounds on the log partition function. IEEE Transactions on Information Theory 51, 2313–2335 (2005)

226. Waterhouse, S.: Classification and Regression using Mixtures of Experts. PhD thesis, Department of Engineering, University of Cambridge (1997)
227. Waterhouse, S., MacKay, D., Robinson, T.: Bayesian Methods for Mixtures of Experts. In: Touretzky, D.S., Mozer, M.C., Hasselmo, M.E. (eds.) Advances in Neural Information Processing Systems 8, pp. 351–357. MIT Press, Cambridge (1996)
228. Watkins, C.J.C.H.: Learning from delayed rewards. PhD thesis, University of Cambridge, Psychology Department (1989)
229. Watkins, C.J.C.H., Dayan, P.: Q-learning. Machine Learning 8(3), 279–292 (1992)
230. Weisstein, E.W.: Banach fixed point theorem. From Mathworld – a Wolfram Web Resource (1999), http://mathworld.wolfram.com/BanachFixedPointTheorem.html
231. Weisstein, E.W.: Jensen's inequality From Mathworld – a Wolfram Web Resource (1999), http://mathworld.wolfram.com/JensensInequality.html
232. Weisstein, E.W.: Relative entropy. From Mathworld – a Wolfram Web Resource (1999), http://mathworld.wolfram.com/RelativeEntropy.html
233. Welch, G., Bishop, G.: An Introduction to the Kalman Filter. Technical Report TR 95-401, University of North Carolina at Chapel Hill, Department of Computer Science (April 2004)
234. Widrow, B., Hoff, M.E.: Adaptive switching circuits. In: IRE WESCON Convention Revord Part IV, pp. 96–104 (1960)
235. Wiegand, R.P., Liles, W.C., De Jong, K.A.: An Empirical Analysis of Collaboration Methods in Cooperative Coevolutionary Algorithms. In: Spector, et al. [201], pp. 1235–1242
236. Wilson, S.W.: ZCS: A zeroth level classifier system. Evolutionary Computation 2(1), 1–18 (1994)
237. Wilson, S.W.: Classifier Fitness Based on Accuracy. Evolutionary Computation 3(2), 149–175 (1995), http://prediction-dynamics.com/
238. Wilson, S.W.: Generalization in the XCS classifier system. In: Koza, J.R., Banzhaf, W., Chellapilla, K., Deb, K., Dorigo, M., Fogel, D.B., Garzon, M.H., Goldberg, D.E., Iba, H., Riolo, R. (eds.) Genetic Programming 1998: Proceedings of the Third Annual Conference, pp. 665–674. Morgan Kaufmann, San Francisco (1998), http://prediction-dynamics.com/
239. Wilson, S.W.: Get real! XCS with continuous-values inputs. In: Lanzi, et al. [145], pp. 209–222
240. Wilson, S.W.: Function Approximation with a Classifier System. In: Spector, et al. [201], pp. 974–981
241. Wilson, S.W.: Classifiers that Approximate Functions. Neural Computing 1(2-3), 211–234 (2002)
242. Xu, L.: BYY harmony learning, structural RPCL, and topological self-organizing on mixture models. Neural Networks 15, 1125–1151 (2002)
243. Xu, L.: Fundamentals, Challenges, and Advances of Statistical Learning for Knowledge Discovery and Problem Solving: A BYY Harmony Perspective. In: Proceedings of International Conference on Neural Networks and Brain, October 2005, vol. 1, pp. 24–55. Publishing House of Electronics Industry, Beijing, China (2005)

Index